Water Resources: Future Perspectives, Challenges, Concepts and Necessities

Water Resources: Future Perspectives, Challenges, Concepts and Necessities

Edited by

Omid Bozorg-Haddad

IWA PUBLISHING

Published by **IWA Publishing**
Unit 104–105, Export Building
1 Clove Crescent
London E14 2BA, UK
Telephone: +44 (0)20 7654 5500
Fax: +44 (0)20 7654 5555
Email: publications@iwap.co.uk
Web: www.iwapublishing.com

First published 2021

British Library Cataloguing in Publication Data
A CIP catalogue record for this book is available from the British Library

ISBN: 9781789062137 (paperback)
ISBN: 9781789062144 (eBook)
ISBN: 9781789062151 (ePub)

Contents

Preface

Now, more than ever, one can feel the overwhelming desire by both public and governing bodies to aspire 'sustainability' in a society as a whole entity. When it comes to water resources, this notion has coincided with the concept of 'integrated water resources planning and management.' Plainly speaking, the idea here is to strive for an equilibrium that could not only meet the demands and needs of a community for water resources but also takes into account, and in turn preserve, these natural resources and the environment that surrounds and depends on it as well. On paper, any practices that follow this guideline can be maintained over a long-time horizon. As it turned out, however, achieving this is easier said than done. The main challenge, in a nutshell, is that water management is, by nature, a multi-dimensional process. As such, a truly integrated and sustainable practice must not only account for technical and engineering challenges, but also addresses the socio-economic, legislative, and political aspect of the problem as well.

Having access to water for drinking and sanitation is a well-established right recognized by the United Nations. One tends to forget, however, that for other consuming sectors, water is essentially an economic commodity. Hence, it needs to be treated as one in order to devise a self-sustaining, long-term water resources planning, and management policy. Thus, knowing these economic-oriented principles and their implication from a water resources management perspective can bring us one step closer to our goal of creating a sustainable future. To that end, in this book, we would be exploring the concepts of water accounting, water economics, water pricing, water markets, and water books from a fresh perspective.

Editor: Omid Bozorg-Haddad
doi: 10.2166/9781789062144_xvii

A comprehensive water resources policy should also accommodate the role of this natural resource from a social and legislative point-of-view. Often, water is a common, shared good and, to some extent, a limited resource. As such, the socio-political aspect of water resources planning and management can play a permanent role in practical, real-world cases. The sheer importance of this matter became obediently clear when one deals with transboundary water bodies. As such, here we tend to take a deep dive into topics like 'rights and laws of transboundary water resources,' 'water policy and governance,' 'shared water resources management,' 'water, culture, civilization, and history,' 'water quality, hygiene, and health,' and 'water and society.'

The book is structured so that each chapter can be seen as a stand-alone lecture on a specific topic. The interested reader, which can be either an experienced and seasoned researcher or an early-career student who wants to get a foot at the door, can explore the said subject area from a somewhat different and fresh perspective. The book can also be seen as a wholesome lecture, and the reader can, and in fact, we encourage the readers to start from the first chapter and go through the entire book. As such, you would be engaged in an engaging, thought-provocative idea that can bring you a fresh perspective on how you perceive sustainability in terms of water resources planning and management.

Omid Bozorg-Haddad

Acknowledgements

The authors thank Iran's National Science Foundation (INSF) for its support of this research.

Chapter 1

New water resources technologies

Omid Bozorg-Haddad[1], Mohammadali Geranmehr[2], Zaher Mundher Yaseen[3] and Vijay P. Singh[4]*

[1]Distinguished Professor, Department of Irrigation & Reclamation Engineering, Faculty of Agricultural Engineering & Technology, College of Agriculture & Natural Resources, University of Tehran, Karaj, Tehran, 31587-77871, Iran. E-mail: OBHaddad@ut.ac.ir

[2]Postdoctoral Researcher, Faculty of Agricultural Engineering and Technology, College of Agriculture and Natural Resources, University of Tehran, Karaj, Alborz, Iran 31587-77871. E-mail: mgeranmehr@gmail.com

[3]Sustainable Developments in Civil Engineering Research Group, Faculty of Civil Engineering, Ton Duc Thang University, Ho Chi Minh City, Vietnam

[4]Distinguished Professor, Regents Professor, Regents Professor, Caroline & William N. Lehrer Distinguished Chair in Water Engineering, Department of Biological and Agricultural Engineering & Zachry Department of Civil Engineering, Texas A&M University, 321 Scoates Hall, 2117 TAMU, College Station, Texas 77843-2117, USA. E-mail: VSingh@tamu.edu

**Corresponding author.*

ABSTRACT

Growing population and increasing water demand, along with limited water resources, pose several challenges to water resources management and planning. Although demand management is one of the critical elements used to reduce water stress, new water resources technologies – such as cloud seeding, fossil water extraction, desalination, extraction of atmospheric water, inter-basin water transfer, and water reuse – can all be effective options to alleviate water scarcity. This chapter discusses and compares each of these technologies with illustrative examples.

Editor: Omid Bozorg-Haddad
doi: 10.2166/9781789062144_0001

Keywords: Cloud seeding, Fossil water, Desalination, Atmospheric water, Inter-basin water transfer, Water reuse.

1.1 INTRODUCTION

Since the beginning of life on Earth, people have used lakes and rivers as surface water sources and taken water from groundwater sources. Uneven distribution of freshwater resources, developments in life quality, and population growth are the main three factors that have led to the need for and development of new water resources (Oki & Kanae, 2006). This forms the motivation for the exploration of new technologies and strategies, and this chapter reviews some of the methodologies being used for the exploration of new water resources. The main concepts of each methodology are discussed briefly and then some practical examples are given. The chapter concludes with a summary, a comparison between methods, and suggestions.

1.2 DEFINITIONS AND TERMINOLOGY

1.2.1 Cloud seeding

Cloud seeding is a process in which adding certain substances into clouds accelerates the growth of water droplets or the formation of ice crystals, increasing precipitation (Keyes *et al.*, 2006). It is a kind of weather modification that triggers and increases the amount of precipitation (i.e., rain or ice) formed over a particular watershed.

1.2.2 Fossil water

Fossil water is water found in an ancient water body that has been contained and undistributed for a long time (such as for millennia), mostly appearing as groundwater confined in an aquifer. Groundwater at great depths (usually more than 1000 meters) is also called fossil water. These waters have been stored deep in the earth for thousands of years and are an almost non-renewable source (Tsur *et al.*, 1989).

1.2.3 Desalination

Desalination is the process of abstraction of mineral elements from saline water. In general, the process of desalination is applied to seawater, oceanic water, or saline groundwater. It is considered to be one of the conventional approaches to freshwater supply (Service, 2006).

1.2.4 Atmospheric water

Producing water from moisture in the air is a common method used to produce renewable water for drinking (Tu *et al.*, 2018).

1.2.5 Inter-basin water transfer

Inter-basin water transfer describes the redistribution of water resources in a region where distribution is uneven, for example where the water per capita in one catchment is much less than in another catchment. Inter-basin water transfer is of the main water resources management approaches (Ghassemi & White, 2007).

1.2.6 Water reuse

Water reuse is the process of wastewater treatment and reusage. Reused or reclaimed water is an approach that is important as a sustainable water source. Water recycling can be used for industry, agriculture, and, in limited cases, even for drinking (Eslamian, 2015).

1.3 BASICS AND SIGNIFICANCE OF THE STUDY

Currently, all available freshwater on Earth is limited. Most of this water, about 97.5%, is in the seas and oceans, whereas only 2.5% is freshwater. About 68.9% of this freshwater is stored as ice, 29.9% is stored in groundwater, and the rest is in lakes, rivers, soil moisture, and the atmosphere (Shiklomanov, 1988). Population growth, increased per capita consumption, and limited water resources are severe challenges for society now and will be in the years ahead. Further, the limited freshwater resources are becoming more and more contaminated by humans.

The world population on Earth is increasing and is expected to reach 8.5 billion (8.5×10^9) by the year 2030, 9.7 billion by 2050, and 11.2 billion by 2100 (McNabb, 2019). This boost in the population will increase the need for freshwater as one of the most critical components of life. Meanwhile, with the development of urbanization and increasing quality of life, the per capita water consumption is increasing day by day. As a result, in many parts of the world, disputes arise over access to water beyond national borders, and access to these resources has become a strategic goal of inter-state interaction. Based on the World Resources Index (WRI), 33 countries face extremely high water stress (Ratio of withdrawal to available water >80%), and 26 countries will have high water stress (Ratio of withdrawal to available water = 40–80%) by 2040 (McNabb, 2015). Due to the limitation of available water resources, two essential solutions need to be considered together: (1) consumption management; and (2) extraction of water from new water sources. The first solution is beyond the scope of the present study. However, the second solution is discussed in more detail in this chapter.

1.4 METHODOLOGY

1.4.1 Cloud seeding

Cloud seeding is a branch of weather modification science and is mainly aimed at combating drought or preventing large hail (Dennis, 1980). Cloud seeding is usually done by adding particular substances into clouds which accelerate the growth of water droplets or the formation of ice crystals. In clouds where ice crystals are not formed or are insufficient, the process of forming ice crystals can be accelerated by injecting silver iodide or dry ice powder. This process is typically called cloud seeding.

There are various methods for seeding clouds (Bruintjes, 1999). One of the conventional methods is the use of ground equipment. In this method, the cloud seeding process is executed by converting the seeding materials into tiny particles and releasing them into the air by terrestrial generators. Another method is to use ground-to-air launch equipment in which a cannon or rocket carrying seeding particles is fired into the clouds. The most commonly used method today is the use of aircraft, depending on the region and the type of cloud. Depending on the aircraft's location, fertilization can take place on the cloud floor inside the cloud, or at the cloud's summit.

In the case of cloud seeding, the following should be considered:

- the process of cloud seeding is a relatively costly method (Sajjadi, 2008);
- based on the uncertain nature of hydrological phenomena, there seems to be no direct and accurate tool to measure the effectiveness of cloud seeding. In this respect, different methods sometimes produce completely contradictory results (Goldreich, 1994);
- although successful experiences in the field have been reported (Ben-Zvi, 1988; Ben-Zvi & Fanar, 1997; Ben-Zvi & Langerman, 1989, 1993; Benjamini & Harpaz, 1986), these artificial changes in the climate may cause irreparable effects;
- there is no particular guarantee that precipitation will occur on the area over which cloud seeding is being done.

1.4.2 Fossil water

Fossil waters are very deep groundwater extracted by oil well drilling systems. Exploration and extraction of these waters are still in the process of identification and research, and no environmental impacts are known. The cost of extracting these waters is relatively high. Meanwhile, the water source is almost non-renewable. Therefore, this water source should be treated with caution (Issar, 1994).

1.4.3 Desalination

Since much of the water on Earth is saline, the desalination process is fundamental. There are generally two types of methods for desalting saltwater. The first category comprises thermal methods, which are based on the principle of distillation and require high energy. The second category consists of membrane-based methods. Reverse osmosis (RO) is one of the most common tools in this category (Ehteram *et al.*, 2020).

One of the disadvantages of desalination is the production of highly saline wastewater, and its proper disposal is of great importance. A study of desalination in Spain showed that it costs 0.7 euros per cubic meter of water (Lapuente, 2012).

1.4.4 Atmospheric water

Air humidity is one of the renewable sources of water supply with no environmental damage. This method is very suitable for areas with sufficient moisture in the air. However, depending on the climate of the region, the water available can be different in several months of the year. However, the amount of water produced is not very high and is often only suitable for drinking. The required energy in this method could be provided from renewable sources like the sun. The state of the art of water supply from the air is the use of Metal-Organic Framework (MOF) crystals, which can produce up to 200 grams of water per day in areas with low humidity, such as in deserts in Arizona, USA (Fathieh *et al.*, 2018).

1.4.5 Inter-basin water transfer

Inter-basin water transfer is a way of balancing the uneven distribution of water resources. Water is usually transferred from a rich basin to a water-deficient basin. However, transfer will have unavoidable economic, political, security, social, cultural, environmental, and other impacts on the source and destination basins. The challenge is, therefore, very complex and requires extensive study and evaluation. It is tough to come up with a balanced solution to solve the problem of transferring water from basin to basin (Getches, 2005). An inter-basin water transfer survey in Spain showed that it costs 0.5 euros per cubic meter of water (Pouget *et al.*, 2012).

1.4.6 Water reuse

Due to the limitations of water resources, recycling and reuse are almost essential. Recycling can be classified into three categories: (1) wastewater treatment; (2) greywater; and (3) stormwater harvesting, as described below.

1.4.6.1 Wastewater treatment

One of the sustainable water sources is wastewater treatment. The water obtained this way can be used for industry, agriculture, and groundwater recharge. For example, 26% of the water required for agriculture is supplied by wastewater in Pakistan, whilst this figure is 80% in Hanoi in Vietnam (Jiménez-Cisneros, 2014). Considering the possibility of contamination of soil and plants, it is necessary to investigate the amount of heavy metals present in the wastewater. Although most wastewater treatment plants are somewhat similar, each treatment plant will be unique in terms of the type of wastewater processed. In this regard, due to health standards and the quality of wastewater, the performance of the treatment plant must always be monitored (Bixio *et al.*, 2006).

1.4.6.2 Greywater

Greywater is part of the sewage produced in homes, including from showers, laundry, dishwashers, and sometimes kitchen sinks (Pepper, 2016). The amount of greywater is about 30% of the total water consumed (De Gisi *et al.*, 2017). After preliminary treatment, this water can be used for some applications, such as subsurface irrigation and toilet flushing. Extensive use of greywater, especially in arid regions, saves drinking water and, at the same time, provides nutrients for plant growth (Pepper, 2016).

1.4.6.3 Stormwater

Rainfall runoff can be a key but unstable source of water. Depending on the seasonality of rainfall, storage tanks can be used to ensure water availability in the months with low rainfall or even with no precipitation (Goonetilleke *et al.*, 2017). Treated stormwater can be used for irrigating green spaces, toilet flushing, clothes washing, fire hydrants, and cooling towers (Ngo Huu *et al.*, 2016).

1.5 ILLUSTRATIVE EXAMPLES

1.5.1 Cloud seeding

An aircraft-assisted cloud seeding program in northern Jiangxi province in eastern China took place from 2008 to 2014 (Wang *et al.*, 2019). Silver iodide (AgI) was used in this project to seed supercooled clouds. A total of 71 seeding operations were carried out, with each operation releasing 80 to 120 grams of AgI in the clouds. The results of this seven-year project show that average winter precipitation in the target area and downwind domain increased by 17.3% and 21.7%, respectively (Wang *et al.*, 2019).

1.5.2 Fossil water

Fossil water was first discovered accidentally in 1953 during oil drilling in Libya. The volume of water found was about 150,000 m^3, which remains a vital resource for Libya If this water had not been used, no area of Libya would have been inhabited today. This water has been transferred to various parts of Libya under the Great Manmade River Project, which has expanded agriculture. However, in recent years, due to the increasing demand for water and its limited supply, there have been many problems in Libya. In this regard, combined supply and demand management has been proposed (Wheida & Verhoeven, 2007).

1.5.3 Desalination

The Kingdom of Saudi Arabia (KSA) is one of the most arid countries in the world with no permanent rivers. One of the primary water resources in KSA is seawater desalination. Almost 50% of Saudi Arabia's drinking water comes from desalination, 40% from non-renewable groundwater, and 10% from surface water (Tlili *et al.*, 2019). Based on the high energy requirement for desalination, it is proposed that by 2050 all Saudi Arabian desalination plants will use renewable energy (Caldera *et al.*, 2018).

1.5.4 Atmospheric water

India, especially in the northern states of the country, is experiencing a shortage of drinking water due to its large population. The eastern states of India have great potential for producing water from the atmosphere. In these areas, the average temperature is 20–30 °C, and the average humidity is 75–85%. Under these conditions, it is possible to produce up to 3 kg of water per hour at an approximate cost of 3 rupees per kg of water produced (Patel *et al.*, 2020).

1.5.5 Inter-basin water transfer

Behesht Abad is an inter-basin water transfer project in Iran that transfers water from the Karun basin to the Gavkhuni basin. The transmission line is 65 kilometers long and 6 meters in diameter, designed to transport one billion m^3 of water. Nevertheless, after pressure from experts in the source basin, the quantity transported has fallen to 580 million m^3.

The sustainability review of this plan shows that inter-basin water transfer may improve stability, while water resources management is appropriately implemented (Kefayati *et al.*, 2018). A MODSIM-based optimization model, coupled with game theory, has recently been proposed to resolve disagreements between different stakeholders (MODSIM is a generalized river basin network flow model). Results show that with an average annual payment of $250.66 million as compensation from the Gavkhuni basin to the Karun basin, both basins would be in an optimal and balanced condition (Ahmadi *et al.*, 2019).

1.5.6 Water reuse

Three water reuse examples are described below, including of wastewater treatment, greywater, and stormwater harvesting:

- in 2014, it was estimated that about 45 million m^3 of municipal sewage were discharged daily in the United States, which could be a sustainable source of water. For example, three wastewater treatment plants in South Florida recycle up to 90% of the water consumed (Mohan *et al.*, 2014);
- the greywater reuse in a student dormitory in Turkey shows that collecting and treating bath and washing wastewater provides up to a 25% daily water-saving and supplies the water needed for toilet flushing (Giresunlu & Beler Baykal, 2016).
- a project in an Australian city to collect 69 million liters of stormwater per year was investigated. This water is used for the irrigation of a city park after initial treatment and disinfection with UV (Furlong *et al.*, 2017).

1.6 SUMMARY

Increasing population and water demand, along with water resource limitations, are some of the most critical challenges faced today and will continue to be so in the future. Some new

Table 1.1 Requirements and limitations of new water resources.

Water Resource	Requirements	Limitations
Cloud seeding	Access to essential technologies and raw materials	There must be suitable clouds in the area
Fossil water	Access to deep drilling and extraction technologies	It does not exist in all regions
Desalination	Economical in terms of energy consumption	It is suitable for coastal areas
Atmospheric water	Uses advanced technologies	It is suitable for areas with sufficient moisture
Inter-basin water transfer	Construction of water transfer structures	It is suitable for basins with low levels of water adjacent to basins with very high water levels
Water reuse: wastewater treatment	Construction of a sewage collection system and a treatment plant with continuous monitoring	Some religious, cultural, and social views oppose the recycling of sewage
Water reuse: greywater	Modification or construction of new buildings with a greywater reuse approach	Can cause adverse effects on plants and health problems
Water reuse: stormwater	Construction of stormwater collection facilities	Not suitable for dry areas with low rainfall

Table 1.2 Advantages and disadvantages of new water resources.

Water Resource	Advantages	Disadvantages
Cloud seeding	Natural water supply with increased rainfall, resulting in increased surface runoff	Uncertainty of the method performance; possibility of creating unknown climatic effects
Fossil water	A high volume of water supply in susceptible areas	A non-renewable resource; unknown environmental impacts
Desalination	No restrictions on water supply	High energy consumption; the need for proper disposal of highly saline wastewater
Atmospheric water	No environmental impacts	Initial investment; low volume of water supply
Inter-basin water transfer	Balances the distribution of water resources	Causing multiple problems in the source basin
Water reuse: wastewater treatment	A sustainable source with the potential to produce methane gas as an energy source	People's Health Problems; the need for constant water quality control
Water reuse: greywater	Saves water and reduces municipal wastewater volume	Insufficient supervision for quality control
Water reuse: stormwater	A significant volume of water can be reused	An unstable source; limitations in arid areas

Table 1.3 Applications of water produced from different resources.

Water Resources	Drinking	Industry	Agriculture	Groundwater Recharge	Environmental Demand	Other
Cloud seeding	•	•	•	•	•	–
Fossil water	•	•	•	–	–	–
Desalination	•	•	•	–	–	–
Atmospheric water	•	–	–	–	–	–
Inter-basin water transfer	•	•	•	–	•	–
Water reuse: wastewater treatment	–	•	•	•	•	–
Water reuse: greywater	–	–	•	–	•	•
Water reuse: stormwater	–	•	•	•	•	•

• Used; Not Used.

Table 1.4 Cost and water quantity supplied for each water resource.

Water Resource	Cost	Water Supply
Cloud seeding	\$0.5–19 M/year (Bruintjes, 1999)	About 5–30% increase in precipitation (Wu *et al.*, 2018)
Fossil water	More than \$0.4/m^3, depends on the depth (Hellegers *et al.*, 2011)	1.5×10^{10} to 6×10^{13} m^3 (Tsur *et al.*, 1989)
Desalination	€0.63–0.72/m^3 (Lapuente, 2012)	Limited to the potential of the desalination system
Atmospheric water	\$0.2–0.4/L (Elashmawy, 2019)	Up to 200 g/day (Fathieh *et al.*, 2018)
Inter-basin water transfer	€0.5/m^3 (Pouget *et al.*, 2012)	Limited to the potential of the source basin and the capacity of the hydraulic structures
Water reuse: wastewater treatment	€0.3–0.37/m^3 (Miller, 2006)	Limited to the potential of the wastewater treatment plant and the population of the city
Water reuse: greywater	About \$1/m^3 when the design flow is 9 m^3/h (Friedler, 2008)	About 30% of the total water consumed (De Gisi *et al.*, 2017)
Water reuse: stormwater	Depends on the volume of the precipitation and the catchment area	Limited to the precipitation and the potential of the catchment area; up to 4700 ML/year, (Furlong *et al.*, 2017)

water sources are briefly explored in this study. As a comparison, the requirements and the limitations of each method are presented in Table 1.1. Some methods, such as cloud seeding, fossil water, and water extraction from humid air, require the use of up-to-date technologies. However, suitable facilities must be constructed and utilized to use some other methods, such as inter-basin water transfer and water reuse. There are natural limitations to some of the methods. Methods, such as fossil water extraction and desalination, are only specific to particular areas. Also, cloud seeding, water extraction from humid air, and stormwater reuse are not suitable for all areas.

The advantages and disadvantages of each method are presented in Table 1.2. In addition to the limitations and requirements of each method, it is essential to pay attention to the volume of water produced, the sustainability of the source, the environmental effects, the initial costs, and the energy consumed.

In Table 1.3, the uses of water produced by each method are categorized. In general, it is recommended to use the atmospheric water and desalination methods (if available) for drinking water supply. In addition, inter-basin water and wastewater treatment can be used for agriculture and industry. Greywater and stormwater are also good options for artificial groundwater recharge, environmental demand, and irrigation.

In general, based on the limitations of most methods, it seems reasonable to use all methods simultaneously. As a summary, the following suggestions are generally recommended:

- the use of fossil water (if available) must be limited to only necessary conditions due to its non-renewability;
- the method of cloud seeding is suggested as a complementary method, if sufficient studies, proper efficiency, and economic justification are available;
- the cloud seeding method gives more and wider efficient alternatives of water usage owing to its natural phenomena. Hence, it is preferable in comparison to the other methods;
- despite the availability of these various water resources, the need to pay attention to consumption management should not be forgotten.

Further, the amount of water supply and the economic evaluation of each method depends on the environmental conditions of the region. Therefore, it is not possible to compare the costs and the water supply volumes of different methods, except in a general way. Hence, Table 1.4 reports an estimation of the costs and water supply volumes for each resource.

It should be noted that the use of different methods is highly dependent on the conditions of the region. Some specific methods may be of higher priority for a particular region in comparison with other methods.

REFERENCES

Ahmadi A., Zolfagharipoor M. A. and Afzali A. A. (2019). Stability analysis of stakeholders' cooperation in inter-basin water transfer projects: a case study. *Water Resources Management*, **33**(1), 1–18, doi: 10.1007/s11269-018-2065-7

Ben-Zvi A. (1988). Enhancement of runoff from a small watershed by cloud seeding. *Journal of Hydrology*, **101**(1), 291–303, doi: 10.1016/0022-1694(88)90041-8

Ben-Zvi A. and Fanar A. (1997). The effect of randomized cloud seeding on runoff depth from a small watershed in central Israel. *Journal of Hydrology*, **191**(1), 161–178, doi: 10.1016/S0022-1694(96)03073-9

Ben-Zvi A. and Langerman M. (1989). Estimation of flow enhancement in the Kinneret watershed due to cloud seeding. *Stochastic Hydrology and Hydraulics*, **3**(3), 203–216, doi: 10.1007/BF01543860

Ben-Zvi A. and Langerman M. (1993). Assessment of runoff enhancement by randomized cloud seeding in case of a carry-over flow. *Journal of Hydrology*, **142**(1), 391–408, doi: 10.1016/0022-1694(93)90020-A

Benjamini Y. and Harpaz Y. (1986). Observational rainfall – runoff analysis for estimating effects of cloud seeding on water resources in northern Israel. *Journal of Hydrology*, **83**(3), 299–306, doi: 10.1016/0022-1694(86)90158-7

Bixio D., Wintgens T., Ravazzini A., Thoeye C., Aharoni A., Koning J. D. and Melin T. (2006). Water reuse. In: Wastewater Quality Monitoring and Treatment, P. Quevauviller, O. Thomas and A. V. D. Beken (eds), Wiley, UK, pp. 329–349.

Bruintjes R. T. (1999). A review of cloud seeding experiments to enhance precipitation and some new prospects. *Bulletin of the American Meteorological Society*, **80**(5), 805–820, doi: 10.1175/1520-0477(1999)080<0805:AROCSE>2.0.CO;2

Caldera U., Bogdanov D., Afanasyeva S. and Breyer C. (2018). Role of seawater desalination in the management of an integrated water and 100% renewable energy based power sector in Saudi Arabia. *Water*, **10**(1), doi: 10.3390/w10010003

De Gisi S., Casella P. and Notarnicola M. (2017). Grey water. In: Encyclopedia of Sustainable Technologies, M. A Abraham (ed.), Elsevier, Oxford, pp. 77–89.

Dennis A. S. (ed.) (1980). Weather Modification by Cloud Seeding. International Geophysics Series 24. Elsevier, ScienceDirect.

Ehteram M., Salih S. Q. and Yaseen Z. M. (2020). Efficiency evaluation of reverse osmosis desalination plant using hybridized multilayer perceptron with particle swarm optimization. *Environmental Science and Pollution Research*, **27**(13), 15278–15291, doi: 10.1007/s11356-020-08023-9

Elashmawy M. (2019). Experimental study on water extraction from atmospheric air using tubular solar still. *Journal of Cleaner Production*, **119322**, doi: 10.1016/j.jclepro.2019.119322

Eslamian S. (2015). Urban Water Reuse Handbook. CRC Press, Boca Raton, USA.

Fathieh F., Kalmutzki M. J., Kapustin E. A., Waller P. J., Yang J. and Yaghi O. M. (2018). Practical water production from desert air. *Science Advances*, **4**(6), eaat3198, doi: 10.1126/sciadv.aat3198

Friedler E. (2008). The water saving potential and the socio-economic feasibility of greywater reuse within the urban sector – Israel as a case study. *International Journal of Environmental Studies*, **65**(1), 57–69, doi: 10.1080/00207230701846697

Furlong C., De Silva S., Gan K., Guthrie L. and Considine R. (2017). Risk management, financial evaluation and funding for wastewater and stormwater reuse projects. *Journal of Environmental Management*, **191**, 83–95, doi: 10.1016/j.jenvman.2017.01.007

Getches D. H. (2005). Interbasin Water Transfers in the Western United States: Issues and Lessons. Water Conservation, Reuse, and Recycling: Proceedings of an Iranian–American Workshop, the National Academies Press, Washington, DC.

Ghassemi F. and White I. (2007). Inter-Basin Water Transfer: Case Studies from Australia, United States, Canada, China and India. Cambridge University Press, UK.

Giresunlu E. and Beler Baykal B. (2016). A case study of the conversion of grey water to a flush water source in a turkish student residence hall. *Water Supply*, **16**(6), 1659–1667, doi: 10.2166/ws.2016.078

Goldreich Y. (1994). Applications of spatial analysis techniques for assessing the contribution of cloud seeding to rainfall in Israel. *Physical Geography*, **15**(6), 573–593, doi: 10.1080/02723646.1994.10642536

Goonetilleke A., Liu A., Managi S., Wilson C., Gardner T., Bandala E. R. and Rajapaksa D. (2017). Stormwater reuse, a viable option: fact or fiction? *Economic Analysis and Policy*, **56**, 14–17, doi: 10.1016/j.eap.2017.08.001

Hellegers P., Perry C. J. and Al-Aulaqi N. (2011). Incentives to reduce groundwater consumption in Yemen. *Irrigation and Drainage*, **60**, 93–102, doi: 10.1002/ird.538

Issar A. S. (1994). Fossil water under the deserts of the Middle East: a safeguard against strife. *Studies in Environmental Science*, **58**, 351–361.

Jiménez-Cisneros B. (2014). 3.15: water reuse and recycling. In: Comprehensive Water Quality and Purification, S. Ahuja (ed.), Elsevier, Waltham, pp. 296–323.

Kefayati M., Saghafian B., Ahmadi A. and Babazadeh H. (2018). Empirical evaluation of river basin sustainability affected by inter-basin water transfer using composite indicators. *Water and Environment Journal*, **32**(1), 104–111, doi: 10. 1111/wej.12304

Keyes C. G., Jr, Bomar G. W., DeFelice T. P., Griffith D. A. and Langerud D. W. (2006). Guidelines for Cloud Seeding to Augment Precipitation. American Society of Civil Engineers, Reston, VA.

Lapuente E. (2012). Full cost in desalination. A case study of the segura river basin. *Desalination*, **300**, 40–45, doi: 10.1016/j.desal.2012.06.002

McNabb D. E. (2015). Aqueduct Projected Water Stress Country Rankings, Technical Note. World Resources Institute, Washington, DC.

McNabb D. E. (2019). The population growth barrier. In: Global Pathways to Water Sustainability, D. E. McNabb (ed.), Springer International Publishing, Cham, Switzerland, pp. 67–68.

Miller W. G. (2006). Integrated concepts in water reuse: managing global water needs. *Desalination*, **187**(1), 65–75, doi: 10.1016/j.desal.2005.04.068

Mohan G. R., Speth T. F., Murray D. and Garland J. L. (2014). Municipal wastewater: a rediscovered resource for sustainable water reuse. In: Potable Water: Emerging Global Problems and Solutions, T. Younos and C. A. Grady (eds), Springer International Publishing, Cham, Switzerland, pp. 153–179.

Ngo Huu H., Guo W., Surampalli Rao Y. and Zhang Tian C. (2016). Green Technologies for Sustainable Water Management. American Society of Civil Engineers, Reston, VA, USA.

Oki T. and Kanae S. (2006). Global hydrological cycles and world water resources. *Science*, **313**, 1068–1072.

Patel K., Patel J. and Raval H. (2020). Potential study of atmospheric water generator (AWG) for humid climatic conditions of eastern states in India. In: Renewable Energy and Climate Change, D. Deb, A. Dixit and L. Chandra (eds), Springer, Singapore, pp. 239–248.

Pepper I. (2016). Graywater reuse for irrigation: benefits and potential hazards. In: Sowing Seeds in the City: Ecosystem and Municipal Services, S. Brown, K. McIvor and E. Hodges Snyder (eds), Springer, Dordrecht, The Netherlands, pp. 69–77.

Pouget L., Escaler I., Guiu R., Mc Ennis S. and Versini P.-A. (2012). Global change adaptation in water resources management: the water change project. *Science of the Total Environment*, **440**, 186–193, doi: 10.1016/j.scitotenv.2012.07.036

Sajjadi S. A. (2008). The precipitable water and determination of cloud seeding threshold through thermodynamic graphs and MODIS satellite in Tabriz. *Journal of Applied Sciences*, **8**(7), 1299–1304, doi: 10.3923/jas.2008.1299.1304

Service R. F. (2006). Desalination freshens up. *Science*, **313**(5790), 1088–1090, doi: 10.1126/science.313.5790.1088

Shiklomanov I. A. (1988). World Water Resources: a New Appraisal and Assessment for the 21st Century. UNESCO, Paris, France.

Tlili I., Alkanhal T. A., Othman M., Dara R. N. and Shafee A. (2019). Water management and desalination in KSA view 2030. *Journal of Thermal Analysis and Calorimetry*, **139**(6), 3745–3756, doi: 10.1007/s10973-019-08700-z

Tsur Y., Park H. and Issar A. (1989). Fossil groundwater resources as a basis for arid zone development? *International Journal of Water Resources Development*, **5**(3), 191–201, doi: 10.1080/07900628908722433

Tu Y., Wang R., Zhang Y. and Wang J. (2018). Progress and expectation of atmospheric water harvesting. *Joule*, **2**(8), 1452–1475, doi: 10.1016/j.joule.2018.07.015

Wang W., Yao Z., Guo J., Tan C., Jia S., Zhao W. and Gao L. (2019). The extra-area effect in 71 cloud seeding operations during winters of 2008–14 over Jiangxi Province, East China. *Journal of Meteorological Research*, **33**(3), 528–539, doi: 10.1007/s13351-019-8122-1

Wheida E. and Verhoeven R. (2007). An alternative solution of the water shortage problem in Libya. *Water Resources Management*, **21**(6), 961–982, doi: 10.1007/s11269-006-9067-6

Wu X., Yan N., Yu H., Niu S., Meng F., Liu W. and Sun H. (2018). Advances in the evaluation of cloud seeding: statistical evidence for the enhancement of precipitation. *Earth and Space Science*, **5**(9), 425–439, doi: 10.1029/2018EA000424

Chapter 2

Inter-basin water transfer

Shima Kheirinejad[1], Omid Bozorg-Haddad[2] and David Quanrud[3]*

[1]M.Sc. Student, Department of Irrigation & Reclamation Engineering, Faculty of Agricultural Engineering & Technology, College of Agriculture & Natural Resources, University of Tehran, Karaj, Tehran, Iran. E-mail: Shima.kheyri20@ut.ac.ir
[2]Distinguished Professor, Department of Irrigation & Reclamation Engineering, Faculty of Agricultural Engineering & Technology, College of Agriculture & Natural Resources, University of Tehran, Karaj, Tehran, 31587-77871, Iran. E-mail: OBHaddad@ut.ac.ir
[3]Research Scientist, School of Natural Resources and the Environment, The University of Arizona, Tucson, Arizona, USA. E-mail: quanrud@email.arizona.edu
**Corresponding author.*

ABSTRACT

Providing water to satisfy human demands is one of the greatest challenges of the present century and has been the source of many changes in the world. Since the amount of renewable water the Earth receives is only equal to, or even less than, that received many years ago, renewable water per capita decreases as human population increases. Inter-basin water transfer is a method to augment supply in water-scarce regions.

Increasing demand for water in some areas sometimes justifies the transfer of water from one basin to another. Water consumption in densely populated cities and in large industrial and agricultural operations produces an economic value that justifies the cost of expensive water transport equipment and infrastructure such as transmission channels, pipelines, pumping stations and dams. Under these conditions, water that is not in high demand in one basin is transferred to another basin for use. These projects are supply-driven engineering solutions to major social challenges. Inter-basin water transmission is carried

Editor: Omid Bozorg-Haddad
doi: 10.2166/9781789062144_0015

out across local, regional, national, and international boundaries to overcome water scarcity, to meet demand in the agricultural, industrial and domestic sectors as well as to achieve economic and social development.

Inter-basin transfer (IBT) is an important water resource management tool that can have significant impacts on the donor and recipient basins. This chapter aims to provide insights into the concepts, logics, methods and tools used to evaluate inter-basin transfer projects. Challenges that may arise with implementation of such projects and management methods to overcome these challenges are reviewed. Several case study examples of existing projects are presented to provide insight into how to better manage such projects in the future.

Keywords: Inter-basin water transfer, Water resources.

2.1 INTRODUCTION AND LITERATURE REVIEW

This chapter provides an overview of studies in this field conducted to date and highlights lessons to be learned from developed IBT projects that can serve to guide future planning. It begins with a review of previous literature on the topic of IBT.

Davies *et al.* (1992) assessed IBTs that existed prior to 1992 and examined case studies from three regions: Southeastern Australia, South Africa, and Central and Southwestern United States. Importantly, they noted a need for environmental impact studies on IBT on a global scale. Changes in water quality, extinction of native plant and animal species, climate change, increasing disease, changes in hydrological regimes, and other environmental impacts were noted as information gaps, prompting Davies *et al.* (1992) to propose an international meeting on IBTs to assess the extent of the expansion, geographical distribution, and ecological and sociological impacts and consequences.

Significant advances in engineering over time have permitted larger water transfer volumes and longer travel distances, making IBTs an increasingly important water supply management tool internationally (Snaddon *et al.*, 1999). However, evaluation of IBT projects based solely on their effects on water systems and human communities is not sufficient. Snaddon *et al.* (1999) emphasized that evaluation of IBT projects should consider two important issues: (1) increasing our knowledge of the distribution, operation and impact of IBTs at national, international (especially Australian and American) and global scales; and (2) understanding the ecological (physical, chemical, and biological), socio-political and socio-economic effects of IBTs. Specifically, the environmental impacts of an IBT project depend on several factors, including the types of donor basin, recipient basin and transmission route. In all of these projects, water transfer causes changes such as in water quality, temperature regime, and mixing of organisms in the basins involved.

Snaddon *et al.* (1999) also provided an overview of IBT-related studies, in particular their effects on freshwater ecosystems, with an emphasis on IBTs located in South Africa, the United States and Australia. They also highlighted problems and conflicts that have arisen through IBT projects around the world and provided recommendations for planning and managing these projects.

Ghassemi and White (2007) examined IBT projects constructed under different political, social, economic, and geographical conditions in Canada, United States, Australia, China, and India. They assessed strengths and weaknesses of these projects and provided recommendations for future projects, concluding that it is essential to have a comprehensive vision and a thorough understanding of the situation and potential of the area, and to also consider the reasons and conditions of successful and unsuccessful projects.

Gupta and van der Zaag (2008) evaluated IBT projects from a multidisciplinary perspective using the concept of integrated water resources management (IWRM). They reviewed recommendations developed by international commissions, policy societies, and scientists to formulate a coherent set of evaluation criteria for inter-basin transmission plans.

Several factors, including differing stakeholder interests, system physical constraints and socio-economic criteria complicate the optimal allocation of shared water resources in water transmission projects. Sadegh *et al.* (2010) developed a new approach for optimizing water resource allocations between basins based on the method of crisp and fuzzy Shapley games and assessed the effectiveness of this approach in a large case study of water transfer from the Karoon River Basin in southern Iran to the Rafsanjan Plain in central Iran. They found that their proposed method can rationally allocate water, and benefits water users in IBT projects. A set of water allocation times for all water consumers and related ancillary payments are also provided according to the criteria of justice and efficiency.

Rezaei and Basirzadeh (2011) examined challenges facing IBT schemes from the perspective of land use planning and IWRM. Their main purpose was to investigate the role of these plans in sustainable development in a case study in Khuzestan, Iran. They found that food security, energy security, border security and national security are all closely linked, concluding that IBT has broad economic and social implications and can affect national security, and also that many potential tensions could be avoided with a win-win scenario.

In a review of IBT projects in India, China and Australia, Boddu *et al.* (2011) described project specifications and operating conditions, assessing their advantages and disadvantages. They found that while IBT projects provide many benefits, they may also cause negative environmental effects. For example, adverse effects of an IBT project on the Snowy River in Australia include reduction in water flow, clogging of the river channel with silt, disruption of fish migration and salt water infiltration into estuaries.

Maknoon *et al.* (2012) examined three proposed protocols in water resources management (in the face of climate change) in an IBT scheme from Dez to Qomrood Basin in Iran. Based on economic analysis and a project efficiency assessment, the Fix Downstream (FD) allocation protocol was found to best provide the water required by a larger population at a fixed cost.

Karakaya *et al.* (2014), examining IBT projects in Turkey, found it important to consider socio-economic and environmental challenges and the spatial and temporal scale of IBT projects' implementation. They also emphasized considering available management alternatives for providing water supply and managing demand, and that qualitative and quantitative assessment indicators of transferred water should both be considered in water resources management decision making.

In a study on IBT environmental impacts, Zeng *et al.* (2015) investigated physical, chemical, biological and hydrological changes in basins receiving water transferred 1267 km from south to north China. The effects of 13 process management scenarios on the water quality variables nitrogen (N), phosphorus (P), and chlorophyll a (Chl a) were examined. Results indicated that in order to avoid impacts on N, P and Chl a, IBT projects should be designed to maintain the flow uniformity of transferred water as much as possible and to also minimize the flow rate of transferred water.

To fully consider all potential positive and negative aspects of IBT projects, several methods, algorithms and approaches have been proposed to optimize their design. Peng *et al.* (2015) proposed three multi-core parallel particle swarm optimization (PPSO) algorithms (PPSO_ring, PPSO_star, and PPSO_share) to optimize a joint operation model for inter-basin water transfer systems. Their results show that the PPSO_share algorithm can increase computational efficiency and accuracy of the collaborative operation model solution, and demonstrate its feasibility and potential validity for future multi-reservoir system operations.

Based on participatory game theory and the concept of virtual water, Manshadi *et al.* (2015) developed a modeling approach considering economic, justice and environmental criteria to assess how well IBT projects can lead to sustainable development. They applied their model to assess an IBT project from Solakan to Rafsanjan Basin in Iran and showed that the use of game theory doubled the net profit of the project in the case of initial allocation.

The imbalance between distribution of and demand for water resources across human society have made water transfers inevitable. Zhuang (2016) noted that basins involved in IBTs may experience positive or negative environmental impacts. Positive effects in receiving basins include water cycle facilitation, improved meteorological conditions, reduced environmental water shortages, improved environmental system conditions and conservation of endangered wildlife and plants. Negative effects in donor basins include salinity changes and environmental damage. Negative effects in receiving basins include increased water consumption and spread of diseases. Because IBTs may pose significant environmental risk, it is necessary to analyze the water balance relationship between the basins and to coordinate potential differences and environmental quality problems across regions. In addition, there are some effective alternatives to IBTs, such as demand management, improving water use efficiency, freshwater production from seawater, rainwater harvesting, water reuse, and more.

Basin water transfers may significantly impact freshwater environments. Berbel-Filho *et al.* (2016) examined effects of inter-basin transmission on fish species in the Katina freshwater coastal area of northeast Brazil located between an inter-basin water diversion connecting water of the Sao Francisco coast with the four main northeast basins of Caitina. They found significant morphological differences in freshwater fish among basins and concluded variable environments and probable economic developmental damage may result from artificial river connection.

Cooperation of a multi-reservoir system in a water transfer project is a challenging problem due to the complex structure and policy of the cooperation. Wan *et al.* (2018) developed a joint optimization operational model with respect to water supply and transmission and proposed the concept of 'shape constraints'. An IBT project in northeast China, consisting of three tracks and eight reservoirs, was selected as a case study. Progressive Reservoir Algorithm-Particle Swarm Optimization (PRA-PSO) was used to obtain the solution, and the results show that this method is more promising than other optimization techniques. This case study also shows that the new optimization method can effectively increase the chances of jumping from local optimized points, thus looking for better solutions.

Alahdin *et al.* (2019) believed that water allocation is the most important cause of disputes (whether between independent nations, different user groups, or neighboring basins) in the field of water resources, and in IBT schemes the recipient basin usually benefits and the donor basin suffers from problems. Since there is no single solution that optimizes all goals simultaneously in a multi-objective optimization problem, decision makers are concerned about finding the only compromise solution that balances the conflicting goals in a socially acceptable way. Because game theory can identify and interpret the behavior of parties in water resources, Alahdin *et al.*'s (2019) paper is used in conjunction with optimization. In their study, an inter-basin water transfer project from Zohreh basin (multi-reservoir and multi-purpose) in southwestern Iran to the coastal region of the Persian Gulf was investigated using this system.

To create more rapid development of water supply chain management in China, the eastern and central routes of the South-North Water Transmission Project are considered as two water supply sources. Du *et al.* (2019) developed a chain-to-chain competition model based on a two-stage Stackelberg game in three alternate scenarios, including wholesale price contracts, two-part pricing contracts, and mixed pricing contracts, achieving optimal pricing and ordering.

Although groundwater supplies fresh water for drinking and irrigation and maintains ecosystem health, serious consequences of unstable groundwater discharge have been widely reported, and limiting pumping in watersheds requires the identification of alternative sources. It is important to determine how much water can be made available to prevent groundwater crises, and to formulate rules for allocating water to achieve regional water sustainability. Yao *et al.* (2019) reviewed and identified shortcomings of the groundwater pumping control policies of the South China North Water Transmission Project and suggested an integrated quantification framework and a policy reform approach.

The dynamic interactions between multiple stakeholders in the supply chain in IBT from the perspective of social welfare maximization (SWM) have not been adequately studied in the past. Chen *et al.* (2019) created, analyzed, and compared IBT equilibrium and coordination decision models with respect to concerns about fairness in scenarios with or without the goal of SWM.

2.2 DEFINITIONS

Inter-basin transfer (IBT): an IBT can be defined as the physical transfer of water from one basin to another basin, during which displacement a catchment loses (origin) and another catchment receives water (destination). From the point of view of political divisions, the source and destination watersheds may be located in one province, more than one province or more than one country. Any transfer of surface, underground or sewage water to the outside of the river basin at the source of the transmission, if it crosses the boundaries of a basin, shall constitute an IBT. IBT involves the removal of surface or groundwater from one basin to another to provide beneficial uses.

Other terms that describe the transfer of water from an area of surplus to an area where demand exceeds or will soon exceed available resources include the following: remote water transmission, interregional water transmission, interstate transmission, large-scale water transfer. Quinn (1981), in guidelines developed to identify IBTs in Canada, proposed two main criteria:

(1) the diverted flow does not return to the stream of origin (or parent stream) within 24 km of the point of withdrawal; and
(2) the annual average flow transferred shall not be less than 0.71 m^3 per second or 1,233,500 m^3.

Since the concept of sustainable water resources has been used extensively in the definitions of standardized IBT, this concept should also be defined. Hence, Sustainable water resources:

(1) maintain, improve and create the right life for every individual in society;
(2) preserve, improve and sanitize the environment; and
(3) maintain, improve and create a healthy economy in which equity, self-reliance and adequate production are met to meet needs.

A basin is an area of land in which all water drains to a single outlet at the lowest elevation, due to the slope and shape of the land. The land is called the basin from the beginning of the branches to the end of the lowlands. A donor basin is a basin that donates surplus water to a basin that lacks sufficient water, based on expert reviews. A recipient basin is a basin that receives water transferred from a donor basin, based on expert reviews.

2.3 BASICS AND LOGIC

2.3.1 UNESCO criteria for inter-basin water transfer

In 1987, a report entitled 'Our Common Future' was presented to the United Nations Global Environment and Development Group by Dr Gro Harlem Brundtland (United Nations, 1987) in which the concept of environmental sustainability was defined as follows: 'Providing the needs of present and future generations by preserving their rights while maintaining the cultural, ecological and hydrological integrity of communities.'

Subsequently, at the UNESCO Workshop on Interbasin Water Transfer in April 1999 in Paris, the following specific criteria were identified as essential for the implementation of water transport projects (Cox, 1999):

(1) the destination area, despite the use of alternative sources of water supply and all reasonable measures to reduce its demand, is still severely deficient in meeting current and projected needs;
(2) the recipient basins are either currently or in the near future affected by a water crisis, or the development of the region in various respects is dependent on water supply through transmission. Also, the transfer of water from donor basins does not present a serious crisis in the present or future development;
(3) the project should not have significant and irreparable damage to the environment. A comprehensive assessment of the environmental consequences must show that there is a reasonable level of certainty that the transmission plan does not substantially impair the environmental quality in the recipient or donor basin. However, if environmental compensation costs are provided, the transmission plan may be justified;
(4) a comprehensive assessment of social and cultural consequences must show that there is a reasonable level of certainty that the transfer plan will not cause significant social and cultural disruption in the recipient or donor basin. If the project has a lot of social issues in the recipient or donor basin or along the route, it should not be implemented;
(5) the net benefits arising from the implementation of the plan should be distributed equitably among the recipient and donor basins. The plan should have an economic return, but the benefits should not be spent solely on the project executor or on the recipient basin.

Because sustainability is a function of various economic, environmental, social, and physical goals, water resource management must address multi-objective issues through multidisciplinary and multi-participatory decision-making approaches, and water resource management systems must be programmed in a way that water and related land resources are not destroyed over time. As a result, restrictions should be applied at every stage of development, from project planning to final operation and management in its overall social and technical system (Loucks, 2000). Therefore, in order to maintain the sustainability of IBT projects, the following aspects should be considered, so that the best plan can be implemented:

(1) project infrastructure in the area;
(2) the impact of the project on controlling running waters, especially cross-border waters;
(3) dependence of industries, agriculture and drinking water on transitional water in recipient or donor basins;
(4) potential to improve agricultural lands, and increase productivity, entrepreneurship and job creation, and prevent unemployment.
(5) investigation of the potential of the donor basins;
(6) evaluation of water quality in the donor basin from source to downstream and environmental requirements of discharge including sea, wetland, etc;
(7) investigation of changes in water level in the donor basin and land subsidence if groundwater levels fall;
(8) assessment of the effects of the project on the elimination or destruction of tourist attractions;
(9) establishment of social justice in the allocation of basin water and the need to eliminate deprivation in the recipient or donor basins;
(10) occurrence of adverse social impacts and irregular migration from the donor basin and focus on recipient basins due to the presence of water, and the integrated role of water management in the country;
(11) project impact on land use planning;
(12) people's satisfaction;
(13) social security;
(14) social welfare;
(15) environmental protection;
(16) sustainable operation;
(17) economic productivity.

Other considerations for sustainable water transfer schemes include indicators such as efficiency, effectiveness, adaptability, reversibility and vulnerability, and these must be taken into account to ensure project sustainability over the long term (Loucks, 2000).

2.4 CHALLENGES FACING INTER-BASIN WATER TRANSPORT

Inter-basin water transfer projects have significant impacts on economic, social, environmental, technical and administrative contexts. Feasibility studies are necessary to achieve successful water transfer between basins and will facilitate more efficient managerial decision making. In order to optimize decisions with greater awareness and confidence, Cox (1999; Section 4.1) outlined five principles for evaluating the feasibility of inter-basin water transmission projects: the first and second principles are related to economic issues, the third is related to environmental issues, the fourth is related to social issues and the fifth is related to the equitable distribution of profits from project implementation in two basins.

Due to the extent of the affected areas and the need for a comprehensive and long-term approach to estimating water transfer benefits, costs and risks, significant challenges are created for planners and managers. To this end, Cox (1999), in collaboration with UNESCO, proposed criteria for the above-mentioned factors and the distribution of benefits in these areas:

(1) Technical feasibility of the project. Project feasibility studies should consider the length of transmission route, water losses along the route, volume of useful water received, groundwater loss in the donor basin, stability of transmission channel walls, liquefaction potential of sand, required height of the probable dam, and ultimately the system simulation and modeling to determine the appropriate options. Studies show that the failure of many IBT projects is related to insufficient planning. One of the important issues to recognize during planning is that environmental assessments should not be a function of technical and economic aspects (Davies *et al.*, 1992; Greer, 1983);

(2) Economic impacts. Costs and benefits (tangible and intangible) are considered in the economic aspects of projects. IBT projects incur significant capital costs and the benefits must be shared among donor and recipient basins. The most important issues to consider in the economic evaluation are damages, including loss of homes, farmland, and people's livelihoods, that are caused during implementation of water transfer projects, demolition and loss of land to build transmission canals, income losses associated with water removed from the donor basin, and income losses due to land use and cultivation pattern changes in both donor and recipient basins.

(3) Social effects. Social impacts are often the most significant negative effect of an IBT and a project cannot be implemented unless communities are socially calm and stable. The most important societal adverse effects include joblessness due to loss of arable land and pastures leading to rural migration to cities. One way to counteract this is to create sustainable jobs in an affected region. Obtaining social acceptance is no easy task but can be achieved with some consideration. One of the actions that can be taken to attract social acceptance is to distribute the profits from the water transfer to both the recipient and donor basins;

(4) Environmental effects. Environmental impacts must be considered before implementing an inter-basin water transfer. There are many examples of IBT projects that have caused irreparable environmental damage. In recent years, researchers have examined impacts in donor and recipient basins from a variety of perspectives, especially environmental considerations (Ben Fraj *et al.*, 2019; Tien Bui *et al.*, 2020; Wu *et al.*, 2020). Taken together, the results of these studies show that implementation of IBT projects can present some challenges, especially in donor basins. Important potential environmental impacts of inter-basin water transport include:

- aquifers: an inter-basin transfer of surface water may impose a negative impact on hydraulically connected groundwater in the donor basin. If natural aquifer recharge is provided by a river, then a decrease in the flow of that river will reduce the volume of groundwater recharge;

- water quality: Transferring water from one river to another river means moving a physico-chemical regime from one region to another, causing changes in various parameters affecting water quality. Water transfer can cause fundamental changes in hydrological balance because there is a close interaction between quantity and water quality (Das, 2006). This may have adverse impacts including sedimentation as a result of reduced discharge in the donor basin or as a result of water transferred across saline or contaminated soils (from mines), leading to water quality decline and lack of suitability for the end user. Excessive groundwater recharge by unintended channel leakage may cause a rise in groundwater level, which may lead to soil salinization and swampiness problems in a large area along the transmission route (Shao *et al.*, 2003);
- aquatic habitat: inter-basin water transfers can alter the volume and timing of river flow, potentially impacting aquatic habitats. Changes in water temperature or dissolved oxygen may stress or threaten certain aquatic species. Creation of dams may make it impossible for fish to migrate (Das, 2006; Snaddon *et al.*, 1998). Collectively, these changes can result in loss of biodiversity and loss of sustainable water resource conditions.

In arid regions, negative impacts of inter-basin water transfers can be more severe. Therefore, greater scrutiny by expert studies is warranted to identify their necessity and the extent of exploitation. Potential negative impacts of improper inter-basin water transmission in arid areas include the following:

(1) hydrological balance disruption. Improper inter-basin water transfer can disrupt the hydrological balances in both donor and recipient basins, leading to adverse effects such as dam filling, flooding, drought, loss of wetlands and salinity changes;
(2) lack of permanent water flow in small areas of arid regions. In arid regions, there are usually no permanent rivers, which are the most important source of energy, surface water and groundwater for agriculture and domestic use, and support ecosystems. Most water resources in arid areas are shallow or deep aquifers and unmanaged transfer of groundwater from an arid basin may reduce or eliminate natural groundwater discharge, destroying dependent aquatic or riparian habitat and leading to devastating ecosystem effects;
(3) increased hydraulic gradient of groundwater flow. With increasing hydraulic gradient in the basin after untreated water transfer, the upstream lands of the basin rapidly lose their water and the effects of drought and scarcity are evident throughout the basin.

2.5 SOLUTION AND MANAGEMENT OF PROBLEMS CAUSED BY INTER-BASIN WATER TRANSFER

2.5.1 Integrated water resources management of inter-basin water transfers

The combined effects of a rapid rate of urbanization, increased prosperity and public health, industrial growth, and global population growth (expected to reach 8 billion (8×10^9) by

2030) are resulting in increased demand for water, food and energy, which all impact on water resources. In addition to augmenting water supplies through increased utilization of groundwater and inter-basin water transfer, demand management is also becoming increasingly important. Organizations are being forced to shift to demand management and water saving programs rather than modifying water consumption patterns and regulating the behavior of water users, and so on, as low-cost, sustainable and reliable solutions. In this regard, IWRM has been proposed in response to increasing pressures on water resource systems. By prioritizing water as a cornerstone of sustainable development, IWRM is important in influencing and communicating with different sectors and applying a variety of participatory perspectives in water management, allowing for coherent water planning. Operationally, IWRM involves applying knowledge in a variety of ways, and taking into account the views of different stakeholders for effective management and implementation, as well as providing rational and equitable solutions to water and development issues. IWRM is thus a way of balancing social and economic needs and ensuring the protection of ecosystems for the next generation. Cosgrove and Rijsberman (2000) defined IWRM as a philosophy that addresses the issue of water in an all-encompassing perspective, balancing the different needs of agriculture, industry, municipalities, and the environment. In IWRM, the management of water resources and services should reflect the relationships and interactions between different water uses and should therefore be coordinated with different sectors. A fair, efficient and sustainable situation will arise if there is a good coherence in the management of water resources and services.

Inter-basin transmission plans affect the status of both the donor and recipient basins, and inadequate management may cause irreparable detrimental effects on the socio-economic and environmental conditions of both. As a result, it is impossible to make a decision without thoroughly identifying the plan. The philosophy of IBT plans should be to transfer only surplus water and to prioritize current and future needs of the donor basin. In addition, to ensure a sustainable pattern of consumption and ensure that water consumers in a donor basin do not oppose its exploitation, the groundwater and surface water resources of a donor basin should not be negatively affected, and transferred water should only be used to meet current consumption requirements of a recipient basin. The transferred water should never be used to support land or agricultural development.

Therefore, in light of the above, observing the basic principles of IWRM in inter-basin water transfers can justify those plans and sustain the development and compliance of all stakeholders.

2.5.2 Comprehensive land use planning in inter-basin water transfers

As the water resource crisis is very serious and undeniable, short-term thinking to meet temporary needs should be avoided and comprehensive planning, based on sustainable land use plans, should be implemented – otherwise, unsustainable water transmission between basins will be made not on the basis of scientific principles, natural geographical conditions or for economic and social justifications, but sometimes for propaganda purposes. When a province or a watershed obtains more power, they may opt to receive transferred water at any cost, including the cost of environmental degradation, or at the expense of water shortage in the donor basin. Developing and adhering to a comprehensive land-use plan can prevent resource degradation, contribute to sustainability of the land, and prevent the loss of natural and land-based assets, as well as losses of social and human capital.

2.5.3 A systematic approach to inter-basin water transfers

A systematic approach considers all components and their relevance to the system, examining and analyzing real-world problems with appropriate precision in detail, by modeling and identifying the causes of the problem. Since the environment is created by both living and non-living factors that interact with each other and form a system, these relationships are essential for the survival of the environment. It is necessary to integrate all stages (policy, planning, implementation, monitoring, and evaluation) in order to prevent environmental vulnerability and to help adapt to the needs of the community.

Given the central role of water as one of the most important components of the environment and sustainable development, a systematic approach to all natural factors must be the basis of any decision-making on inter-basin water transport. It is important to accurately assess the relationship between different environmental factors (such as water, soil, plants, humans, and other living creatures) in the donor and recipient basins. After completing the necessary assessments, the relationship between IBT and governance, economy, culture and environmental law should be carefully studied and even modeled systematically, to ensure a sustainable approach.

2.5.4 Achieving a balance

Citizens in donor basins typically resist water transfer plans because of inevitable economic and environmental impacts. Achieving an equitable balance to meet the needs of donor and recipient basins is difficult. The influence, political and economic power of a recipient basin is typically greater than that of a donor basin, which usually lacks sufficient political power to prevent or modify water transfer plans. It is therefore necessary to develop methods for achieving an equitable solution that balances positive and negative impacts for both the donor and recipient basins.

One way to overcome the challenges of inter-basin water transfer schemes is to employ a win-win strategy, taken from game theory, as has been used in developed countries for many years. Although experts have utilized this approach in other contexts, this method should also be institutionalized in water resource development plans, particularly in inter-basin water transfer plans.

The usual outcome of an IBT project is an increase of economic activity in the recipient basin and a decrease in economic activity in the donor basin. Therefore, water allocation and dispute resolution is one of the most important issues in achieving a successful water transfer scheme. Game theory techniques provide an effective and accurate language for discussing specific water resource conflicts and take into account the preferences, disagreements, relative power of decision makers, and reallocation among claimants. A decision-making process involving more than one person is accompanied by different and often conflicting goals, views and priorities. Differences in views and priorities in decision-making pose problems. The final decision in such a situation must be taken in such a way that all the preferences, disagreements, and relative power of the decision makers (players) are properly taken into account.

The methods used to solve this type of problem are known as conflict resolution models and one of these methods is based on game theory. A game is a metaphor of the logical behaviors of multiple actors in a reciprocal or interdependent state, such as cooperation or coalition, conflict, competition, coexistence, etc. (Wei & Gnauck, 2007).

In conflicts over scarce water resources, with many stakeholders having conflicting interests, game theory is a useful tool to determine and identify behaviors of different water resource segments and to describe relationships between them. Quantitative and cooperative methods of game theory include: Shapely value; Nuclear; Nucleolus; Kalai-Smorodinski; Nash bargaining solution; Bankruptcy Model; Minimum Costs Remaining Savings (MCRS); and Separable Costs Remaining Benefits (SCRB). Uncooperative methods of game theory include: Non-Myopic Stability; Limited-Move Stability; Sequential Stability (SEQ); Symmetric Metarationality (SMR); and General Metarationality (GMR).

Game theory methods are effective for resolving conflicts when the actions of actors (individuals or groups) interact and the choices of all actors influence the outcome (Scharpf, 1997). A country, a region, a group, a person, an organism or even nature can all be an actor. The essence of game theory is to examine the interaction, strategies, and balance of different actors (Wei *et al.*, 2010). A game can be defined as the following set (2.1):

$$G_T(N, A, V, I, O, E) \tag{2.1}$$

but a form game (or strategic game) is generally described as set (2.2):

$$G(N, S, V) \tag{2.2}$$

where G_T is a general symbol for all kinds of games, G is a normal form game (or strategic game), N is a set of players, A is actions (moves), S is a strategy profile, V is payoffs (or utility), I is the information set, O is outcomes, and E is equilibrium or equilibria. As an example, a water conflict problem in China was simulated by Wei *et al.* (2010) as a normal (or strategic) form game with two levels, including one main game and 4 sub-games. The aim of this study was to investigate and resolve conflicts arising from the allocation of water in the Middle Route of the South-to-North Water Transfer Project (in China) taking into account environmental issues (reduction of nitrogen). In the main game, the benefits and losses of different main players (Beijing municipality, and the six cities in the provinces of Shaanxi, He'nan and Hubei in the upper basin of Hanjiang River), given their cooperation and non-cooperation, are compared. In the first sub-game, the sub-players (industry, households and agriculture of every main player) decide how to get their water. The other three sub-games are classified as total nitrogen (TN) reduction games, where sub-players decide whether to reduce the amount of TN discharge into the river. The results of this study show that cooperation results in maximizing collective positive benefits, although some players may suffer losses. Wei *et al.* (2010) concluded that providing economic incentives for affected players will increase cooperation and maximize net benefits whilst considering all players.

2.6 CASE STUDIES

There has always been a debate about whether inter-basin water transfers are justified, and their implementation requires the consideration of many factors. A review of existing projects can provide insights on positive and negative effects.

The design and implementation of major IBT projects in Western industrialized and developed countries dates back to the 1960s. At that time, large-scale plans for inter-basin transfer of water to dry areas were also first proposed in the Soviet Union. The main focus of these proposals was to showcase engineering capabilities and economic growth in industrialized countries. The ambitions of many proposed IBT projects were constrained in the 1980s due to strong opposition by donor basins, the high economic costs of these projects, and environmental considerations. In some countries, including the United States, the former Soviet Union, and China, the transfer of water from one basin to another has been a common method used to increase water resources in arid watersheds. In this context, the city of Denver's water supply from the Colorado River can be cited as an example.

Well-known long-distance IBT projects include the Amodaria Water Transmission Plan to the Turkmenistan Irrigation Network in the former Soviet Union, the Triple Water Transmission Plans from the Western, Eastern and Southern Sections of the Beijing Basin (capacity of 20,000,000,000 m^3/yr) in China, the Water Transfer Plan from Lesotho to South Africa, and the Tokyo Water Transmission Plan in Japan.

The inter-basin transfers of large volumes of water to cities such as Paris, Marseille, Athens, Helsinki, New York, Los Angeles and Tokyo have accelerated population growth, centralized urban structures in these areas, and hence exacerbated long-term water supply challenges for recipient basins. These IBT projects utilize water from distant sources and have been met with varying levels of resistance from local people and donor communities impacted by them.

To illustrate the experiences and consequences of an IBT project, a case study in India is explored here in more detail. Freshwater resources in India account for only 4% of the world's total and less than 1% of India's water resources are available for human consumption. Floods and droughts occur periodically in India, and water availability is a major socio-economic concern.

India has many large and small rivers that provide a livelihood for many people, and which are heavily influenced by monsoons. The seven main rivers of India are the Narmada, Tapi, Krishna, Indus, Brahmaputra, Godavari and Mahanadi. River basins comprise approximately 80% of the country's total geographic area and are classified as either Himalayan or Peninsula. The rivers passing through the Himalayas are the Ganga, Indus and Brahmaputra, whilst the major rivers in the peninsula are the Narmada, Tapi, Krishna, Cauvery and Mahanadi. India also has several beautiful lakes that provide freshwater and also play an important role in aquifer recharge as well as providing for flood control and assimilating pollutants from sewage.

India is one of the largest food producers in the world and its agricultural infrastructure relies heavily on groundwater. Future agricultural productivity in India is threatened by increasing water scarcity associated with climate change. Zaveri *et al.* (2016) used a multidisciplinary approach to assess impacts of climate change on the Indian agricultural system. They concluded that the effectiveness of large-scale water infrastructure projects designed to address these challenges, such as the proposed large National River Linking Project (NRLP), would have limited capacity to reduce groundwater stress.

The Indian government is studying about 30 inter-basin water transfer projects to link 37 rivers to implement a river confluence plan in the country. The overall NRLP aims to provide multiple water resource benefits including increased access to irrigation water, increased recharge of groundwater, employment generation and improvements in irrigation, power, education, water availability, water quality, agriculture, family income, industry, flood status, drainage systems, tourism, soil, fishing, public health status and other economic aspects. Kumar and Verma (2020) studied this river linkage scheme designed to improve water management and flood control in India and to meet the water needs of all basins. Since floods and droughts occur periodically in India, they suggested that monsoon flood waters in excess of that utilized within a basin should be captured and transferred to water-stressed areas. They also concluded that the NRLP in India, with its proper water distribution and proper flood and drought management, would meet the country's water resource needs.

The national water policies in India address the importance of water resource management planning and include several recommendations for water conservation and management, including the beneficial role provided by large-scale IBT projects for proper management of floods and droughts in the country.

2.7 SUMMARY

Implementation of IBT projects in basins that do not have the potential for population or economic growth will result in the waste of natural, economic and human capacities. These capacities should instead be applied towards achieving the best possible social and economic efficiency. Decision making for IBT investments requires knowledge of a region's geographic potential and proper execution of land using planning. Integrated water resources planning is a necessary factor for sustainable development. Decision making in water resources management should be conducted in a systematic approach since the effect of a decision on one factor can result in negative consequences to related factors. To achieve a balance among all economic, social and environmental considerations, an IWRM approach should be used throughout the planning and implementation of an IBT project.

Before proceeding with an IBT project, the sum total of positive and negative effects in the donor and recipient basins should be determined in all aspects (economic, social, political and environmental) and the plan implemented only if there is a net positive effect and if negative effects can be compensated.

Finally, since IBT is a method to manage water resources through augmentation of water supplies, it is also worth mentioning that other management tools should also be considered to achieve the best solution for a given situation. Development of water reuse/recycling programs and methods to improve water use efficiency and reduce water waste should also be given consideration. If these methods cannot sufficiently alleviate water stress and if the considered basin has investment potential and value, the transfer of water from other basins should then be considered.

REFERENCES

Alahdin S., Ghafouri H. R. and Haghighi A. (2019). Multi-reservoir system operation in drought periods with balancing multiple groups of objectives. *KSCE Journal of Civil Engineering*, **23**(2), 914–922.

Ben Fraj W., Elloumi M. and Molle F. (2019). The politics of interbasin transfers: socio-environmental impacts and actor strategies in Tunisia. *Natural Resources Forum*, **43**(1), 17–30. Blackwell Publishing Ltd, Oxford, UK.

Berbel-Filho W. M., Martinez P. A., Ramos T. P., Torres R. A. and Lima S. M. (2016). Inter-and intra-basin phenotypic variation in two riverine cichlids from northeastern Brazil: potential eco-evolutionary damages of Sao Francisco interbasin water transfer. *Hydrobiologia*, **766**(1), 43–56.

Boddu M., Gaayam T. and Annamdas V. G. M. (2011). A Review on Inter Basin Transfer of Water. IPWE 2011, Proceedings of 4th International Perspective on Water Resources & the Environment, January 4–6, National University of Singapore, Singapore.

Chen Z., Su S. I. I. and Wang H. (2019). Inter-basin water transfer supply chain equilibrium and coordination: a social welfare maximization perspective. *Water Resources Management*, **33**(7), 2577–2598.

Cox W. (1999). Determining when interbasin water transfer is justified: criteria for evaluation. Interbasin Water Transfer, Proceedings, International Workshop on Interbasin Water Transfer, April 1999, UNESCO, Paris, pp. 173–178.

Cosgrove W. J. and Rijsberman F. R. (2000). Making Water Everybody's Business. World Water Vision. The World Water Council, Earthscan Ltd, London.

Das D. K. (2006). Environmental impact of inter-basin water transfer projects: some evidence from Canada. *Economic and Political Weekly*, **41**(17), 1703–1707.

Davies B. R., Thoms M. and Meador M. (1992). An assessment of the ecological impacts of inter-basin water transfers, and their threats to river basin integrity and conservation. *Aquatic Conservation: Marine and Freshwater Ecosystems*, **2**(4), 325–349.

Du W., Fan Y., Liu X., Park S. C. and Tang X. (2019). A game-based production operation model for water resource management: an analysis of the south-to-north water transfer project in China. *Journal of Cleaner Production*, **228**, 1482–1493.

Ghassemi F. and White I. (2007). Inter-Basin Water Transfer: Case Studies from Australia, United States, Canada, China and India. Cambridge University Press, Cambridge, UK.

Greer R. (1983). The Texas water system: implications for environmental assessment in planning for interbasin water transfers. In: Long-Distance Water Transfer: A Chinese Case Study and International Experiences, A. K. Biswas, J. E. Nickum, L. Changming and Z. Dakang (eds), UNU and Tycooly International Publishing Ltd, Tokyo.

Gupta J. and van der Zaag P. (2008). Interbasin water transfers and integrated water resources management: where engineering, science and politics interlock. *Physics and Chemistry of the Earth, Parts A/B/C*, **33**(1–2), 28–40.

Karakaya N., Evrendilek F. and Gonenc I. E. (2014). Interbasin water transfer practices in Turkey. *Journal of Ecosystem and Ecography*, **4**(2), 149.

Kumar N. and Verma A. (2020). Inter-basin water transfer and policies of water resource management. In: Environmental Concerns and Sustainable Development, V. Shukla and N. Kumar (eds), Springer, Singapore, pp. 257–274.

Loucks D. P. (2000). Sustainable water resources management. *Water International*, **25**(1), 3–10.

Maknoon R., Kazem M. and Hasanzadeh M. (2012). Inter-basin water transfer projects and climate change: the role of allocation protocols in economic efficiency of the project. Case study: Dez to Qomrood Inter-Basin Water Transmission Project (Iran). *Journal of Water Resource and Protection*, **4**(09), 750.

Manshadi H. D., Niksokhan M. H. and Ardestani M. (2015). A quantity-quality model for inter-basin water transfer system using game theoretic and virtual water approaches. *Water Resources Management*, **29**(13), 4573–4588.

Peng Y., Chu J., Peng A. and Zhou H. (2015). Optimization operation model coupled with improving water-transfer rules and hedging rules for inter-basin water transfer-supply systems. *Water Resources Management*, **29**(10), 3787–3806.

Quinn F. (1981). Water transfers—Canadian style. *Canadian Water Resources Journal*, **6**(1), 64–76.

Rezaei M. and Basirzadeh H. A. (2011). Evaluation of water transfer project aspect of national security with an emphasis on sustainable development and land use planning, river engineering. Proceeding of Ninth International Seminar, Shahid Chamran University, Ahvaz, Iran [in Persian].

Sadegh M., Mahjouri N. and Kerachian R. (2010). Optimal inter-basin water allocation using crisp and fuzzy shapley games. *Water Resources Management*, **24**(10), 2291–2310.

Scharpf F. W. (1997). Games Real Actors Play. Actor-Centered Institutionalism in Policy Research. Routledge, New York, USA.

Shao X., Wang H. and Wang Z. (2003). Interbasin transfer projects and their implications: a China case study. *International Journal of River Basin Management*, **1**(1), 5–14.

Snaddon C. D., Wishart M. J. and Davies B. R. (1998). Some implications of inter-basin water transfers for river ecosystem functioning and water resources management in Southern Africa. *Aquatic Ecosystem Health and Management*, **1**(2), 159–182.

Snaddon C. D., Davies B. R., Wishart M. J., Meador M. E. and Thoms M. C. (1999). A global overview of inter-basin water transfer schemes, with an appraisal of their ecological, socio-economic and socio-political implications, and recommendations for their management. Water Research Commission Report No. TT120/00, Water Research Commission, Pretoria.

Tien Bui D., Talebpour Asl D., Ghanavati E., Al-Ansari N., Khezri S., Chapi K., Amini A. and Thai Pham B. (2020). Effects of inter-basin water transfer on water flow condition of destination basin. *Sustainability*, **12**(1), 338.

United Nations (1987). Our Common Future. Report of the World Commission on Environment and Development. Annex to Document A/42/427 – Development and International Cooperation: Environment. United Nations, New York, USA.

Wan W., Guo X., Lei X., Jiang Y. and Wang H. (2018). A novel optimization method for multi-reservoir operation policy derivation in complex inter-basin water transfer system. *Water Resources Management*, **32**(1), 31–51.

Wei S. and Gnauck A. (2007). Game theory based water quality models for reservoir management. In: Proceedings of EnviroInfo Warsaw 2007: Environmental Informatics and Systems Research, Volume 1: Plenary and session papers, Warsaw, Poland, O. Hryniewicz, J. Studziński and M. Maciej Romaniuk (eds), Shaker Verlag, Aachen, pp. 363–370.

Wei S., Yang H., Abbaspour K., Mousavi J. and Gnauck A. (2010). Game theory based models to analyze water conflicts in the middle route of the south-to-north water transfer project in China. *Water Research*, **44**(8), 2499–2516.

Wu L., Bai T. and Huang Q. (2020). Tradeoff analysis between economic and ecological benefits of the inter basin water transfer project under changing environment and its operation rules. *Journal of Cleaner Production*, **248**, 119294.

Yao Y., Zheng C., Andrews C., He X., Zhang A. and Liu J. (2019). Integration of groundwater into China's south–north water transfer strategy. *Science of the Total Environment*, **658**, 550–557.

Zaveri E., Grogan D. S., Fisher-Vanden K., Frolking S., Lammers R. B., Wrenn D. H., Prusevich A. and Nicholas R. E. (2016). Invisible water, visible impact: groundwater use and Indian agriculture under climate change. *Environmental Research Letters*, **11**(8), 084005.

Zeng Q., Qin L. and Li X. (2015). The potential impact of an inter-basin water transfer project on nutrients (nitrogen and phosphorous) and chlorophyll a of the receiving water system. *Science of the Total Environment*, **536**, 675–686.

Zhuang W. (2016). Eco-environmental impact of inter-basin water transfer projects: a review. *Environmental Science and Pollution Research*, **23**(13), 12867–12879.

Chapter 3

Nanotechnology application in water resource management

Arezoo Boroomandnia[1], *Omid Bozorg-Haddad*[2*], *Jimmy Yu*[3] *and Mariam Darestani*[4]

[1]*M.Sc. Graduate, Department of Irrigation & Reclamation Engineering, Faculty of Agricultural Engineering & Technology, College of Agriculture & Natural Resources, University of Tehran, Karaj, Tehran, 31587-77871, Iran.*
E-mail: aboroomandnia@ut.ac.ir
[2]*Distinguished Professor, Department of Irrigation & Reclamation Engineering, Faculty of Agricultural Engineering & Technology, College of Agriculture & Natural Resources, University of Tehran, Karaj, Tehran, 31587-77871, Iran.*
E-mail: OBHaddad@ut.ac.ir
[3]*Senior Lecturer, School of Engineering and Built Environment, Civil and Environmental Engineering, Griffith University, N79 3.18A, Nathan Campus, Brisbane, Australia. E-mail: jimmy.yu@griffith.edu.au*
[4]*Senior Lecturer, School of Engineering, Western Sydney University, Room XB.3.34, Building XB, 56 Second Avenue, Kingswood, 2747, Sydney, Australia.*
Email: m.darestani@westernsydney.edu.au
**Corresponding author.*

ABSTRACT

Fast-growing water demand, population growth, global climate change, and water quality deterioration all drive scientists to apply novel approaches to water resource management. Nanotechnology is one of the state-of-the-art tools in scientists' hands which they can use to meet human water needs via reuse of water and utilizing unconventional water resources. Additionally, monitoring water supply systems using new nanomaterials provides more efficient water distribution networks. In this chapter, we consider the generic concepts of nanotechnology and its effects on water resources management

Editor: Omid Bozorg-Haddad
doi: 10.2166/9781789062144_0033

strategies. A wide range of nanomaterials and nanotechnologies, including nano-adsorbents, nano-photocatalysts, and nano-membranes, are introduced to explain the role of nanotechnology in providing new water resources to meet growing demand. Also, nanomaterial application as a water alternative in industry, reducing water demand in the industrial sector, is presented. Another revolution made by nanomaterials, also discussed in this chapter, is their use in water supply systems for monitoring probable leakage and leakage reduction. Finally, we present case studies that clarify the influence of nanotechnology on water resources and their management strategies. These case studies prove the importance and inevitable application of nanotechnology to satisfy the rising water demand in the modern world, and show the necessity of nanotechnology awareness for today's water experts.

Keywords: Nanotechnology, Water resource management.

3.1 INTRODUCTION

The prefix 'nano' (from the Greek, meaning something very small) is used in 'nanometer' to represent a one thousand millionth of a meter (1×10^{-9} m). The characteristics of all substances at a nanoscale size are enormously different in comparison to those at the macroscale. Mechanical, electrical, optical, and magnetic properties are part of the wide range of size-dependent characteristics which change significantly. The large specific surface area is the most well-known reason for nanomaterials' unique characteristics. Many nanomaterials also typically have characteristics of high reactivity, catalysis, and adsorption capacity which have made them the subject of much research and development in recent years.

To understand the evolution of nanotechnology over recent decades, a summary of chronological progress in nanotechnology is first reviewed in this chapter. Then the importance and benefits of applying nanotechnology to water resource management are explained. Nanotechnology applications in water-related areas are assessed in Section 3.4, which is followed by a more detailed explanation of nanomaterials and technologies. In the last two sections, the influence of applying nanotechnology in water management strategies is covered, a detailed review of three case studies is presented. After studying this chapter, it is hoped that readers will have a good knowledge of nanotechnology application in water science and of the improvements provided by nanotechnology in water management strategies.

3.2 HISTORY OF NANOTECHNOLOGY DEVELOPMENT

The concept of nanotechnology was first introduced in 1959 by Richard Feynman at the annual meeting of the American Physical Society. Feynman is considered the father of modern nanotechnology. The term 'nanotechnology', however, was first used and defined in 1974 by Norio Taniguchi, the Japanese scientists. With the invention of the scanning tunneling microscope (STM) in 1981, scientists could finally see materials at an atomic level and two years later the first STM image of the topmost layer of silicon atoms was released. The STM was followed by development of the atomic force microscope (AFM) and scanning probe microscopes (SPM), the main instruments now applied in nanotechnology research. In the same decade, the first book on nanotechnology, '*Engines of Creation: The Coming Era of Nanotechnology*', was published by K. Eric Drexler (Drexler, 1986). This book improved public awareness and made the theory of 'molecular engineering' more common. It is noteworthy that during this time, powerful computers played a complementary role in large scale simulations of nanoscale material structures and their properties (Bayda *et al.*, 2019).

In the late 1990s and early 2000s, nanotechnology activities were spread almost all over the industrialized world. A breakthrough in 1991 was the discovery of the carbon nanotube, resulting from the discovery of the buckyball, a molecule which consists of 60 carbon atoms. Carbon nanotubes remain one of the most promising areas of nanotechnology in many fields, such as catalysis, medicine, sensing, and the environment. Graphene and carbon dots are two other nanostructure materials that were discovered in 2004 and are still used in various nano-technological applications (Bayda *et al.*, 2019). At the beginning of the 21st century, the rising interest in nanotechnology has affected methods of environmental protection, including source emission control and elimination.

3.3 NANOTECHNOLOGY'S PLACE AND IMPORTANCE IN WATER RESOURCE SCIENCE

Before the industrial revolution period, which resulted in a dramatic increase in new pollution emissions to water resources, simple sand filtration was used for water treatment. In 1826, Henry Doulton initiated water filtration in Great Britain and also invented the concept of the ceramic candle water filter. Another breakthrough was chlorine usage as an effective water disinfectant in the USA in 1908. With the industrial revolution and wide emission of new man-made chemicals and products to the environment, humans faced serious water resource pollution and an urgent need to address this problem. To end the increasing pollution of water sources which were being threatened by a broad range of new contaminants, chlorine alone had not been sufficient. Therefore, in the 1970s scientists developed novel treatment techniques such as membranes for reverse osmosis filtration and ozonation. Also, a combination of different methods including sedimentation, filtration through various structures, and chlorination was found to be an effective and beneficial method of water treatment. Filtration methods have experienced continuous improvement, particularly with the emergence of nanotechnology. The strong adsorption capacities, high reactivity, and high mobility of nanomaterials in solution, all of which results from their large specific surface areas, leads to the growing application of nanomaterials in water treatment. Numerous research projects have proven the effectiveness of nanomaterials in removing various pollutants in water, such as heavy metals, bacteria, organic pollutants, and inorganic anions. For example, iron nanoparticles are utilized to degrade chlorinated organic molecules from groundwater resources.

With a significant rise in the world population and consequent water demand, utilizing reclaimed water and unconventional water sources is now an important part of sustainable water management. However, the existence of various noxious pollutants such as heavy metals, radionuclides, nitrates, pathogens, viruses, and metalloids in wastewater and even freshwater resources (due to sewage discharge to surface and groundwater resources) requires the utilization of novel treatment approaches for the purpose of water reuse. While conventional treatment methods were not functional, effective treatment techniques like reverse osmosis, electroflotation, ion exchange, adsorption, membrane separation, and nanotechnology have succeeded in the elimination of toxic chemicals. Among these techniques, nanotechnology-based water treatment techniques have the utmost efficiency and viability, which has led to their enormous growth (Sahu *et al.*, 2019).

Applying functional nanomaterials improves the efficiency of purification and waste treatment systems by combining the separation and degradation of retention processes. Moreover, among different methodologies that have been applied to separate oil from oil-contaminated water, nanotechnology has proven to be the most cost-effective technique so far. Indeed, high separation efficiency, environmental friendliness, good recyclability, and easy fabrication are the properties distinguishing nanomaterials from other methods (Machado *et al.*, 2019). Additionally, the formation of byproducts is another important factor in the selection of an appropriate method (these byproducts are produced when chemical disinfectants react with organic materials and inorganic ions present in water or a sewer). The use of antimicrobial nanomaterials is another example of nanomaterials improving water treatment, as the weak oxidants do not produce toxic byproducts.

Desalination is another essential source of supplying water, and the effectiveness of conventional methods – including precipitation, oxidation, reduction, ion exchange, membrane filtration, and adsorption – are not as high as nanotechnology in extracting salt and mineral components from diverse sources of water (including seawater, brackish water, and reclaimed water). For instance, reverse osmosis membranes need high pressure as well as producing substantial quantities of liquid waste, while utilization of nanomaterials requires less pressure and results in low energy consumption. Nanomaterials' high salt rejection has led to their application in sustainable membrane desalination, increasing the membrane lifecycle (Dhakras, 2011).

Nanotechnology application in water reuse can also decrease the energy footprint of water on account of its availability for point-of-entry (POE) and point-of-use (POU) applications. Indeed, nanofiltration membranes provide decentralized water treatment, including POE and POU units, which eliminate the energy cost of diversion and distribution for wastewater and reclaimed water. POE systems contain small treatment units at the point of entry of the water supply for apartment complexes, and POU treatment units are fixed at the water dispensing points (e.g., specific tap/faucet). Application of nanotechnology in POE and POU systems improves the cost-efficiency of supplying potable water in remote areas (Dhakras, 2011).

Another application of nanomaterial which leads to energy saving is their use in repairing leakages and modifying inefficient delivery processes. Annually, 126 billion (126×10^9) m^3 of water is lost through leakage in the water supply systems around the world, with an estimated value of around 39 billion USD (Liemberger & Wyatt, 2019). However, new self-healing nanoscale coatings can be applied to repair leakage in a water delivery system (Zhai *et al.*, 2020). Moreover, nanomaterials can be applied as sensors for real-time leakage control. Therefore, nanomaterials play an active role in saving both water and energy in water distribution systems (NTSC Committee on Technology, 2016). Sometimes nanotechnology-enhanced fluids are also used as a proper alternative for water as a heat transfer fluid, which reduces water consumption in industry, and particularly in thermal power plants (Kim, 2013).

Therefore, the high level of effectiveness of nanomaterials in water treatment in comparison to other methods, their low production cost (in some nanomaterials, not all), lower energy consumption over the treatment process, less production of by-product

Table 3.1 The chronological trend of nanotechnology achievement in water industry.

Application	Nanomaterial/Technology	First Application
Adsorption	Metal based nano-adsorbents	1970s
	Polymeric nano-adsorbents	2001
	Carbon nanotube	2004
Membranes and membrane processes	Nanocomposite membranes	2001
	Nanofiber membranes	2004
	Thin film nanocomposite (TFN) membranes	2007
	Biologically inspired membranes	2007
Photocatalysis disinfection and microbial control	Fullerene derivatives	1995
	Nano-Ag	1997
	Nano-TiO_2	2001
	Carbon nanotubes	2010
Self-healing nanomaterials	Nano-clay, SiO_2, TiO_2, ZrO_2 and CNTs	1981
Water replacements	Al_2O_3, TiO_2, CNT, CuO, SiO_2, SiC, MgO, ZnO, Ag	1995
Sensing and monitoring	Noble metal nanoparticles	2003
	Dye-doped silica nanoparticles	2007
	Quantum dots	2007
	Carbon nanotubes	2008
	Magnetic nanoparticles	2013

pollution, point-of-use applications, uniqueness in treating particular contaminants, control and reduction of leaky pipes, and many other reasons, make nanotechnology usage necessary to meet the growing demand for high quality water.

Since there is much research into nanotechnology application in the water industry, the chronological trend of research in this field is summarized in Table 3.1 (Qu *et al.*, 2013a). These different applications are explained in detail in Section 3.4 below.

3.4 NANOSTRUCTURED MATERIALS AND TECHNOLOGIES APPLIED IN WATER SCIENCE

ISO/TS 80004 defines a nanomaterial as a ‘material with any external dimension in the nanoscale or having internal structure or surface structure in the nanoscale’. Nanomaterials divide into two categories: nano-objects and nanostructured materials. Discrete pieces of material of nanometer dimensions are nano-objects; however, nanostructured materials have internal or surface structures on the nanoscale.

Based on the number of an object’s dimensions which fall in the nanoscale, nano-objects are classified as nanoparticles, nanofibers, or nanoplates. Nanoparticles have all three external dimensions in the nanoscale and there is no significant difference between its longest and shortest axes. If an object has two nanoscale dimensions, it is considered as a nanofiber. Hollow nanofibers are known as nanotubes and solid nanofibers are classified as nanorods. By decreasing the nanoscaled external dimension to only one axis, a nanoplate is formed. A nanoribbon is a kind of nanoplate with a significant difference between the two larger dimensions. For nanofibers and nanoplates, one dimension – such as length or width – is larger, and might be in nanoscale or micro scale, while one dimension at least is in nanoscale. The image of various nano-objects under a scanning electron microscope (SEM) is illustrated in Figure 3.1.

Some bulk materials contain features on the nanoscale: nanocomposites, nanostructured films, nanocrystalline materials, and nanotextured surfaces are examples of nanostructured materials. Nanocomposites contain at least one physically or chemically distinct region which has at least one nanoscale dimension. There are organic and inorganic nanocomposites with numerous applications: their application in water purification is explained in this chapter. A thin film (e.g., graphene) is an example of a nanostructured film consisting of crystalline materials with a two-dimensional single layer of atoms. Nanostructured materials are employed in a wide range of, manufacturing processes, and as final products such as paints, filters, insulation, and lubricant additives. In the water industry, high-quality filters produced using nanostructures can remove particulate as small as a virus.

Nanotechnology and nanomaterial applications can be classified into four main categories: water and wastewater treatment; desalination; water delivery and consumption; and water monitoring and sensing. The first two categories, providing new water resources, are considerably more prevalent than the others. In the following subsections, features of various nanomaterial and technologies are expounded under three main applications: water availability enhancement, water delivery and consumption improvement, monitoring.

3.4.1 Water availability enhancement

Undoubtedly, freshwater resources are scarce, and in the face of growing water demand and climate change, water reuse and desalination (of seawater and brackish water) are sustainable options to overcome water shortage. The World Health Organization anticipates that around 50% of the world's population will settle in water-stressed regions, and wastewater treatment and desalination in coastal areas has risen over the last decades. There are numerous forms of pollutants released to water resources or even naturally existing which must be eliminated at various levels before usage in different sectors. As many of these contaminants in water and wastewater cannot be treated via identical methods, nanotechnology is employed to reach water purity standards cost-effectively. For instance, the activated sludge process which is used for the treatment of municipal wastewater, cannot degrade toxic and recalcitrant pollutants in an industrial sewer. To guarantee the elimination of biological and chemical contaminants, nanotechnology has attracted scientists' attention. In wastewater treatment also, a combination of conventional and novel approaches (like aerobic degradation and nanoparticle usage) provides treatment on a large scale. All in all, nano-absorbents, nano-catalysts, nano-minerals, or nano-membranes are all being applied according to the characteristics of the contaminants and impurities being treated, desired purification levels, budget limitations, etc., to provide new water resources. The main nano-agents in water treatment and their features are classified in Table 3.2, and their complementary explanation is given in the following subsections (Machado *et al.*, 2019).

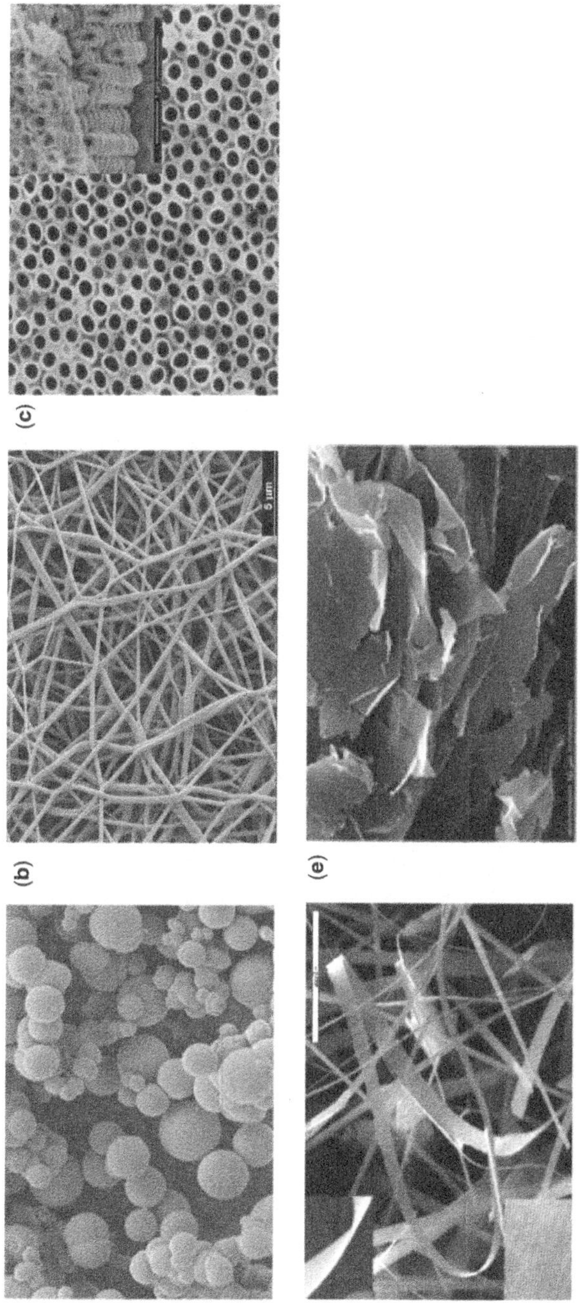

Figure 3.1 SEM images of various nano-objects: (a) nanoparticle; (b) nanofiber; (c) nanotube (d) nanoribbon; and (e) nanoplate.

Table 3.2 Nanomaterial classification based on treatment technique (Gehrke *et al.*, 2015).

Treatment Agent	Advantages/Disadvantages	Pollutants Removed	Nanomaterial Example
Nano-adsorbents	+ High specific surface, higher adsorption rates, small footprint – High production costs	Bacteria, organic and inorganic contaminants	Nanostructured mixed oxides, magnetic nanoparticles, metallic oxide nanoparticles, and carbon based nano-adsorbents
Nano photocatalysts	+ Short intraparticle diffusion distance, compressible, abrasion-resistant, magnetic, high interaction with light – Less reusable	Bacteria and organic substances	Inorganic materials such as semiconductors and metal oxides
Nano-membrane	+ Reliable, largely automated process – High energy consumption	Dyes, heavy metals, minerals and other contaminants	Nanotubes, nanoribbons, and nanofibers

3.4.2 Nano-adsorbents

Nano-adsorbents, with the capability of removing organic and inorganic contaminants in water and wastewater, are considerably more effective than conventional adsorbents which are limited by surface area or active sites, a lack of selectivity, and by adsorption kinetics. The prominent features of nano-adsorbents are their high specific surface area and associated sorption sites, short intraparticle diffusion distance, controllable pore size and surface chemistry. Nano-absorbents are divide into two groups, carbon-based and metal-based, and their treatment mechanisms and features are described below (Qu *et al.*, 2013a).

3.4.2.1 Carbon-based nano-adsorbents

Carbon-based nanomaterials are divided into two groups: graphene/graphene oxide nanomaterials, and carbon nanotubes (CNTs). Large specific surface area, more accessible sorption sites and their different contaminant interactions make the adsorption efficiency of CNTs for various organic chemicals higher than that of activated carbon. In the aqueous phase, loose bundles resulted from the hydrophobicity of the graphitic are formed over the external surface of nanotubes. Additionally, CNT aggregates contain interstitial tunnels which are high adsorption energy sites for polar organic compounds. Generally, a broad range of organic compounds interact with CNTs through diverse methods, including hydrophobic effects, pep interactions, hydrogen bonding, covalent bonding, and electrostatic interactions (Qu *et al.*, 2013a).

CNTs also perform efficiently when absorbing bacteria, inorganic elements, and heavy metal pollutions. The surface functional groups of CNTs absorb metal ions by electrostatic attraction and chemical bonding. Also, their highly accessible adsorption sites and short intraparticle diffusion distance result in fast adsorption kinetics. Although activated carbon adsorbs an expansive range of heavy metal pollutants, CNTs can have unique applications in targeting specific contaminants. The small quantity of required materials for these applications makes CNT application cost effective. Graphite oxide is another form of carbon-based absorbent; produced at low cost it outcompetes activated carbon in removing Hg^{2+} and bulky dye molecules, such as azo dye and xanthene dye, efficiently (Gehrke *et al.*, 2015).

Some CNT features, such as their good thermal and electrical conductivity, hydrophilicity, and high breaking strength, make their performance acceptable for use in desalination. The multi-walled carbon nanotube is one of the common materials used in electrocatalytically active conductive membranes for desalination (Anis *et al.*, 2019).

There are also numerous studies on heavy metals and organic contaminants removal by graphene and graphene oxide. For example, graphene-based hydrogel and aerogel are used for removal of Cu2+, Pb(II), Cr(VI), and organic dye from water. Modification of graphene by polymer or organic molecules indicates improvement in its pollution selectivity (Bilal *et al.*, 2020).

3.4.2.2 Metal based nano-adsorbents

Iron oxide, titanium dioxide, and alumina (aluminium oxide) are well-known metal oxides which effectively adsorb radionuclides and heavy metals, including arsenic, lead, mercury, copper, cadmium, chromium, and nickel. In addition to effectively adsorbing organic infections and inorganic contaminants, the outstanding feature of these oxides (such as nano-magnetite and TiO_2) is their ability in arsenic removal, for which activated carbon has limitations. Magnetic nanomaterials also act efficiently in desalination (Anis *et al.*, 2019). The higher specific surface area, shorter intraparticle diffusion distance and larger number of surface reaction sites in metal oxide nanoparticles leads to their higher adsorption capacity and faster kinetics. By reducing the particle size below 20 nm, adsorption capacity rises radically.

Besides high adsorption capacity, some iron oxide nanoparticles have the capability of converting to being superparamagnetic when their size decreases to below 40 nm. Becoming superparamagnetic provides easy separation and recovery within a low-gradient magnetic field. The magnetic nanoparticles can act as adsorbents. They can also be placed as the core of a core-shell nanoparticle structure, facilitating nanoparticle regeneration and reuse while the shell adsorbs the pollutants. Fine powders and porous pellets are commercial forms of metal oxide nanocrystals used in industry. The most prevalent metal absorbents are described in detail below (Qu *et al.*, 2013a).

- **Silver nanoparticles:** Silver nanoparticles (Ag NPs) are known as an efficient material in bacterial pollutions degradation. The degradation process starts when the Ag NPs adhere to the bacterial cell walls and subsequently penetrate them. This process changes the structure of the cell membranes and consequently increases their permeability. Additionally, contact between Ag NPs and bacteria leads to the generation of free radicals with the ability to damage the cell membranes and leading to the death of cells. Another destructive factor is the interaction between Ag NPs and abundant sulfur and phosphorus elements in DNA (Lu *et al.*, 2016). On top of that, antimicrobial $Ag+$ ions, released from the dissolution of Ag NPs, can interact with the thiol groups of many vital enzymes and disrupt cell function. It was also found that the efficiency of Ag NPs in filters with higher porosity in bacteria removal is higher than those with lower porosity (Gehrke *et al.*, 2015).
- **MgO nanoparticles:** Magnesium oxide nanoparticles (MgO NPs) have been receiving scientists' attention recently owing to their significant ability in toxic gas removal, and toxic waste remediation. Formation of O_2 on MgO NPs' surfaces stops the growth of microorganisms or adsorbs the negatively charged bacteria. MgO NPs have an antimicrobial performance which is affected by parameters including particle specific surface area, interaction time, the alkaline condition of the particle surface, and presence of active oxygen on the surface (Sahu *et al.*, 2019).
- **ZnO nanoparticles:** Zinc oxide nanoparticles (ZnO NPs) are antimicrobial nanoparticles which degrade both bacteria and a few endospores. To bind bacterial cells on the surface of ZnO NPs, electrostatic forces play a vital role. Also, research has revealed that the larger surface area and higher concentration of ZnO NPs empower the antibacterial activity. It is noteworthy that the antimicrobial performance of ZnO NPs is more efficient than Ag NPs, while they have the advantage of low cost over TiO_2 NPs. Yet, ZnO applications in drinking water treatment are restricted on account of its harmful effect on aquatic organisms if it enters natural water bodies (Dhakras 2011; Lu *et al.*, 2016; Sahu *et al.*, 2019). A combination of ZnO with graphene oxide which are embedded into polysulfone (PSF) UF membranes, used for desalination, have shown *Escherichia coli* (*E. coli*) growth prevention (Chung *et al.*, 2017).

- **Iron nanoparticles:** Zerovalent iron (nZVI) nanoparticles have the highest rate of assessment amongst zero-valent metal nanoparticles in water treatment studies on account of their low cost, adsorption properties, and precipitation and oxidation (in the presence of dissolved oxygen, DO). Significant properties such as their being non-toxic, abundant, inexpensive, and easy to produce have accelerated their application for toxic materials removal. The contaminants removal process is based on directional electron transfer from nZVI to the contaminants. Adsorption, reduction, precipitation, and oxidation (in the presence of DO), are chemical reactions between nZVI and the contaminants, all of which finally lead to degradation of a large range of contaminants including halogenated organic compounds, nitroaromatic compounds, organic dyes, phenols, heavy metals, inorganic anions such as phosphates and nitrates, metalloids, and radioactive elements or isotopes (Sahu *et al.*, 2019). nZVI is used in water contaminated field sites as well as in soil remediation, distinguishing it from other nanoparticles. nZVI is also widely used in groundwater remediation. Rapid, highly effective, and low-cost use are the reasons for the large-scale applications of nZVI. nZVI is also effective in textile dye removal, with even 100% removal proven to be achievable, though its application efficiency depends on the pH, doses of nZVI, contaminant load, and volume of dye. One of the successful applications of iron oxide nanoparticles in desalination has been its combination with CNTs to fabricate novel nanocomposite membranes with high hydrophilicity and fouling resistance (Fernandez-Gonzalez *et al.*, 2017). Despite its many advantages, however, aggregation, oxidation, and separation difficulties from the degraded system are regarded as nZVI drawbacks. As an example, to enhance the reactivity of nZVI, it is doped with other metals. Surface coating and conjugation with supports are also used to prevent aggregation of nZVI (Gehrke *et al.*, 2015).
- **Polymeric nano-adsorbents (dendrimers):** Both organics and heavy metals can be properly absorbed by dendrimers. These polymers adsorb organic compounds via the hydrophobic feature of their interior shells while their exterior branches adsorb heavy metals. Complexation, electrostatic interactions, the hydrophobic effect, and hydrogen bonding take place during the sorption process. A successful example of a dendrimer application is in Cu^{2+} removal, in which it is combined with an ultrafiltration system for the dendrimers' recovery. Biopolymer adsorbents also perform efficiently in anionic compound removal, such as for dyes from textile wastewater (Gehrke *et al.*, 2015; Qu *et al.*, 2013a).
- **Natural minerals:** The physiochemical properties of some natural minerals like zeolite and clay, as well as their abundant availability and cost-effectiveness, have led to their wide applications in pollution and salt removal from water bodies. Natural zeolite, built of crystalline aluminosilicates with porous structure, removes organic micropollutants, cations and heavy metals through ion-exchange and sorption properties (Choi *et al.*, 2016; Jiang *et al.*, 2018). Natural zeolite with a porous structure is an inexpensive mineral which is applied solely or in combination with other materials or technologies in various sizes, from nanometers to centimeters. High silica zeolite powder effectively eliminates organic micropollutants such as nitrosamines. To absorb anions in wastewater or salty water via zeolite, numerous modification methods are successfully assessed to increase the treatment efficiency of zeolite. However, pH value, initial concentration of contaminates, temperature and

contact time determine the removal rate (Margeta *et al.*, 2013). Like zeolite, clay minerals with small molecular size are highly porous and have a large surface area, which leads to great bonding power with contaminants, and these features make them suitable for physical and chemical reactions. Nano-clays are known as good adsorbents for heavy metals, antibiotics, dyes, inorganics, and biocidal compounds in wastewater. Additionally, they remove bad odor and taste from water whilst simultaneously softening it (Awasthi *et al.*, 2019).

3.4.3 Nano photocatalysts

Some hazardous and non-biodegradable contaminants can be made fragile through photocatalytic oxidation. Photocatalysis as a pretreatment process treats recalcitrant organic compounds. Although it is a low-cost, environmentally friendly, and sustainable water treatment technology, its disadvantages are the slow kinetics stemming from limited light availability and photocatalytic activity. The most widely used photocatalyst in water/wastewater treatment is TiO_2. The high application of this semiconductor is due to its low toxicity, chemical stability, low cost, and abundance as a raw material (Gehrke *et al.*, 2015). Although various photocatalysts like WO_3, Fe_2O_3, CuS, ZnO are used in desalination, TiO_2 is the most predominant substance used due to its great absorption in the UV spectrum and significant chemical stability. Additionally, TiO_2 is used for membrane modification in desalination on account of its hydrophilicity (Anis *et al.*, 2019). Indeed, TiO_2 increases the flux rate of the membrane.

UV photons provoke the electrons in TiO_2, which is the reason for consequent pollution degradation. Much research has been carried out to improve the photoactivity of Nano TiO_2 by optimizing particle size and shape, reducing electron/hole recombination by noble metal doping, maximizing reactive facets, and surface treatment to enhance contaminant adsorption. Photocatalytic water treatment is carried out in a photo-reactor, in which various parameters impact on the water quality, including TiO_2 loading, dissolved oxygen, pH, temperature, contaminant type and concentration, light wavelength and intensity. This method is used on a commercial scale with a 2 million gallons (7570.8 m^3) per day capacity and operates with fairly low specific power consumption of about 4 kWh/m^3 (Qu *et al.*, 2013a).

3.4.4 Nano-membranes

Membranes work by removing undesired constituents from water via a physical barrier designed based on the contaminants' size. Their advantages are their high level of automation, less land occupation and chemical use, and flexible design. However, the high energy demand needed to supply pressure-driven membrane processes, as well as membrane fouling, are considered as their major challenges. Thanks to the application of nanomaterials in membranes, the new generation benefits from improved permeability, fouling resistance, mechanical and thermal stability, and self-cleaning (Ramsden, 2009). Various types of nano-membranes and their features are listed in Table 3.3, and extra explanation is presented in the following subsections (Gehrke *et al.*, 2015).

3.4.4.1 Nanofiber membranes

After the invention of electrospinning, producing nanofibers from diverse materials has become simple, low-cost, and effective, and has improved filtration technology.

Table 3.3 Nano-membrane types and features.

Nano-membrane Type	Significant Features	Substances Used
Nanofiber membranes	• Simple, low-cost and efficient membrane produced by electrospinning • High porosity and bactericidal	Polymers, ceramics, metals for example, TiO_2
Nanocomposite membranes	• Low fouling rate and high water permeability • Better mechanical and thermal stability • High bacterial inactivation	Hydrophilic metal oxide nanoparticles (Al_2O_3, TiO_2, zeolite, silica), antimicrobial nanoparticles (Ag-NPs and CNTs), catalytic nanomaterials (TiO_2, zero-valent iron)
Thin film nanocomposite membranes	• Improved permeability and selectivity	nano-zeolites, nano-Ag, nano-TiO_2, and CNTs
Biologically inspired membranes	• Highly ionic selective and permeable	Zwitterionic polyelectrolyte nanoparticles

In addition to high specific surface area and porosity which are the intrinsic benefits of nanofibers independent from their production method, the other advantages of electrospun nanofibers are the easy manipulation of their diameter, their morphology, composition, secondary structure, and the spatial alignment of the nanofibers. Nanofiber membranes are proficient in removing micron-sized elements dissolved or suspended in an aqueous phase at a high rejection rate and low fouling rate, which makes them appropriate for utilizing as pretreatment steps in ultrafiltration or reverse osmosis processes (Suwaileha *et al.*, 2020). In addition, a multifunctional nanofiber membrane can be produced either directly from electrospinning intrinsically multifunctional materials like TiO_2 or by doping functional materials on the electrospun nanofibers. Indeed, nanofibers can make a scaffold which is able to accept a variety of capture agents such ceramic nanomaterials for the purpose of removing a wide range of heavy metals or organic pollutants (Tlili & Alkanhal, 2019). For instance, Ag/TiO_2 nanofiber membranes are able to degrade organic pathogen pollutants from water. Hollow fiber membranes are also widely used in desalination and in removing ammonia from irrigation water (Anis *et al.*, 2019).

3.4.4.2 Nanocomposite membranes

To reduce the fouling rate, scientists have assessed the effect of an increase in hydrophilicity by adding nanomaterials to polymeric or inorganic membranes. The analyzed nanomaterials for this purpose include hydrophilic metal oxide nanoparticles, as well as antimicrobial and photocatalytic nanomaterials. Investigations show that by adding metal oxide nanoparticles to polymeric ultrafiltration membranes, not only do the membrane surface hydrophilicity and water permeability increase, but their mechanical and thermal stability is also enhanced. Adding antimicrobial nanomaterials also can reduce membrane biofouling but they do not perform efficiently for the long-term. Moreover, photocatalytic nanoparticles reduce fouling by a physical separation function, as well as by contaminant degradation via a chemical reaction. For instance, nZVI serves as an electron donor and destroys contaminates through reductive degradation (Qu *et al.*, 2013a).

Nano-chitosan is another nanocomposite with antimicrobial characteristics, embedded in polymer membranes to disinfect drinking water. These nanocomposites have several advantages over contemporary disinfectants, such as lower toxicity to human health as well as an extensive spectrum of activity against bacteria, viruses, and fungi. However, the material's preparation techniques determine its performance efficiency. Also, other factors such as pH, pKa, reaction time, availability of lipids and proteins as interferences, the concentration of chitosan, acid solvents, surface charge, the ionic strength of the matrix, and, lastly, types of bacteria all affect the antimicrobial activity of chitosan (Sahu *et al.*, 2019).

3.4.4.3 Thin film nanocomposite membranes

Nanomaterials play an active role in the development of thin film nanocomposite (TFN) membranes when they are added to the active layer via doping in casting solutions or surface modification. The type, size, and amount of added nanoparticles determine the modification level of membrane permeability and selectivity. Over the last two decades TFN membranes have been the dominant membranes used for reverse osmosis seawater desalination and forward osmosis (Suwaileha *et al.*, 2020). The zeolite TFN membrane is one of the commercial nano-membrane products showing potential in membrane permeability enhancement and better salt rejection. The hydrophilic pores of nano-zeolites create preferential paths for water while they can also carry antimicrobial agents such as Ag. Another nanomaterial which increases the membrane rejection rate while maintaining permeability is nano-TiO_2 (up to 5 wt%). However, detrimental effects on polymeric membrane materials stemming from the close proximity of the photocatalyst and the membrane decrease its long-term efficacy. Application of a small number of CNTs in TFN membranes has resulted in bacteria inactivation while also delaying membrane biofouling (Chung *et al.*, 2019).

3.4.4.4 Biologically inspired membranes

Biological membranes equipped with channel proteins are highly selective and water permeable. Aquaporin-Z from E. coli is an example of a biological material that successfully rejects glucose, glycerol, salt, and urea by channel proteins. The coating of Aquaporin-Z-incorporated lipid bilayers on commercial nanofiltration has been successfully investigated, although it is yet to be applied to industrial desalination (Qu *et al.*, 2013a).

In this section, a comparison has been made of the various applications of nanotechnologies for water treatment, wastewater treatment, and desalination. However, to summarize the broad application of nanotechnology to eliminate undesired substances from water resources, the methods employed and materials used are listed in Table 3.4 by treatment method.

Table 3.4 Nanomaterial classification by treatment method (Chung *et al.*, 2019; Machado *et al.*, 2019).

Pollution type	Treatment Method	Applied Nanomaterial
Heavy metals	Ion Exchange, Chemical Precipitation, Membrane Filtration, Coagulation, Biological or Electrochemical Remediation, Adsorption	• Nanosized metal oxides: iron oxides, manganese oxides, aluminium oxides, and titanium oxides • Carbon based materials: carbon nanotubes, graphene, Graphene oxide • Polymeric nano-adsorbents
Dye	Membrane Filtration, Adsorption Integrated With Microwave Radiation or Ultrasonic Irradiation	• Zero-valent metal nanoparticles (Fe NPs) • Carbon nanotubes, nanoribbons, nanofibers
Oil	Porous Sponge Coated With Nanoparticles	• Hydrophobic nano-silica • TiO_2 nanosheet • Silver nanoparticles • Carbon nanotubes
Salt and minerals	Membrane Filtration, Thin Film	• Nanoporous graphene like Graphene oxide • Nanohydroxyapatite • Thin film nanocomposite membranes doped nano-zeolites
Infection sources	Filtration, Adsorption	• Magnetic nanoparticles (Ag NPs) • Carbon nanotubes and nanoparticle • Nano-catalysts (nano-TiO_2, ZnO) • Chitosan

3.5 IMPROVEMENT OF WATER DELIVERY AND USE

3.5.1 Improving efficiency

Improving the efficiency of water distribution networks is necessary, since freshwater resources are scares and treatment of unconventional water resources put a financial burden on society. Therefore, minimizing water loss through distribution systems is a priority, particularly in water-stressed regions. Moreover, cracks in pipes pose the risk of contaminants entering and growing in the water delivery system. The amount of water lost during the distribution process worldwide is around 126 billion m^3, which is significant. The average water loss during transmission in urban areas is 20–30%; however, this figure can even rise to 50% in old systems without regular maintenance (El-Zahab & Zayed, 2019). Leakage from deficient joints, and holes and cracks in old pipe walls will gradually lead to bursts and a significant amount of water loss (Ahopelto & Vahala, 2020). Therefore, continuous monitoring can prevent probable bursts, while decreasing the leakage of identified cracks with the use of self-healing materials or after network renovation. Also, coating nanomaterials on the inner wall of water transport pipes can reduce friction losses and reduce the energy consumed for water transmission. Moreover, high durability and high strength-to-weight ratios are desirable properties of nanomaterials that can modify a pipe system and protect pipes against corrosion and biofouling.

Nano-clays are one of the types of substances which are shown to improve recovery levels of cracks in water structures built by cement. Nano-clays contain water, and supply extra hydration to the cracks (Wu *et al.*, 2012). SiO_2, TiO_2, ZrO_2 and CNT nanostructures are other common materials applied as nanofillers in concrete (Zhang *et al.*, 2020).

Another effective measure for reducing water demand is finding alternatives for water in industrial applications. For instance, water is widely used in thermoelectric power generation. Nanotechnology-enhanced fluids can replace water in power generation systems and heat exchangers. These processes consume large volumes of water as a heat transfer medium. Yet, the emergence of nanomaterials with comparatively better intrinsic characteristics, such as thermal conductivity, density, specific heat, and viscosity, not only improve the heat transfer efficiency but also decrease the water consumption. Some of the nanomaterials (in the form of suspended particles) used as heat transfer include Al_2O_3, TiO_2, CNT, CuO, SiO_2, SiC, MgO, ZnO, and Ag. These nanomaterials are used in combination with a liquid such as water or ethylene glycol (Perarasu *et al.*, 2013).

3.5.2 Monitoring water systems

Nanotechnology is applied for both pollution monitoring and leakage monitoring in water systems. Nanosensors with high sensitivity and selectivity have been developed to ensure that traces of pollutants, certain contaminants, and pathogens in water are detected. These sensors are made of three main parts: recognition agents (responsible for detecting special pollutants), nanomaterials, and signal transduction tools (for indicating the pollutants' presence or absence). Nanomaterials have improved sensitivity and detection speeds, because of their electrochemical, optical, and magnetic properties. Magnetic nanoparticles, noble metals, quantum dots, CNTs, and dye-doped nanoparticles are broadly used in pollution sensing and monitoring (Hossain, 2019). Nanosensors are categorized into three major groups: optical, electrochemical, and mechanical (or vibrational). In optical nanosensors, electromagnetic radiation is used to generate analytical signals in transduction elements. Any changes in optical properties indicate the presence of a specific impurity. In electrochemical nanosensors, chemical reactions affect the electrical current, potential, conductance, and impedance of the electrodes. Mechanical nanosensors such as thin and flexible rods can detect changes at the molecular scale (Graboski *et al.*, 2020).

For water leakage quantity monitoring, one example of use is the nanosensors fabricated by Yu-Feng Lin and Chang Liu, sponsored by the Illinois State Water Survey, for monitoring leakage in a water pipeline. The nanosensors developed detect leakage through flow and pressure changes along the PVC pipe length. This group also developed a temperature nanosensor for water quality detection (Lin & Liu, 2009). However, further investigations are required to assess their real-life application, including manufacturing and maintenance costs, fouling examination, real-time data communication software, and long-term stability.

3.6 NANOTECHNOLOGY IN WATER RESOURCE MANAGEMENT

Freshwater makes up just 1% of the world's total water resources, while climate change and population growth put more pressure on freshwater resources. The United Nation's anticipate that by 2025 around 2.7 billion people in developing countries will face severe water scarcity. Numerous strategies have been taken to increase water availability using water reuse, including desalination. This approach converts unconventional resources such as wastewater, seawater, and fossil water to usable water resources. Therefore, not only does water reuse decrease the pressure on freshwater resources, but this approach also reduces the level of pollution in water resources, since effluent is not discharged to the water bodies. The first benefit has resulted in reclaimed water now having a 25% share of total supplied water in Australia, and the second benefit has motivated European authorities to use reclaimed water to meet 30% of their total demand (Paul *et al.*, 2019). Although water reuse can be an alternative to traditional clean water demand, there are some constraints. The most important limiting factor is the quality of the treated water. The presence of pathogens such as bacteria, viruses, and parasites, salt, heavy metals, and emerging contaminants in water reused for any purpose considerably limits the water reuse approach. Even though secondary treatment (traditional activated sludge or trickling filter/aerobic treatment) is sufficient for some applications, tertiary treatment is essential where human contact is involved, like for watering lawns and gardening. Therefore, to decrease the risk of public health as well as surface and groundwater pollution, soil contamination, and risk to aquatic life, nanotechnology is being utilized to advance treatment as well as to monitor the water quality technologies. According to water quality standards in various reuse categories, treatment by nano-agents can either be added to conventional treatment methods or applied alone. The different categories of utilizing reclaimed water (except overall water demand reduction, which is the most important advantage) are listed in Table 3.5 and explained in the following subsections.

Nanotechnology application in treating and monitoring water and wastewater at different quality levels for different purposes is shown in Figure 3.2 (Qu *et al.*, 2013b). These nanotechnology-enabled modules can be similarly applied to drinking water treatment.

In the water recycling process, producing high-quality reclaimed water for non-potable reuse is a waste of time and money. As shown in Figure 3.2, providing such high-quality water is not necessary for non-potable water uses. Appropriate technologies should be applied to implement 'fit for purpose' water reuse.

3.6.1 Impact of nanotechnology on agricultural water demand

Due to the presence of emerging pollutants in wastewater, their treatment for reuse in agricultural purposes is essential. Based on the quality of wastewater, advanced treatment may be essential. Nanomaterials for disinfection and nano-adsorbents are effective in both treating harmful contaminates of wastewater and decreasing the salinity of drainage effluents, which convert them to suitable new water resources for reuse in irrigation.

As far as the economic aspect is concerned, the level of the cost-effectiveness of reuse projects depends on the volume of reclaimed water used and the quality of water (Urkiaga *et al.*, 2008). Indeed, the more water used, the more cost-effective the project. Therefore,

Table 3.5 Details of water reuse in various sectors (Urkiaga *et al.*, 2008).

Consumption Sector	Application
Agricultural irrigation	Commercial nurseries Crop irrigation
Industrial recycling and reuse	Cooling Process water Boiler feed Heavy construction
Urban uses (landscape irrigation, non-potable and potable reuse)	Parks Greenbelts School yards Residential uses Cemeteries Freeway medians (central reservations) Golf courses Fire protection Air conditioning Toilet flushing Blending in water supply reservoir Pipe to pipe water supply
Recreational/environmental uses	Fisheries Snowmaking Lakes, ponds and marsh enhancement Stream flow augmentation Groundwater replenishment Salt water intrusion control

reclaimed water utilization for irrigation is the most cost-effective approach in comparison to other applications on account of the high volume of water required and lower quality needed, in comparison to industry and potable usage. Its cost-effectivity can also be higher if the reclaimed water is used for irrigating non-food agricultural products. Furthermore, not only does it decrease freshwater demand, but also reduces diversion costs stemming from reduced demand. Another advantage of reclaimed water is the saving in fertilizer costs. The presence of nitrate and phosphate in treated water can enrich the soil and improve yield and productivity (Urkiaga *et al.*, 2008).

3.6.2 Impact of nanotechnology on industrial water demand

Nanomaterials have broad application in industrial effluent treatment which is accompanied by numerous advantages. The obvious benefits of nanotechnology applications in industrial water treatment are the reduction in raw materials, reagents, and water or heat consumption, as well as reduced environmental penalties resulting from wastewater discharge (Urkiaga *et al.*, 2008). Applied nanomaterials in various industries are listed in Table 3.6.

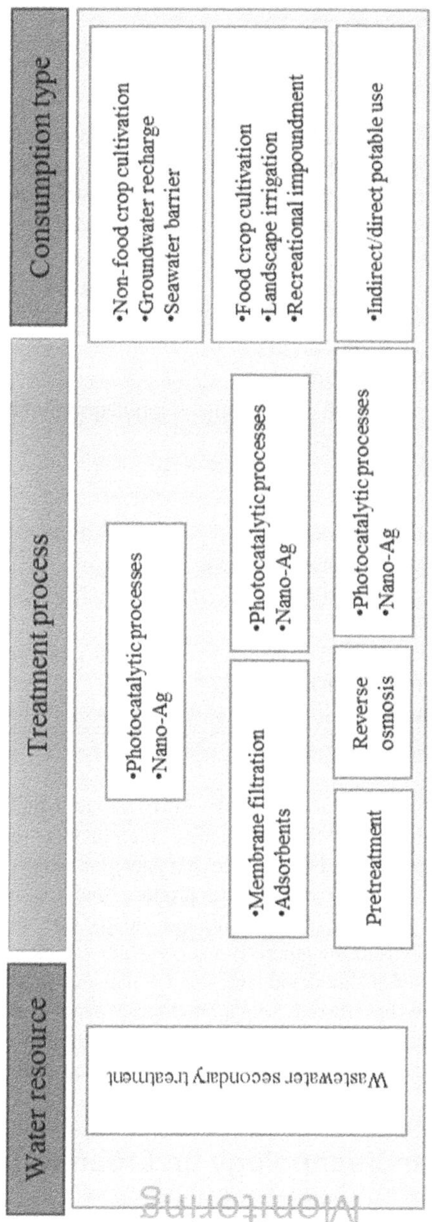

Figure 3.2 wastewater tertiary treatment systems with nanotechnology-enabled modular design for differential water reuse scenarios.

Table 3.6 Nanomaterials employed in various industrial wastewater.

Industry Type	Pollutants	Nanotechnology Applied
Steel and coke	Cyanides, thiocyanate, phenol, ammonia, oil, and grease	Nanofiltration membranes
Pharmaceutical	Drug residues with high BOD, COD, toxic substances, surfactants, and volatile organic compounds	Nano-adsorbents (e.g., CNT, magnetic nanocomposite), nano-photocatalysts (e.g., TiO_2 nanofibers), nanomembrane
Paper and pulp	Ligneous material, primarily resin acids and fatty acids, inorganic compounds containing sodium	Nanofiltration
Leather	Toxic metals like chromium, arsenic, and organic compounds such as organic dyes	Nanofiltration membranes
Petroleum refinery	Polycyclic aromatic compounds, sulfur, nitrogen, heavy metals, and other chemicals	Nanofiltration membranes
Textile	Suspended solids, nitrogen, heavy metals, dyes, COD, color, acidity	Nanofiltration membranes

Almost all the industries mentioned in Table 3.6 consume a high volume of water; however, thanks to membrane-based technologies, high-quality reusable water from wastewater streams can be produced. For instance, the paper industry is a water-consuming industry in which $150\,m^3$ water is required to produce one ton of paper and allied products. By applying a downstream nanofiltration loop, safe reusable water at a reasonably high level is produced. Singapore is a successful example of water reuse in the industry sector, with 32% of its total industrial water demand recycled and reused. A shortage of water resources has forced Singapore to use nanotechnologies to improve water availability. Generally, industrial wastewater treatment through nanotechnology is a common solution to combat water shortage and environmental pollutions in developed countries.

3.6.3 Impact of nanotechnology on urban water demand

Urban water demand is divided into potable and non-potable sectors. To reach quality standards, some or all of the nanotechnology treatments including disinfection, media filtration, and nanomembranes should be implemented. Harvested rainwater, urban stormwater, and residential and commercial effluent are graywater resources in urban areas

Table 3.7 Relationship between treatment cost, volume and quality of treated water.

Treatment System Type and Capacity	Energy Intensity of Treatment Process (kWh/kL)		
	5 (Mlit/day)	5–200 (Mlit/day)	Greater than 200 (Mlit/day)
Centralised potable	0.75–2.00	0.20–0.90	less than 0.80
Centralised non-potable	0.48–2.00	0.25–0.75	less than 0.55

Mlit/day: million liters/day.

that are treated for potable and non-potable reuse. In coastal cities, seawater desalination is also another important source of water which is usually introduced as a backup in arid and populated areas. Nanotechnology treatments are utilized in centralized or decentralized form in cities. In decentralized treatment systems, the volume of effluent is lower than in centralized systems. Although the cost of treatment on a small scale is higher than for centralized treatment, the cost of consumed energy for diversion and distribution in the centralized systems is significant. Also, the final treatment cost depends on the volume and quality of the treated water, which indirectly affects the consumed energy for treatment and its related cost. This concept is shown in Table 3.7.

Centralized treatment systems in the USA and Australia supply 5% and 30% of their urban water needs through water reuse, respectively. The non-potable centralized system requires a dual-pipe system that has high pumping energy consumption due to the long distance between treatment plant and users. However, reclaimed water of a potable standard could be injected directly to the distribution network, or enter potable water supply resources indirectly. In other words, treated water could be either released to surface streams or recharged groundwater in the vicinity of treatment plants to eliminate transport costs. Another reason for the indirect use of reclaimed water for potable purposes is the public acceptance of water reuse (Setty *et al.*, 2019). Generally, nanotechnologies provide a wide range of options for satisfying part of urban water demand through reclaimed water, which can be useful in a wide range of circumstances.

3.6.4 Impact of nanotechnology on the environment and recreational water demand

On the one hand, contaminants in wastewater are regarded as a serious threat to the environment. On the other hand, severe water withdrawal from limited freshwater resources puts pressure on the environment. To solve both problems, nanotechnologies are employed. Reuse of treated water by nanotechnology brings the following benefits:

- increases river flow quality and quantity for downstream users and improves river aesthetics;
- reduces environmental impacts (such as wetland and lake droughts), due to less consumption of fresh water resources;
- reduces impacts on fisheries and aquatic life;

- improves the quality of seawater and beaches;
- reduces the risk of salinity intrusion in aquifers
- reduces the risk of aquifer subsidence.

Additionally, reused water for environmental purposes improves recreational values of waterways and helps recreational activities flourish. Further, better public health and an increase in fauna and flora species due to the creation of new recreational zones, parks, gardens, and green areas are subsequent positive effects.

Generally, a closed water consumption cycle lessens the effects of competition between humans and wildlife over limited freshwater sources. Despite the broad advantages of water reuse mentioned above, the potential for water reuse and recycling has not yet been fully exploited in the world. To achieve sustainable water management, authorities should allocate funds for water reuse and for further research of nanotechnology applications in water reuse (Machado *et al.*, 2019). Although nanotechnology performs effectively in contaminant removal from water resources, more research on the nanomaterials' impacts on the environment is required. Due to the probable harmful effects which threaten organisms exposed to nanomaterials, the factors affecting these impacts – such as concentration, presence of organic or inorganic materials, the lifetime of the nanomaterials – should all be assessed. Coming to an informed conclusion needs more research about the advantages and disadvantages of nanotechnology applications in terms of environmental-related issues (Gehrke *et al.*, 2015).

3.7 CASE STUDIES

In this section, the application of nanotechnology in water reuse in three real cases is presented. These cases differ in various aspects, including economic conditions, population, and geographical features. Employing nanotechnology in water reuse and desalination in numerous environments reveals its wide application capability, with some limitations.

3.7.1 Namibia

Namibia is an arid country in Southern Africa. Windhoek, the capital city of Namibia, is the most populated city in the country with escalating water demand. At the time of the last census in 2012, around 300,000 people were living in Windhoek using 27 million m^3 of water. Windhoek experiences severe water scarcity: less than 360 mm/year rainfall, evaporation up to 50% of the dam storage capacity, and no perennial rivers. Surface water resources, including three dams located 60–200 km away from the city, and aquifers supplying 17 Mm^3 and up to 5 Mm^3 water, respectively, on average (Lafforgue & Lenouvel, 2015). The remaining demand is supplied through reclaimed water. The establishment of a closed water loop dates back to the 1960s when authorities embarked on a water resources expansion. Since the surface water resources depend heavily on meteorological conditions, water demand has been successfully met by emphasizing water reuse. The different water sources and their relationships are depicted in Figures 3.3 and 3.4 (Lafforgue & Lenouvel, 2015).

To reuse effluent, two water reclamation plants treat the wastewater. Activated sludge and biofilter systems are pretreatment steps before the effluent is sent to the treatment plants. The

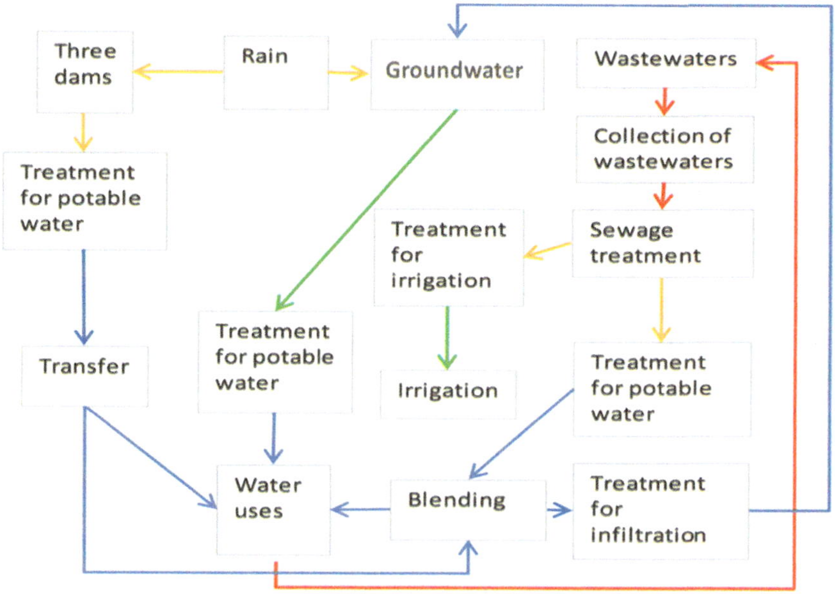

Figure 3.3 The urban water cycle in Windhoek (red: wastewater, yellow: raw water, green: treated water for non-potable uses, blue: drinking water).

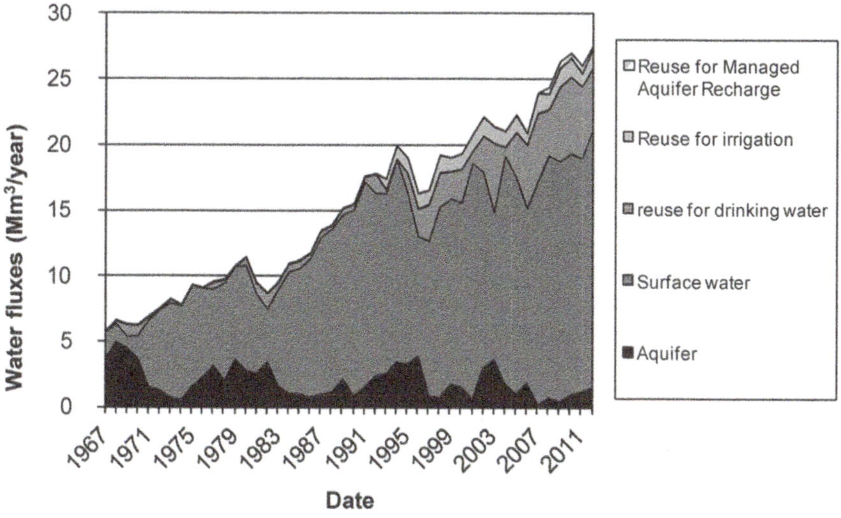

Figure 3.4 trends in water production in Windhoek as a function of supply source.

Table 3.8 Windhoek's treatment plants properties.

Name	Establishment/ Development Date	Capacity (m^3/day)	Reclaimed Water Quality	Use Sector	Portion of Total Annual Demand
Old Goreangab	1968/1997	7500	Semi-purified	Irrigation of the city's parks and gardens	8%
New Goreangab	2002	21,000	High quality	Potable water, aquifer recharge	26%

treatment processes and other features of treatment plants are summarized in Table 3.8 (Lafforgue & Lenouvel, 2015).

To produce high-quality water for supplying potable water as well as for recharging the aquifer in the southern part of Windhoek, the following process is carried out in which nanotechnology also takes part: pre-ozonation, coagulation-flocculation, dissolved air flotation, dual media filtration, ozonation, biological active carbon filtration, adsorption, ultrafiltration, and chlorination. In the ultrafiltration step, nanofiltration membrane filters are applied to guarantee disinfection and to control salinity, nitrate, and dissolved organic carbon (DOC) levels. Due to the high percentage of evaporation in Windhoek, the surplus reclaimed water is injected to boreholes located in the southern part of the city for managed aquifer recharge (MAR) which has offered storage possibilities of up to 11 Mm^3 per year. Water management and energy assessments show that MAR is the optimal solution with relatively low cost (Lafforgue & Lenouvel, 2015).

3.7.2 Singapore

Singapore with its 5.4 million inhabitants is a densely populated island city-state. At the time of Singapore's independence in 1965, authorities wished to avoid dependency on Malaysia's water resources. On the one hand, the continuous economic and population growth led to a rise in total water demand (Lafforgue & Lenouvel, 2015). On the other hand, the limitations of the water resources motivated the authorities to find alternatives to improving water consumption per capita. Therefore, they focused on reducing water demand and increasing the availability of water. Incentive-based billing, communication campaigns, and water-saving devices were all used to reduce water demand, while a final goal was set of reaching around 140 liters per capita per day by 2030. As far as reducing water losses has been concerned, the efficiency of the water distribution network has been increased to approximately 95%, while in most developed countries this figure was 65–85% in 2006. Singapore's water demand is met by four main sources of water (Figure 3.5) (Lafforgue & Lenouvel, 2015).

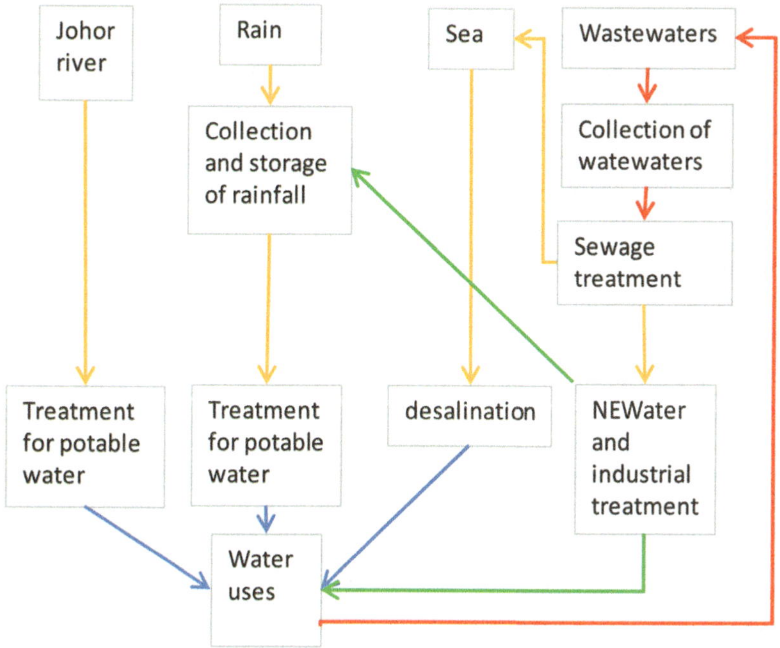

Figure 3.5 The urban water cycle in Singapore (red: wastewater, yellow: raw water, green: treated water for non-potable uses, blue: drinking water).

Imported water from the Johor river supplies about 40% of Singapore's annual water demand (250 Mm^3), while the target is to minimize this share by 2061. Since the average annual rainfall in Singapore is considerable (2400 mm), rainwater harvesting has become the second source of water supply. To use the collected rainwater, some major urban planning changes have been implemented, including building new water treatment plants that employ nanofiltration membranes to treat water. Rainwater has been harvested from two-thirds of the total watershed area, while the long-term goal is to raise this to 90% of the catchment area (Lafforgue & Lenouvel, 2015).

The third water resource is reclaimed water, called NEWater. To improve the reclaimed water, quality nanofiltration, reverse osmosis, and UV treatment are applied. In 2012, 110 Mm^3 – equal to 18% of the total water demand of the country – was supplied by NEWater. Around a quarter of reclaimed water was produced from municipal wastewater and used by the petrochemical industry; 32% of industrial and commercial needs were met by this source, and 2% of the reclaimed water was combined with harvested rainwater to supply drinking water.

The fourth source of water is seawater desalination, which is used as a backup source due to its high energy costs. Nanofiltration and reverse osmosis technologies are employed in two desalination plants in Singapore, with a 165 Mm^3 cumulative capacity (Lafforgue & Lenouvel, 2015).

One benefit of using nanotechnologies in Singapore water treatment is the reduction in the energy footprint of water production. Energy consumption in treatment plants has been reduced by replacing the activated sludge process (i.e., aeration and final clarification) with a membrane-based process (microfiltration or ultrafiltration) followed by a nanofiltration membrane bioreactor.

3.7.3 Saudi Arabia

Saudi Arabia is one of the largest arid countries in the tropical and subtropical desert region of the Middle East without perennial rivers or lakes. Seasonal rivers created by flash flooding provide roughly 0.45 Mm^3 runoff annually, which is captured in more than 200 dams across the country. The temperature in some areas rises to more than 50°C, while the long-term average rainfall varies between 100 and 200 mm annually with an average of 114 mm per year (DeNicola *et al.*, 2015). Although Saudi Arabia is one of the wealthiest nations in terms of oil resources, it suffers severely from a shortage of natural renewable water resources. In addition, mismanagement in agricultural water consumption as well as a water-consuming affluent lifestyle (250 L per capita daily), resulting from a low tariff structure for water (US$0.03/$m^3$), have deteriorated its water-shortage condition and depleted precious groundwater sources. Considerable population growth has also increased annual water demand by 8.8%. In 2010, renewable sources supplied 34% of water consumed, whilst the remaining requirements were supplied by nonrenewable groundwater, known as 'fossil' water. All these factors have forced authorities in Saudi Arabia to look for alternatives for improving water availability and therefore desalination technologies as well as water reuse have replaced dwindling groundwater resources in supplying domestic water demand. Also, to reduce water consumption in the agriculture sector, outsourcing of food crops has been introduced (DeNicola *et al.*, 2015).

At present, desalination supplies 60% of water demand in Saudi Arabia, the largest producer of desalinated water in the world. To be more precise, more than 70% of the country's drinking water is supplied by desalination that utilizes nanotechnology. As there are not any other water resources, authorities endure the high capital cost and energy consumption of both the water production process and transmission from desalination plants. Around half of the domestic oil production of Saudi Arabia is consumed by desalination plants. Although this new source of water is vital for the country, it will cause a serious economic crisis if their growing water consumption trend does not change in the future (DeNicola *et al.*, 2015).

Water recycling and reuse is another efficient way used by Saudi Arabia to curb water demand. Presently, roughly 20% of the total wastewater is collected, treated, and reused. The reclaimed water is used for restricted irrigation (limited to fodder and industrial crops, cereals, trees, and pastures) in Saudi Arabia and has resulted in increased yield and profit for some products owing to the nutrients in treated water. Saudi Arabia plans to take third place in sewage collection and treatment in the world by investing US$23 billion in this sector. This huge investment is necessary to assure the complete removal of contaminants from wastewater through novel technologies like nanomaterials (DeNicola *et al.*, 2015).

Saudi Arabia is a great example of how traditional poor water management strategies put pressure on water availability and obliged authorities to take sustainable measures. Now, Saudi Arabia serves as an example of water-scarcity management by developing desalination plants and expanding water recycling processes and infrastructure.

Table 3.9 Comparison of the case studies.

Case Study	Climate	Available Water Resource	Economic Condition	Population Density	Sector in Which Nanotechnology is Applied
Namibia	Arid	Groundwater, Surface water	Poor	Low	Rainwater and wastewater treatment for reuse
Singapore	Humid	Groundwater, surface water, seawater	Wealthy	High	Rainwater and wastewater treatment for reuse, desalination
Saudi Arabia	Arid	Groundwater, surface water, seawater	Wealthy	Low	Municipal wastewater treatment for irrigation, desalination

3.7.4 Case studies comparison

The case studies presented above show the capabilities and necessity of nanotechnology applications in sustainable water resource management in a wide range of economic, geographic, social, and environmental conditions. The case studies are compared in Table 3.9.

In all three case studies, nanotechnologies have been applied to improve water availability, while they were accompanied by other effective strategies in some cases. For instance, water authorities in Namibia and Singapore strive to reduce household water consumption through pricing policy. In contrast, low water prices and high water demand in Saudi Arabia threaten its future.

3.8 SUMMARY

Population growth, climate change, and water resource scarcity guide decision-makers toward sustainable water resource management strategies. In a sustainable approach, water demand is managed within a loop which necessitates water reuse. On account of the presence of numerous contaminants in wastewater and sometimes in freshwater resources in the modern world, more efficient treatment technologies are necessary. Nanotechnologies with the capability of removing a wide range of contaminants and minerals not only make water reuse for different purposes feasible but also provide a new source of water by effective desalination. Nowadays, desalination by nanotechnology is an important source of water in many coastal cities in arid regions around the world. Generally, applications of nanotechnology in water reuse can be summarized as disinfection, filtration, and monitoring. Yet, further research is essential to propagate the application of nanoscience in real-time water quality and quantity monitoring. Other than that, nanomaterials can also be used to improve the supply network through leakage reduction and stronger pipelines. However, the use of nanotechnology for self-healing in PVC pipes still requires further investigation. Also, employing nanoparticles as working fluids instead of pure water has slightly decreased industrial water consumption. In conclusion, the impact of nanotechnology on the water industry at present is beyond question, while more research should be carried out to make this novel science more cost-effective.

REFERENCES

Ahopelto S. and Vahala R. (2020). Cost–benefit analysis of leakage reduction methods in water supply networks. *Water*, **12**(195).

Anis Sh. F., Hashaikeh R and Hilal N. (2019). Functional materials in desalination: a review. *Desalination*, **468**(114077).

Awasthi A., Jadhao P. and Kumari K. (2019). Clay nano-adsorbent: structures, applications and mechanism for water treatment. *SN Applied Sciences*, **1**(9), 1076.

Bayda S., Adeel M., Tuccinardi T., Cordani M. and Rizzolio F. (2019). The history of nanoscience and nanotechnology: from chemical-physical applications to nanomedicine. *Molecules*, **25**(1), 112.

Bilal M., Rasheed T., Mehmood S., Tang H., Fernando L., Naresh Bharagava R. and Iqbal H. M. N. (2020). Mitigation of environmentally-related hazardous pollutants from water matrices using nanostructured materials: a review. *Chemosphere*, 126770.

Choi H. J., Yu S. W. and Kim K. H. (2016). Efficient use of Mg-modified zeolite in the treatment of aqueous solution contaminated with heavy metal toxic ions. *Journal of the Taiwan Institute of Chemical Engineers*, **63**, 482–489.

Chung Y. T., Mahmoudi E., Mohammad A. W., Benamor A., Johnson D. and Hilal N. (2017). Development of polysulfone-nanohybrid membranes using ZnO-GO composite for enhanced antifouling and antibacterial control. *Desalination*, **402**, 123–132.

Chung T.-S., Zhao D., Gao J., Lu K., Wan C., Weber M. and Maletzko C. (2019). Emerging R&D on membranes and systems for water reuse and desalination. *Chinese Journal of Chemical Engineering*, **27**(7), 1578–1585.

DeNicola E., Aburizaiza O. S., Siddique A., Khwaja H. and Carpenter D. O. (2015). climate change and water scarcity: the case of Saudi Arabia. *Annals of Global Health*, **81**(3), 342–353.

Dhakras P. A. (2011). Nanotechnology applications in water purification and waste water treatment: a review. International Conference on Nanoscience, Engineering and Technology (ICONSET 2011), Chennai, India, pp. 285–291.

Drexler K. E. (1986). Engines of Creation: The Coming Era of Nanotechnology. Anchor Press, Doubleday, USA.

El-Zahab S. and Zayed T. (2019). Leak detection in water distribution networks: an introductory overview. *Smart Water*, **4**(1), 1–23.

Fernandez-Gonzalez C., Zhang B., Dominguez-Ramos A., Ibañez R., Irabien A. and Chen Y. (2017). Enhancing fouling resistance of polyethylene anion exchange membranes using carbon nanotubes and iron oxide nanoparticles. *Desalination*, **411**, 19–27.

Gehrke I., Geiser A. and Somborn-Schulz A. (2015). Innovations in nanotechnology for water treatment. *Nanotechnology, Science and Applications*, **8**(1–17). doi: 10. 2147/NSA.S43773

Graboski A. M., Martinazzo J., Ballen S. C., Steffens J. and Steffens C. (2020). Chapter 4: Nanosensors for water quality control. In: Nanotechnology in the Beverage

Industry, A. Amrane, S. Rajendran, T. A. Nguyen, A. A. Assadi and A. M. Sharoba (eds), Elsevier, pp. 115–128. See: https://doi.org/10.1016/B978-0-12-819941-1.00004-3 (accessed 27 May 2021).

M. F. Hossain (ed.). (2019). Sustainable Design and Build. Butterworth-Heinemann, New York, pp. 301–418.

Jiang N., Shang R., Heijman S. G. J. and Rietveld L. C. (2018). High-silica zeolites for adsorption of organic micro-pollutants in water treatment: a review. *Water Research*, **144**, 145–161.

Kim S. (2013). The study of heat transfer for nanofluid with carbon nano particle in an exhaust gas recirculation (EGR) cooler. *Heat and Mass Transfer*, **49**, 1051–1055.

Lafforgue M. and Lenouvel V. (2015). Closing the urban water loop: lessons from Singapore and Windhoek. *Environmental Science: Water Research & Technology*, **1**(5), 622–631.

Liemberger R. and Wyatt A. (2019). Quantifying the global non-revenue water problem. *Water Supply*, **19**, 831–837.

Lin Y.-F. and Liu C. (2009). Smart Pipe: Nanosensors for Monitoring Water Quantity and Quality in Public Water Systems. World Environmental and Water Resources Congress, Great Rivers, Kansas City, Missouri, May 17–21, pp. 1–8.

Lu H., Wang J., Stoller M., Wang T., Bao Y. and Hao H. (2016). An overview of nanomaterials for water and wastewater treatment. *Advances in Materials Science and Engineering*, **2016**, 4964828.

Machado A., Zimmermann G., Feksa L., Berlese D. and Tundisi J.G. (2019). Applications of nanotechnology in water treatment. *Revista Conhecimento Online*, **1**, 3–15.

Margeta K., Zabukovec Logar N., Šiljeg M. and Farkaš A. (2013). Natural zeolites in water treatment – how effective is their use. In: Water Treatment, W. Elshorbagy and R. K. Chowdhury (eds), IntechOpen, pp. 81–112. doi: 10.5772/50738

NTSC Committee on Technology. (2016). *Water Sustainability through Nanotechnology: Nanoscale Solutions for a Global-Scale Challenge*. 12. Published at https://www.nano.gov/ (accessed 27 May 2021).

Paul R., Kenway S. and Mukheibir P. (2019). How scale and technology influence the energy intensity of water recycling systems-an analytical review. *Journal of Cleaner Production*, **215**, 1457–1480.

Perarasu V. T., Arivazhagan M. and Sivashanmugam P. (2013). Heat transfer characteristics of TiO2/water nanofluid in a coiled agitated vessel provided with disk turbine agitator. *Chemical Engineering Communications*, **200**(6), 783–797.

Qu X., Alvarez P. J. J. and Li Q. (2013a). Applications of nanotechnology in water and wastewater treatment. *Water Research*, **47**(12), 3931–3946.

Qu X., Brame J., Li Q. and Alvarez P. J. J. (2013b). Nanotechnology for a safe and sustainable water supply: enabling integrated water treatment and reuse. *Accounts of Chemical Research*, **46**(3), 834–843.

Ramsden J. (2009). In: Nanotechnology Applications for Clean Water: Solutions for Improving Water Quality, N. Savage, M. Diallo, J. Duncan, A. Street and R. Sustich (eds), William Andrew Publishing, Boston.

Sahu J. N., Zabed H., Karri R. R., Shams S. and Qi X. (2019). Applications of nano-biotechnology for sustainable water purification. In: Industrial Applications of

Nanomaterials, S. Sabu Thomas, Y. Grohens and Y. B. Pottathara (eds), Elsevier, pp. 313–340.

Setty K., Loret J.-F., Courtois S., Hammer C. C., Hartemann P., Lafforgue M., Litrico X., Manasfi T., Medema G., Shaheen M., Tesson V. and Bartram J. (2019). Faster and safer: Research priorities in water and health. *International Journal of Hygiene and Environmental Health*, **222**(4), 593–606.

Suwaileha W., Johnson D. and Hilala N. (2020). Membrane desalination and water re-use for agriculture: state of the art and future outlook. *Desalination*, **491**.

Tlili I. and Alkanhal T. A. (2019). Nanotechnology for water purification: electrospun nanofibrous membrane in water and wastewater treatment. *Journal of Water Reuse and Desalination*, **9**(3), 232–248.

Urkiaga A., de las Fuentes L., Bis B., Chiru E., Balasz B. and Hernández F. (2008). Development of analysis tools for social, economic and ecological effects of water reuse. *Desalination*, **218**(1), 81–91.

Wu M., Johannesson B. and Geiker M. (2012). A review: Self-healing in cementitious materials and engineered cementitious composite as a self-healing material. *Construction and Building Materials*, **28**, 571–583.

Zhai L., Narkar A. and Ahn K. (2020). Self-healing polymers with nanomaterials and nanostructures. *Nano Today*, **30**(100826).

Zhang W., Zheng Q., Ashour A. and Han B. (2020). Self-healing concrete composites for sustainable infrastructures: a review. *Composites Part B: Engineering*, **189**, 107892.

Chapter 4

Stormwater management by low impact development practices

Melika Mani[1], Sayed Bateni[2], Omid Bozorg-Haddad[3] and Amanda Cording[4]*

[1]*Department of Civil and Environmental Engineering, Louisiana State University, Patrick F. Taylor Hall, Baton Rouge, LA, 70803-6405, USA*
[2]*Department of Civil and Environmental Engineering and Water Resources Research Center, University of Hawaii at Manoa, Honolulu, USA*
[3]*Distinguished Professor, Department of Irrigation & Reclamation Engineering, Faculty of Agricultural Engineering & Technology, College of Agriculture & Natural Resources, University of Tehran, Karaj, Tehran, 31587-77871, Iran.*
E-mail: OBHaddad@ut.ac.ir
[4]*Water Resources Research Center, University of Hawaii at Manoa, Honolulu, USA*
**Corresponding author.*

ABSTRACT

Due to the effects of climate change, urban and suburban expansion, and urban pollutants on runoff quality and quantity, applying contemporary stormwater management approaches in urban areas have become more critical. Low impact development (LID) practices are environmentally friendly stormwater management methods, seeking to replicate the natural hydrologic regimes in urban areas. They have become popular methods to reduce/prevent adverse stormwater runoff impacts in urban catchments, mainly by improving on-site infiltration or harvesting and reusing runoff. This study introduces LID practices and the importance of using them. Thereafter, the structure, benefits, and limitations of common LID practices are explained to help water resource engineers and urban planners have a better understanding of these practices, and choose the most suitable LID practice based on the needs of the project and features of the site.

Editor: Omid Bozorg-Haddad
doi: 10.2166/9781789062144_0063

Keywords: Porous/permeable pavement, Vegetated/bio-swale, Rain garden, Bioretention cell, Infiltration trench, Green roof, Rainwater harvesting system, Rain barrel, Best management practices, Low impact development.

4.1 INTRODUCTION

Expanding urban, suburban and rural development has resulted in a decrease in permeable soils and vegetated land cover, and in an increase in impermeable surfaces such as asphalt for roads and parking areas. Impermeable surfaces increase stormwater peak flow and volume, due to reduced infiltration into the soil profile. The resulting discharge rates can be very high, resulting in erosion and flooding downstream when not properly managed. In addition, runoff carries contamination from impermeable surfaces such as streets and parking lots to receiving waters – which can be considered as one of the main polluting sources of receiving waters (Davis *et al.*, 2001). Furthermore, increasing air temperature due to global warming can result in rainfall intensification, thereby enhancing the risk of flooding in urban areas (Hettiarachchi *et al.*, 2018). Owing to these problems, stormwater management has become an important issue.

When stormwater is conveyed into conventional infrastructure such as concrete channels or pipes, the pollutants in the runoff become more concentrated, and can have ecological impacts on the receiving water bodies. Although the goal of conventional infrastructure is to prevent localized flooding, centralizing stormwater can actually increase flood risk and erosion downstream (Barkdoll *et al.*, 2016).

Low impact development (LID) – also known as water sensitive urban design (WSUD), and sustainable urban drainage system (SUDS) – is an approach to development that aims to mimic predevelopment hydrology and uses ecological engineering techniques such as green stormwater infrastructure (GSI) to remove pollutants in stormwater prior to re-use or replenishment of local groundwater supplies. LID can be structural or non-structural. Non-structural LID techniques include the preservation of open vegetated space within a development plan, lowering overall freshwater usage by installing low flow fixtures or incorporating recycled water for irrigation, and/or the selection of native or drought tolerant vegetation. Structural LID techniques are increasingly called GSI and include practices which are either introduced as a retrofit design to improve the performance of an existing urban drainage system or can be applied to new developments. Examples of GSI include permeable materials such as permeable asphalt, permeable concrete, or permeable pavers, vegetated roofs known as 'green roofs', bioretention systems also known as biofilters, rain gardens, vegetated swales, and rainwater harvesting. These techniques each function somewhat differently, but are all designed to slow the movement of stormwater from a site, reduce the total volume, and remove pollutants being carried, releasing cleaner water either to existing municipal infrastructure, groundwater, or storage for future re-use. LID techniques generally are intended to maximize evapotranspiration and pollutant uptake processes in vegetation, and increase residence times necessary for physical, chemical, and biological pollutant reductions. They techniques try to mimic the natural hydrologic conditions of the area. LID practices are often stormwater runoff control devices that reduce, purify, and sometimes reuse urban runoff. Instead of transferring runoff to other locations, these practices create conditions for runoff storage, evaporation, and infiltration. Application of LID controls can have many benefits, including: reduction of flooding and pressure on sewage treatment systems (where stormwater is connected to the sewer network); reduction of nutrient, bacteria, and sediment pollution reaching receiving waters; moderating the effects of climate change on stormwater runoff in urban areas; providing for additional freshwater sources in times of soft drought (i.e., rain water harvesting); replenishment of groundwater supplies; reduction in construction costs of urban infrastructures; and providing habitat for wildlife (Chang *et al.*, 2018; Chau, 2009; Fletcher *et al.*, 2013; Leimgruber *et al.*, 2019; Mooers *et al.*, 2018). Many experimental studies have evaluated the performance of LID practices, and shown their practical use in urban areas (Cording *et al.*, 2018; Davis, 2008; Dietz & Clausen, 2008; Emerson *et al.*, 2010; Fach *et al.*, 2011; Huang *et al.*, 2016; Jiang *et al.*, 2020; Kamali *et al.*, 2017; Lucke & Nichols, 2015; Ma *et al.*, 2019; Wang *et al.*, 2019).

A number of works have investigated the capability of LID practices to mitigate the adverse effects of urban development on urban runoff quantity and quality, and introduced computational models for LID practice simulations (Ahiablame *et al.*, 2012; Cording *et al.*, 2018; Dietz 2007; Eckart *et al.*, 2017; Hager *et al.*, 2019; Li *et al.*, 2019; Shafique & Kim 2015; Yang *et al.*, 2020; Zhu *et al.*, 2019). Overall, the literature indicates that LID controls are useful for qualitative and quantitative control of urban runoff, but they do need to be designed, implemented, and maintained appropriately. For each location, effective LID practices should be selected based on site characteristics such as land use, soil condition, slope, groundwater level, amount and type of pollutants in the area, precipitation amount and precipitation intensity (Mani *et al.*, 2019).

Of the LID systems reviewed in the literature, the main practices include porous/permeable pavements, vegetated swales, rain gardens, bioretention cells, infiltration trenches, green roofs, and rainwater harvesting systems. The functionality and usability of each of the above LID practices are discussed in this chapter.

4.2 TERMS AND DEFINITIONS

4.2.1 Best management practice (BMP)

Best management practices (BMP) are structural and non-structural practices, techniques, and restrictions to prevent the entrance of pollutions into receiving waters, limit soil erosion impacts and improve agricultural productivity. It is worth mentioning that LID practices are small structural BMPs. Examples of BMPs include applying fertilizers and pesticides wisely, managing household hazardous waste, afforestation and using different types of LID alternatives (Chau, 2009; Mtibaa *et al.*, 2018).

4.2.2 Detention

This approach reduces peak flow and flood damage by storing runoff in detention ponds for a short time, and gradually transferring it to the receiving streams or channels (Johannessen *et al.*, 2018). Detention ponds are dry most of the time except during and after precipitation, and thus they are called dry ponds. Sands and larger silt particles settle down in detention ponds, thereby improving water quality (Laramie County Conservation District, 2016).

4.2.3 First flush

First flush (FF) refers to a higher discharge of pollutants in the first portion of a runoff hydrograph, commonly applied as the first half inch or inch (1.27 or 2.54 cm) of rainfall (Baek *et al.*, 2015; Ma *et al.*, 2002; Stenstrom & Kayhanian, 2005).

4.2.4 Flood

Flood occurs when runoff overflow covers an area due to heavy rainfall. Water can overflow from water bodies, dams, rivers, and streams (Coates *et al.*, 2016).

4.2.5 Green infrastructure

Green infrastructure (GI) is considered as a network of natural and semi-natural areas that is designed and managed to provide a wide range of environmental and social benefits, such as stormwater management, climate adaptation, ecosystem conservation, and air quality enhancement (American Rivers, 2017; Chatzimentor *et al.*, 2020).

4.2.6 Infiltration-based technologies

Infiltration-based technologies restore runoff by recharging it into subsurface flows and groundwater (Fletcher *et al.*, 2013). This process also improves the quality of runoff by filtering it through the soil. Bioretention cells, porous pavements, rain gardens, infiltration trenches, and vegetated swales are infiltration-based LID alternatives (Eckart *et al.*, 2017).

4.2.7 Low impact development

Low impact development (LID) is an approach to development that aims to mimic predevelopment hydrology and uses ecological engineering techniques such as GSI to remove pollutants in stormwater prior to re-use or replenishment of local groundwater supplies. In other words, LID practices preserve the predevelopment hydrologic conditions of watersheds. They are mainly used to manage stormwater runoff and improve water quality. Porous pavement, bioretention cell, rain garden, vegetated swale, green roof, infiltration trench, and rainwater harvesting systems are popular LID practices (Ahiablame *et al.*, 2012; Chau, 2009; Cording *et al.*, 2017).

4.2.8 Maintenance operations

Maintenance operations are actions such as plant replacement, sediment removal, vacuum sweeping, cleaning drainage pipes, and spillway reconstruction to prevent blockage, preserve a desire level of performance and ensure long-term LID operation. These tasks are categorized into short-term (routine and more frequent), long-term (non-routine and less frequent) and major (rare). The various LID alternatives need different maintenance operations (Erickson *et al.*, 2010; Flynn *et al.*, 2012; Kang *et al.*, 2008).

4.2.9 Rainwater harvesting

Rainwater harvesting (RWH) is the collection and storage of runoff for use in irrigation, household water supply, and groundwater recharge (Helmreich & Horn, 2009).

4.2.10 Retention based technologies

Retention based technologies (LID practices) retain stormwater runoff by either attenuation of outflow or/and reduction of it by increasing evapotranspiration and capturing and reusing possibilities (Fletcher *et al.*, 2013; Hager *et al.*, 2019). Wetlands, ponds, green roofs, and rainwater harvesting systems are retention-based technologies (Eckart *et al.*, 2017).

4.2.11 Retention

Retention is a flood control technique that can improve quality of stormwater. A pond retains stormwater for a long period until it is lost due to evaporation and infiltration. Retention ponds are also called wet ponds as they permanently hold water. They remove stormwater pollutants by sedimentation, aquatic plants, and biological mechanisms.

4.2.12 Stormwater management

Stormwater management is an effort to control the quantity and quality of stormwater, particularly in urban areas. Traditional stormwater management practices are concerned with draining stormwater flow through channels, culverts, street gutters, and storm sewers into the downstream, while modern methods emphasize capturing and infiltrating stormwater runoff at local points to mimic the natural hydrology of the site. Traditional and modern methods are complementary to each other, and can be used together for a more efficient stormwater management (Akan & Houghtalen, 2003; Barkdoll *et al.*, 2016; Zhang *et al.*, 2017).

4.2.13 Stormwater

Stormwater is water that comes from rainfall or melting snow. It can infiltrate into the soil and recharge the groundwater, evaporate, or flow away. Less stormwater can infiltrate into the ground in urban areas because of the high level of imperviousness. This leads to an increase in stormwater runoff, ultimately causing flooding and transport of urban pollutants into the receiving waters (Mani *et al.*, 2019; Pekarek *et al.*, 2011).

4.2.14 Urban heat island

This phenomenon occurs when the air temperature in urban areas is 1–5°C (or 2–8°F) higher than its surrounding. The higher temperature in urban areas is mainly due to the higher level of absorption and storage of solar energy by urban structures, which is finally released into the atmosphere, and due to anthropogenic heat released from urban heat sources such as vehicles, power plants and air conditioning (Karamouz *et al.*, 2010; Rizwan *et al.*, 2008).

4.3 FUNDAMENTALS OF USING LID PRACTICES

LID practices are small-scale decentralized retrofit designs for stormwater management (Ahiablame *et al.*, 2012; Coffman, 2000; Newcomer *et al.*, 2014). By increasing the infiltration and evaporation of stormwater runoff through their detention and retention, LID practices reduce stormwater runoff volume, peak flow and mitigate flooding and erosion in urban areas. Also, reducing stormwater runoff volume and peak flow can prevent the overwhelming of the downstream stormwater infrastructure, and consequently reduce the size and cost of these infrastructures (Chang *et al.*, 2018; Gaffield *et al.*, 2003; Taguchi *et al.*, 2020; Zhen *et al.*, 2015).

LID practices can improve the quality of stormwater runoff by local infiltration and filtration (Bae & Lee, 2020; Chui *et al.*, 2016; Dietz & Clausen, 2008; Sadeghi *et al.*, 2018). By chemical sorption, biological processes and passive volatilization, LID practices improve stormwater runoff quality, thereby reducing the transfer of urban pollutants to receiving waters (e.g., streams, rivers, ponds, and lakes). This is vital for preserving coastal and freshwater ecosystems (Ahiablame *et al.*, 2012; Hunt *et al.*, 2010; Newcomer *et al.*, 2014).

Generally, LID practices can be categorized as either infiltration or retention-based technologies (Fletcher *et al.*, 2013; Perez-Pedini *et al.*, 2005). In infiltration-based technologies, soils with high infiltration potential are used at the top of the soil column to enhance runoff infiltration at local points (Ahammed, 2017). Infiltration practices include permeable pavements, bioretention cells, vegetated swales, and infiltration trenches. Retention-based technologies attenuate the stormwater peak flow and volume by capturing and retaining runoff and holding it for a period of time. Some examples of retention-based systems include rain gardens, green roofs and rainwater harvesting (Ahiablame *et al.*, 2012; Chau, 2009; Eckart *et al.*, 2017; Pazwash, 2011). Each has its own efficiency, depending on site features and soil condition, rainfall volume and pattern, amounts and types of pollution and climatic conditions (Hager *et al.*, 2019; Hsieh *et al.*, 2007; Liu *et al.*, 2017; Qin *et al.*, 2013; Sohn *et al.*, 2019; Zhang *et al.*, 2018).

4.4 THE ADVANTAGES OF LID PRACTICES

Due to growing population and urbanization over the past years, impervious areas have increased rapidly, which has increased the volume of stormwater runoff. Furthermore, climate change has led to more frequent extreme rainfalls, which has increased risk of flooding (Hettiarachchi *et al.*, 2018; Zahmatkesh *et al.*, 2014). A wide variety of pollutants (e.g., nitrate, phosphate, metals, hydrocarbons, organic materials, and pesticides) in urban areas originate from car exhausts, tire wear, landscape fertilizers, detergents, trash, fecal waste, atmosphere deposition, and soil erosion (Aryal *et al.*, 2010; Gunawardena *et al.*, 2013; Petrucci *et al.*, 2014). They are carried with stormwater and enter receiving waters (e.g., streams, rivers, ponds, and lakes), posing a major threat to the environment and human health. Hence, the application of modern stormwater management methods has become more important.

In traditional stormwater management practices, stormwater peak flow is reduced by transferring runoff to a nearby water body through gutters, pipes and impervious drainage structures. Alternatively, LID stormwater management practices retain and treat the stormwater, and can be used in concert with traditional practices.

LID practices have advantages over the traditional approaches because they not only reduce the runoff peak flow and volume but also purify runoff pollution via physical filtering, chemical absorption/adsorption, and biodegradation processes such as plant uptake and denitrification. They also offer opportunities to recharge groundwater, save stormwater for other purposes (e.g., irrigation, toilet flushing, terrace cleaning, laundry and car washing), reduce construction costs of urban drainage systems, and to create beautiful landscapes within urban areas.

4.5 DIFFERENT TYPES OF LID PRACTICES

In this section, the structure, advantages, and disadvantages of the most commonly used LID practices are introduced, namely: (1) porous/permeable pavements; (2) vegetated swales; (3) rain gardens; (4) bio-retention cells; (5) infiltration trenches; (6) green roofs, and (7) rainwater harvesting systems.

4.5.1 Porous/permeable pavements

A porous/permeable pavement (PP) looks similar to a traditional pavement, but has pores that allow runoff and snowmelt to infiltrate through it and reach the underlying soil layers. In addition to enhancing localized infiltration and reducing runoff, PPs remove pollutants from road runoff, such as petroleum hydrocarbons, which are broken down by microbial communities in the rock and soil layers below (Pratt *et al.*, 1999). Permeability and total storage capacity of PPs depend on several factors, including the type of pavement and its maintenance and subsurface storage reservoir depth. Permeable materials can be used on sidewalks, parking lots, driveways and roads, although the strength of the materials differ, which needs to be carefully considered when selecting a final product. Permeable materials are a cost-effective solution for populated cities with scarce and expensive land.

There are several types of permeable pavement materials, namely pervious concrete, porous asphalt, permeable interlocking concrete pavers (PICPs), and other options such as plastic and concrete grids (Minnesota Stormwater Steering Committee, 2013). PPs consist of 4 main layers (Figure 4.1): (1) a surface pavement layer; (2) a filter (optional, for additional pollutant removal); (3) a base layer of crushed stone with a mean grain size (d50) of 0.5–1 in (1.27–2.54 cm); and (4) a sub-base layer of larger stones with d50 of 2–3 in (5.08–7.62 cm).

PICPs and some other PP options such as permeable pavers, and concrete and plastic grid pavements generally have a small-sized aggregate bedding layer below the pavement surface to ensure a level surface for the pavers/grids (Eisenberg *et al.*, 2015).

Advantages of PPs include:

(1) extremely effective at reducing runoff volume and peak flow (risk of downstream flooding) at the source of runoff generation;
(2) decrease stormwater runoff pollutant loads;
(3) decrease the risk of cars slipping or skidding in rain;
(4) decrease the urban heat island effect by cooling down the pavement and its surrounding environment through vaporization of water in the pavement pores;
(5) prevent surface ponding and pavement freezing;
(6) no need to provide additional space for implementation outside of the project area – PPs are simple applied to an impervious surface with a deeper excavation for drainage.

Disadvantages of PPs include:

(1) more expensive than conventional pavements;

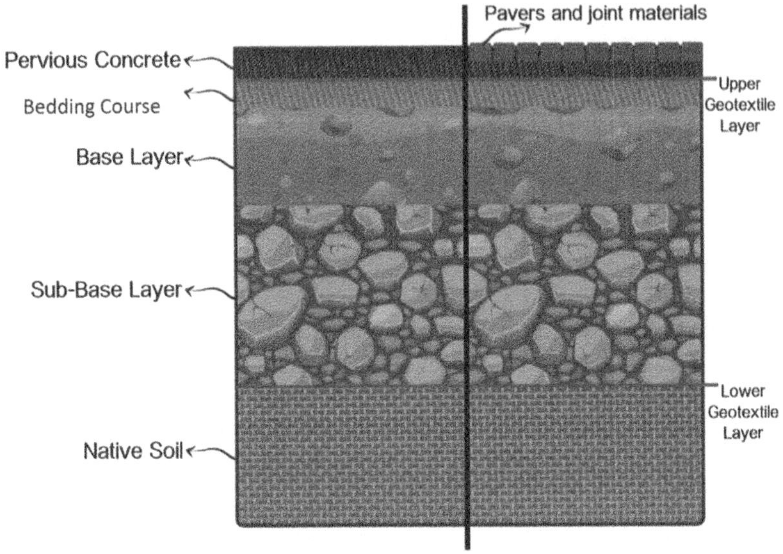

Figure 4.1 A schematic of a porous/permeable pavement.

(2) sensitive to clogging and thus care needs to be taken to prevent run-on of sediments from adjacent areas;
(3) need regular maintenance in the form of sweeping and/or washing to prevent the blockage of pavement pores;
(4) owing to deterioration by runoff, air infiltration, and subsequent oxidation, permeable pavements are less durable than impermeable ones (Scholz & Grabowiecki, 2007).

4.5.2 Vegetated swales

A vegetated swale (VS) is a wide, shallow channel with vegetation (Figure 4.2). It is used to increase infiltration and prevent erosion when runoff is transmitted to a downstream zone. A VS can be used in many applications instead of the traditional curb and gutter system, although care to prevent erosion in high peak flows is critical to the lifespan and usability of these systems. The low slope of a VS and its built-in obstacles promote runoff infiltration and reduce a significant portion of runoff contaminants. It is an effective LID practice for reducing total suspended solids (TSS) in stormwater runoff; however, metals and nutrient removal has been variable, and appears to depend on the VS's vegetation type and density, slope, surface area and length, and on the rainfall pattern (Boger *et al.*, 2018; Leroy *et al.*, 2016; Stagge *et al.*, 2012). Finally, VS runoff can be discharged into a city's drainage system or other LID controls. VSs can be implemented along roads, highways, sidewalks, and parking lots. They are very useful when constructed near highways because they can reduce zinc from zinc contaminants which remain on highways due to cars' braking, which pollutes the runoff (Cahill *et al.*, 2014; Councell *et al.*, 2004).

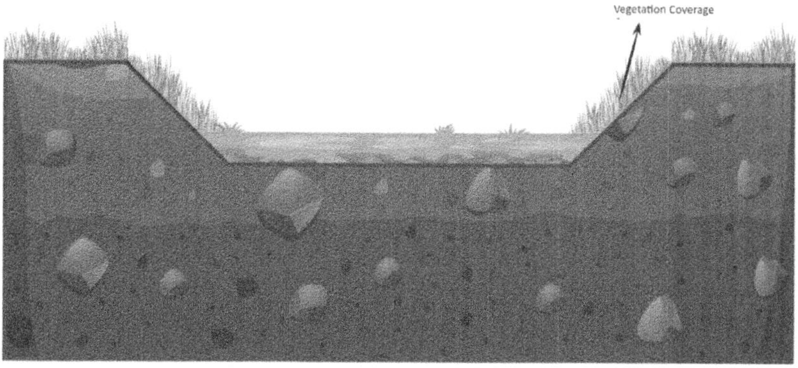

Figure 4.2 A schematic of a vegetated swale.

VSs generally consist of: (1) a shallow channel; (2) erosion and flood resistant vegetation; (3) some barrier to slow down the runoff. The critical parameters for designing a VS are the side and longitudinal slopes, top and bottom width, and the Manning coefficient of roughness of the VS (see Figure 4.2). VS advantages include:

(1) reduces runoff velocity;
(2) decreases stormwater runoff pollutant loads;
(3) less costly than the typical curb and gutter system;
(4) enhances the beauty of cities;
(5) easy to construct and maintain;
(6) applicable in low-conductivity soils.

VS disadvantages include:

(1) lower infiltration rates than other LID controls;
(2) unsuitable for areas with steep slopes because the flow velocity needs to be low (no more than 2 m/s) to prevent erosion and filter pollution;
(3) poor drainage into subsurface soils may lead to standing water and to breeding mosquitoes.

4.5.3 Rain gardens

A rain garden (RG) is a relatively shallow depression which is planted with native vegetation and typically does not include an underdrain. Native vegetation promotes evapotranspiration, takes up pollutants and removes nutrients through nutrient cycling, conserving soil porosity and improving the urban aesthetic. RGs temporarily retain rainwater coming from a roof downspout, driveway, street, or other open area (Figure 4.3). The collected rainwater then slowly evaporates, is taken up by the plants, and infiltrates into the ground. They look like a regular garden, but are typically covered with a bark mulch layer to protect soil from erosion and to increase the microbiological process for pollutant removal. Beneath the planting media is a sand or gravel infiltration bed to provide additional infiltration and storage. Compared to a standard lawn, a rain garden increases rainfall infiltration by about 30% (Freeborn *et al.*, 2012). They are considered to be one of the easiest and most

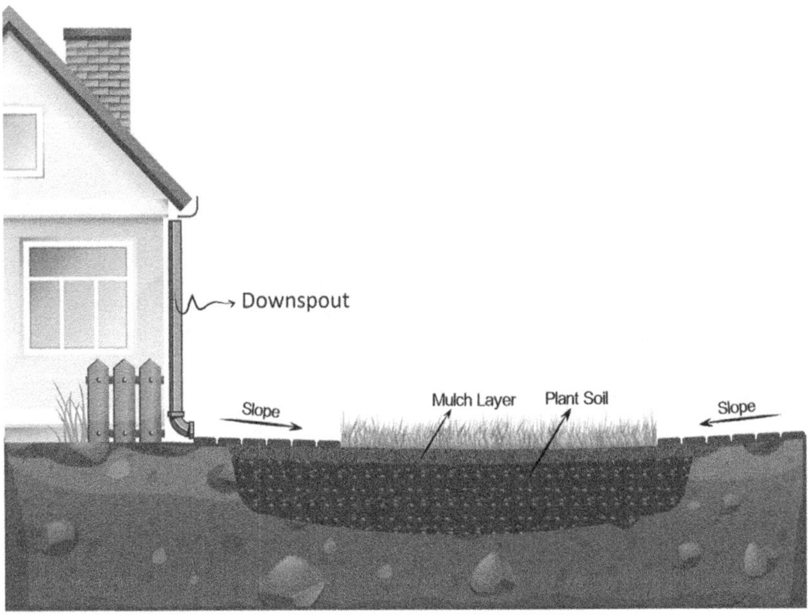

Figure 4.3 A schematic of a rain garden.

cost-effective methods to control runoff pollution at its origin. RGs typically drain stormwater within 12–48 hours, depending on underlaying soil conditions, and thus are not a suitable habitat for mosquitoes (The Groundwater Foundation, 2018).

RGs are particularly effective LID controls for residential areas, since they can be placed in an existing flower bed or lawn. To capture the runoff directly from a roof downspout, the RG should be placed on the downslope portion of the lawn, at least 10 feet away from the building foundations. More complicated RG structures with modified soil and drainage systems are referred to as bioretention.

RGs generally have the following elements: (1) ponding depth (depth of water standing on the surface); (2) an overflow to allow stormwater overflow during intensive rainfall; (3) native vegetation; (3) a mulch layer; (4) appropriate soil for plant growth; and (5) an optional sand or gravel bed (Figure 4.3).

Advantages of RGs include:

(1) reduce runoff volume and peak flow;
(2) decrease stormwater pollutant loads;
(3) recharge the regional groundwater aquifers;
(4) enhance the beauty of cities, providing habitat for butterflies, songbirds and many beneficial insects (US EPA, 2018), and preserve native vegetation;
(5) cost-effective, and easy to maintain;
(6) no fertilizer is needed because native vegetation is planted.

One disadvantage of RGs is that they are unsuitable for steep slope areas.

4.5.4 Bioretention cells

Bioretention cells (BCs) are similar to rain gardens in that, on the surface, they both appear to be a depressed garden with native vegetation. However, a BC is typically a flow-through system, as opposed to a retention system, and is comprised of a number of different soil layers and an underdrain to release water from the system after it has been treated. Native vegetation is used to promote evapotranspiration, enhance soil microbial activities, conserve soil porosity, increase infiltration, and improve uptake of pollutants (Cording *et al.*, 2018; Davis, 2008). BCs are typically used in larger contributing drainage areas than a rain garden is, since they contain an engineered soil layer at the top that allows for rapid infiltration, and a stone reservoir. The bioretention soil media (BSM) used typically consists of soil, sand and added organic matter such as low nutrient compost, coconut coir, biochar, or mulch (Cording *et al.*, 2018; Wardynski & Hunt, 2012) and can be blended locally using native soil excavated from the site, if it is clean and free of pollutants. The soil media is typically covered with a layer of wood mulch, which has been shown to reduce total heavy metals and petroleum-based pollutants through microbiological processes (Lake Superior Streams, 2009). Mulch also slowly breaks down over time, releasing nutrients to the soil and providing carbon for microbial denitrification. One downside of mulch is that it floats and can become a maintenance issue in heavy storm events if the system overflows. Rock mulch is one alternative which also protects underlying soil from erosion if flows are larger than anticipated.

In a BC, the runoff passes through the mulch and permeable soil layers before entering the underlying stone or gravel layer. The gravel layer is used for the storage of infiltrated water and typically contains an underdrain pipe. The underdrain is essential in regions with heavy precipitation and low underlying soil permeability (less than 0.5 inches (1.27 cm) per hour), to ensure proper functioning of the system. When underdrain pipes are elevated using an upturned elbow, as shown in Figure 4.4, they produce an internal water storage zone (IWS), which increases retention time, enhancing nitrate removal via denitrification (Cording *et al.*, 2018; Davis and McCuen 2005; Sun *et al.*, 2011).

Compared to other LID practices, bioretention is one of the most efficient, since it can store large volumes of water in a small footprint and utilizes multiple pollutant removal mechanisms such as sedimentation, infiltration, absorption of pollutants, nitrification, and denitrification (Davis, 2007).

BCs generally have the following elements: (1) ponding depth; (2) surface overflow discharge to convey flow that exceeds the capacity of the BC during heavy rainfall; (3) native vegetation; (4) a surface mulch layer; (5) engineered soil; (6) an underdrain; and (7) a gravel layer.

Advantages of BCs include:

(1) reduce runoff volume and peak flow;
(2) decrease stormwater runoff pollutant loads;
(3) recharge the regional groundwater aquifers;
(4) appropriate for areas even with low soil permeability as they can be equipped with a drainage pipe;
(5) enhance the beauty of cities and prove a suitable habitat for birds and butterflies.

A disadvantage of BCs is that they are unsuitable for steep slope areas.

Figure 4.4 A schematic of a bioretention cell.

4.5.5 Infiltration trenches

An infiltration trench (IT) is a long narrow excavated trench (1–4 m) filled with gravel, sand and a mixture of both (void space 30–40%). Typically, an IT does not have any outlet. Runoff is stored within the pore spaces of rocks and gradually enters the surrounding soil medium. As shown in Figure 4.5, a perforated pipe is often installed inside the trench to increase its storage capacity. In some designs, geotextile or filter fabric is wrapped around the perforated pipe, and along the sides and below the soil layer to prevent clogging. One important design consideration is that, if the underlaying soil is predominantly clay, the geotextile itself may clog. An alternative is to use a graded rock layer (smaller rocks on top of larger rocks above the pipe) in lieu of the geotextile. This also increases storage capacity and decreases maintenance frequency, but may cost more upfront.

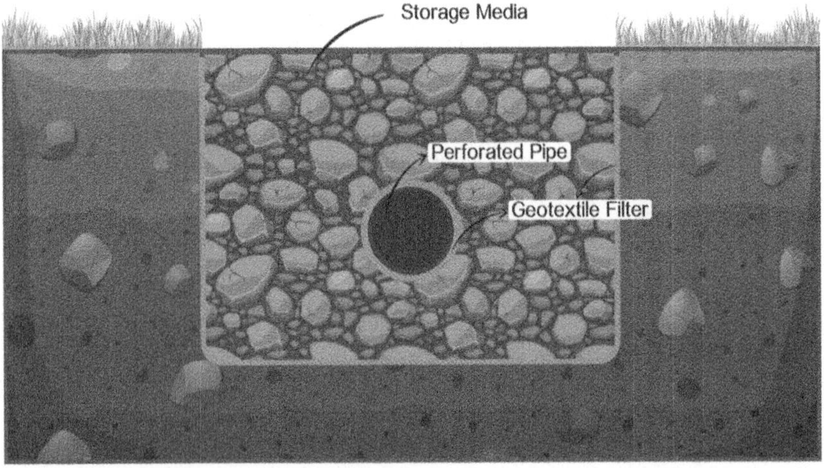

Figure 4.5 A schematic of an infiltration trench.

A positive overflow (bypass) is installed in ITs to convey high flows through the trench (PaDEP, 2006). It is typically embedded in subsurface soil layers to occupy less space, but this makes it difficult to access them and conduct maintenance operations. An observation well is typically also installed in the center of an IT to provide access for monitoring and maintenance operations.

ITs are used in parking lots, residential areas, and/or as part of larger drainage systems. Similar to the other infiltration-based LID controls, ITs are not suitable for areas with high groundwater levels and low soil hydraulic conductivity (Hager *et al.*, 2019; Locatelli *et al.*, 2015). ITs are designed based on runoff volume and soil properties. They can be effective LIDs in reducing many pollutants, such as TSS, biochemical oxygen demand (BOD), chemical oxygen demand (COD), and heavy metals (Hager *et al.*, 2019; Maniquiz *et al.*, 2010).

Maintenance operations should be conducted before and after the rainy seasons to keep up the efficiency of the trench and hinder its clogging. ITs are vulnerable to clogging from sediments in places where sediment loading is not low. In these cases, pre-treatments such as filter strips are needed to remove suspended solids and ensure the efficiency of the ITs (Field & Tafuri, 2006).

ITs generally consist of: (1) a stone-filled trench; (2) a perforated pipe (optional); (3) geotextile or filter fabric; (4) a positive overflow; (5) an observation well.

IT advantages include:

(1) reduce runoff volume and peak flow;
(2) decrease stormwater runoff pollutant loads;
(3) recharge the regional groundwater aquifers;
(4) applicable in areas with limited space because of their narrow dimensions.

IT disadvantages include:

(1) vulnerable to clogging without appropriate and regular maintenance;
(2) risk of groundwater contamination.

4.5.6 Green roofs

A roof of a building that is partially or entirely covered with vegetation is called a green roof (GR). GRs have a thick soil layer for growing vegetation, covered with plants, grass and trees (Dietz, 2007). They also include a drainage layer to ensure the excess water drainage, a filter layer to prevent the clogging of the drainage layer, a waterproofing membrane to prevent water infiltration into the roof structure, and a roof barrier to protect the waterproofing layer from the growing roots (Figure 4.6).

GRs reduce the stormwater runoff from buildings by increasing evapotranspiration. As 40–50% of urban impervious surfaces are generally built roofs, GRs can be an effective way to control urban stormwater (Mentens *et al.*, 2006; Sims *et al.*, 2016). Studies have shown that GRs can efficiently retain about 60–70% of precipitation (Dietz, 2007). The excess water of GRs can be discharged to a BC, a VS, or other LID alternatives for various uses (Barkdoll *et al.*, 2016).

GRs generally have the following elements: (1) soil layer; (2) filter layer; (3) drainage layer; (4) roof barrier; (5) waterproofing membrane.

Figure 4.6 A schematic of a green roof.

Advantages of GRs include:

(1) reduce the runoff volume and peak flow;
(2) enhance the beauty of cities;
(3) applicable in highly urbanized areas;
(4) reduce the heat-island effect in cities;
(5) no need for additional space;
(6) preserve air quality;
(7) prove a suitable habitat for birds and butterflies;
(8) can reduce building energy use for air conditioning due to temperature reduction as a result of the roof;
(9) provide a quieter environment by absorbing sounds.

Disadvantages of GRs include:

(1) more expensive than a conventional roof system to build and maintain;
(2) not applicable for steep roofs;
(3) not recommended for areas with low precipitation;
(4) may need additional watering;
(5) as stormwater remains on GRs, it may damage the roof's waterproof membrane;
(6) can increase the concentration of nutrient pollutants such as nitrogen, phosphorous in stormwater runoff from roof soils, plants, and any fertilizers used (Buffam *et al.*, 2016; Hager *et al.*, 2019; Monterusso *et al.*, 2004; Teemusk & Mander, 2011; Vijayaraghavan & Badavane, 2017).

4.5.7 Rainwater harvesting systems

Rainwater harvesting (RWH) systems are used to capture and retain rainwater from a roof (Figure 4.7). Rain barrels are the simplest form of RWH systems (Figure 4.8). In rain barrels, the rainwater is transferred from the roof to the barrel by eavestroughs (gutters) and downspout pipes. The barrel can be made of plastic, wood, ceramic, fiberglass, or metal. Most barrels have a tap that allows the collected rainwater to be easily used. Complex RWH systems have pumps, tanks, purification systems, and sanitation tanks.

RWH systems generally have the following elements: (1) a catchment area, which is a building rooftop in most cases; (2) a gutter and downspout pipes to convey runoff from the

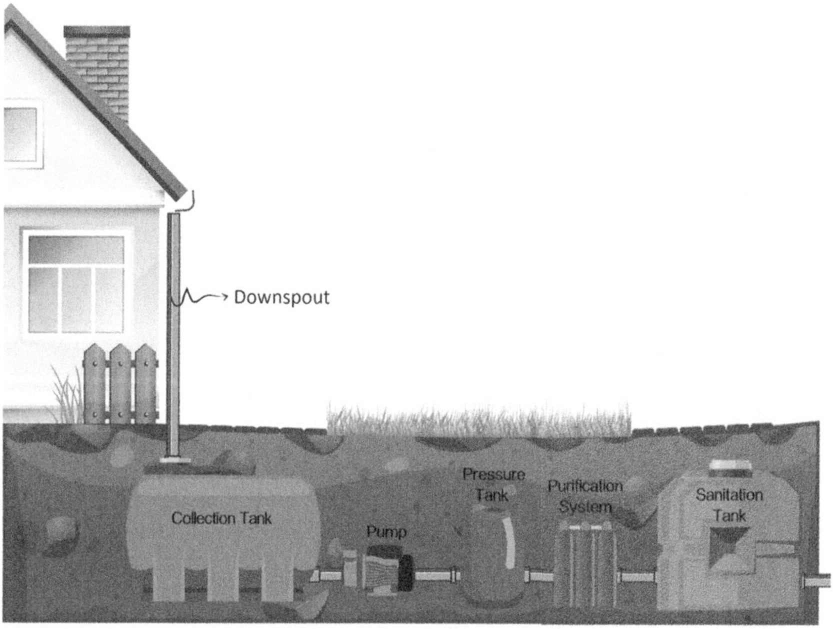

Figure 4.7 A schematic of a rainwater harvesting system.

catchment area to the storage tank; (3) a first flushing diverter to prevent the entrance of runoff water into the storage tank in the early stages of the storm, as that water contains a higher pollutant load; (4) a storage tank to collect runoff for reuse; (5) a water treatment system (optional).

The maintenance operation for RWH systems includes cleaning the roof (catchment), gutter, first flush diverter and storage tank, and checking the pipes and gutter to prevent rusting and leaking (Lee *et al.*, 2016).

Advantages of RWH systems include:

(1) reduce the runoff volume and peak flow;
(2) store stormwater for different purposes such as irrigation, toilet flush tanks, car washing, etc.;
(3) no need for a purification system if the collected water is used for non-drinking purposes.

Disadvantages of RWH systems include:

(1) installation of RWH systems can be expensive;
(2) RWH systems alone are not reliable as a water supply as rainfall is not regular and varies in different seasons;
(3) without a regular maintenance operation, RWH systems can become breeding sites for mosquitoes, rodents, and algae;

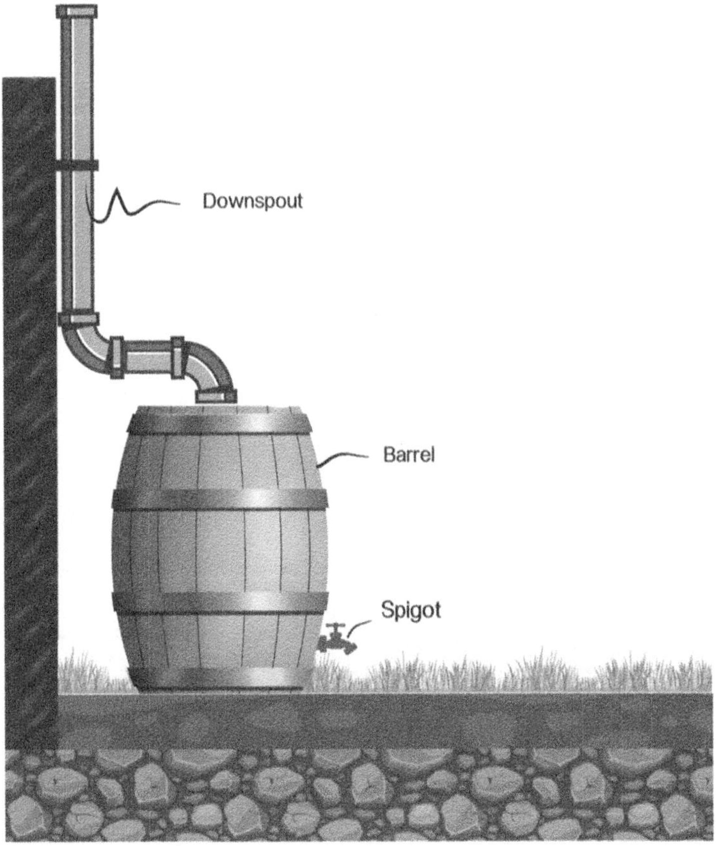

Figure 4.8 A schematic of a rain barrel.

(4) the quality of the collected water is not completely reliable if the RWH system does not have any purification;
(5) they have a limited storage capacity.

4.6 SUMMARY

The proper management of urban runoff via LID practices in cities has become inevitable because of the changes to the rainfall intensity and consequently increasing runoff volume and peak flow, and environmental issues such as rising levels of pollutants.

The LID practices are small-scale stormwater controls in urban areas to slow down, capture, and purify stormwater runoff. The most commonly used LID practices are introduced in this chapter. PP is pavement with permeable surface materials, and it has a base and subbase layer of crushed stone and optional underdrain system. VS is a shallow open channel that can replace traditional curb and gutter to convey, control, and improve stormwater through infiltration, sedimentation, and filtration. RG and BC are landscape depression consisting of vegetation, mulch, and growing media. In BC engineered soil and drainage system used to increase its water treatment and storage capacity. IT is a subsurface trench filled with porous material. GR is a building rooftop that is partially or entirely covered with a thick soil layer to accommodate plants. RWH is a technology to collect and storage of rainwater rather than allowing it to runoff. These LID practices advantages, and disadvantages are explained in this study, and should be selected on a case-by-case basis. Local results may vary, and monitoring may be useful to determine their regional effectiveness (Cording, *et al.*, 2017).

The following factors are crucial for selecting an appropriate LID practice:

(1) climatic conditions, and volume and distribution of rainfall;
(2) soil type, and land use and slope;
(3) groundwater level;
(4) the amount and type of stormwater pollutants, and the need to improve the environmental conditions;
(5) volume of runoff intended to be supplied by LID practices;
(6) the structure of city (urban design, location of infrastructures, and population and construction densities);
(7) costs and economic concerns;
(8) social acceptance;
(9) guidelines of water supply and waste water disposal.

REFERENCES

Ahammed F. (2017). A review of water-sensitive urban design technologies and practices for sustainable stormwater management. *Sustainable Water Resources Management*, **3**(3), 269–282.

Ahiablame L. M., Engel B. A. and Chaubey I. (2012). Effectiveness of low impact development practices: literature review and suggestions for future research. Water, Air, & Soil Pollution, **223**(7), 4253–4273.

Akan A. O. and Houghtalen R. J. (2003). Urban Hydrology, Hydraulics, and Stormwater Quality: Engineering Applications and Computer Modeling. John Wiley & Sons, Inc, Hoboken, NJ.

American Rivers (2017). 'What is green infrastructure?' See: https://www.americanrivers.org (accessed 1 September 2019).

Aryal R., Vigneswaran S., Kandasamy J. and Naidu R. (2010). Urban stormwater quality and treatment. *Korean Journal of Chemical Engineering*, **27**(5), 1343–1359.

Bae C. and Lee D. K. (2020). Effects of low-impact development practices for flood events at the catchment scale in a highly developed urban area. *International Journal of Disaster Risk Reduction*, **44**, 101412.

Baek S. S., Choi D. H., Jung J. W., Lee H. J., Lee H., Yoon K. S. and Cho K. H. (2015). Optimizing low impact development (LID) for stormwater runoff treatment in urban area, Korea: experimental and modeling approach. *Water Research*, **86**, 122–131.

Barkdoll B. D., Kantor C. M., Wesseldyke E. S. and Ghimire S. R. (2016). Stormwater low-impact development: a call to arms for hydraulic engineers. *Journal of Hydraulic Engineering*, **142**(8), 02516002.

Boger A. R., Ahiablame L., Mosase E. and Beck D. (2018). Effectiveness of roadside vegetated filter strips and swales at treating roadway runoff: a tutorial review. *Environmental Science: Water Research and Technology*, **4**(4), 478–486.

Buffam I., Mitchell M. E. and Durtsche R. D. (2016). Environmental drivers of seasonal variation in green roof runoff water quality. *Ecological Engineering*, **91**, 506–514.

Cahill T. A., Barnes D. E. and Spada N. J. (2014). Seasonal variability of ultra-fine metals downwind of a heavily traveled secondary road. *Atmospheric Environment*, **94**, 173–179.

Chang N. B., Lu J. W., Chui T. F. M. and Hartshorn N. (2018). Global policy analysis of low impact development for stormwater management in urban regions. *Land Use Policy*, **70**, 368–383.

Chatzimentor A., Apostolopoulou E. and Mazaris A. D. (2020). A review of green infrastructure research in Europe: challenges and opportunities. *Landscape and Urban Planning*, **198**, 103775.

Chau H. F. (2009). Green infrastructure for Los Angeles: Addressing Urban Runoff and Water Supply through Low Impact Development. Los Angeles, California, USA, pp. 6–34.

Chui T. F. M., Liu X. and Zhan W. (2016). Assessing cost-effectiveness of specific LID practice designs in response to large storm events. *Journal of Hydrology*, **533**, 353–364.

Coates G., Li C., Wright N. G. and Ahilan S. (2016). Investigating the flood responsiveness of small and medium enterprises using agent-based modelling and simulation. In: Flood Risk Management and Response, D. Proverbs and C. A. Brebbia (eds), WIT Press, Southampton, Boston, USA.

Coffman L. S. (2000). Low-impact Development Design: A New Paradigm for Stormwater Management Mimicking and Restoring the Natural Hydrologic Regime an Alternative Stormwater Management Technology. (EPA Report 841-B-00-003). Prince George's County Department of Environmental Resources and US EPA, Maryland County, USA.

Cording A., Hurley S. and Whitney D. (2017). Monitoring methods and designs for evaluating bioretention performance. *Journal of Environmental Engineering*, **143**(12), 05017006.

Cording A., Hurley S. and Adair C. (2018). Influence of critical bioretention design factors and projected increases in precipitation due to climate change on roadside bioretention performance. *Journal of Environmental Engineering*, **144**(9), 04018082.

Councell T. B., Duckenfield K. U., Landa E. R. and Callender E. (2004). Tire-wear particles as a source of zinc to the environment. *Environmental Science and Technology*, **38**(15), 4206–4214.

Davis A. P. (2007). Field performance of bioretention: water quality. *Environmental Engineering Science*, **24**(8), 1048–1064.

Davis A. P. (2008). Field performance of bioretention: hydrology impacts. *Journal of Hydrologic Engineering*, **13**(2), 90–95.

Davis A. P. and McCuen R. H. (2005). Stormwater Management for Smart Growth. Springer Science and Business Media.

Davis A. P., Shokouhian M. and Ni S. (2001). Loading estimates of lead, copper, cadmium, and zinc in urban runoff from specific sources. *Chemosphere*, **44**(5), 997–1009.

Dietz M. E. (2007). Low impact development practices: a review of current research and recommendations for future directions. Water, Air, and Soil Pollution, **186**(1–4), 351–363.

Dietz M. E. and Clausen J. C. (2008). Stormwater runoff and export changes with development in a traditional and low impact subdivision. *Journal of Environmental Management*, **87**(4), 560–566.

Eckart K., McPhee Z. and Bolisetti T. (2017). Performance and implementation of low impact development–a review. *Science of the Total Environment*, **607**, 413–432.

Eisenberg B., Lindow K. C. and Smith D. R. (eds) (2015). Permeable Pavements. American Society of Civil Engineers, Reston, VA, USA.

Emerson C. H., Wadzuk B. M. and Traver R. G. (2010). Hydraulic evolution and total suspended solids capture of an infiltration trench. *Hydrological Processes: An International Journal*, **24**(8), 1008–1014.

Erickson A. J., Gulliver J. S., Kang J. H., Weiss P. T. and Wilson C. B. (2010). Maintenance for stormwater treatment practices. *Journal of Contemporary Water Research and Education*, **146**(1), 75–82.

Fach S., Engelhard C., Wittke N. and Rauch W. (2011). Performance of infiltration swales with regard to operation in winter times in an Alpine region. *Water Science and Technology*, **63**(11), 2658–2665.

Field R. and Tafuri A. N. (eds) (2006). The use of Best Management Practices (BMPs) in Urban Watersheds. DEStech Publications, Inc.

Fletcher T. D., Andrieu H. and Hamel P. (2013). Understanding, management and modelling of urban hydrology and its consequences for receiving waters: a state of the art. *Advances in Water Resources*, **51**, 261–279.

Flynn K. M., Linkous B. W. and Buechter M. T. (2012). Operation and maintenance assessment for structural stormwater BMPs. World Environmental and Water Resources Congress 2012, ASCE, Reston, VA, pp. 3662–3673.

Freeborn J. R., Sample D. J. and Fox L. J. (2012). Residential stormwater: methods for decreasing runoff and increasing stormwater infiltration. *Journal of Green Building*, **7**(2), 15–30.

Gaffield S. J., Goo R. L., Richards L. A. and Jackson R. J. (2003). Public health effects of inadequately managed stormwater runoff. *American Journal of Public Health*, **93**(9), 1527–1533.

Gunawardena J., Egodawatta P., Ayoko G. A. and Goonetilleke A. (2013). Atmospheric deposition as a source of heavy metals in urban stormwater. *Atmospheric Environment*, **68**, 235–242.

Hager J., Hu G., Hewage K. and Sadiq R. (2019). Performance of low-impact development best management practices: a critical review. *Environmental Reviews*, **27**(1), 17–42.

Helmreich B. and Horn H. (2009). Opportunities in rainwater harvesting. *Desalination*, **248**(1–3), 118–124.

Hettiarachchi S., Wasko C. and Sharma A. (2018). Increase in flood risk resulting from climate change in a developed urban watershed-the role of storm temporal patterns. *Hydrology and Earth Science*, **22**(3), 2041–2056.

Hsieh C. H., Davis A. P. and Needelman B. A. (2007). Bioretention column studies of phosphorus removal from urban stormwater runoff. *Water Environment Research*, **79**(2), 177–184.

Huang J., He J., Valeo C. and Chu A. (2016). Temporal evolution modeling of hydraulic and water quality performance of permeable pavements. *Journal of Hydrology*, **533**, 15–27.

Hunt W. F., Traver R. G., Davis A. P., Emerson C. H., Collins K. A. and Stagge J. H. (2010). Low impact development practices: designing to infiltrate in urban environments. In: Effects of Urbanization on Groundwater: An Engineering Case-based Approach for Sustainable Development, N. Chang (ed.), America Society of Civil Engineers, Environmental Water Resources Institute, Reston, VA, pp. 308–343.

Jiang C., Li J., Li H., Li Y. and Zhang Z. (2020). Low-impact development facilities for stormwater runoff treatment: field monitoring and assessment in Xi'an area, China. *Journal of Hydrology*, **585**, 124803.

Johannessen B. G., Muthanna T. M. and Braskerud B. C. (2018). Detention and retention behavior of four extensive green roofs in three nordic climate zones. *Water*, **10**(6), 671.

Kamali M., Delkash M. and Tajrishy M. (2017). Evaluation of permeable pavement responses to urban surface runoff. *Journal of Environmental Management*, **187**, 43–53.

Kang J. H., Weiss P. T., Gulliver J. S. and Wilson B. C. (2008). Maintenance of stormwater BMPs frequency, effort, and cost. *Stormwater*, **8**, 18–28.

Karamouz M., Moridi A. and Nazif S. (2010). Urban Water Engineering and Management. CRC Press, New York.

Lake Superior Streams (2009). Lake Superior Streams: Community Partnerships for Understanding Water Quality and Stormwater Impacts at the Head of the Great Lakes. See: http://www.lakesuperiorstreams.org (accessed 18 July 2019). University of Minnesota-Duluth, Duluth, Minnesota.

Laramie County Conservation District (2016). Best management practices for stormwater runoff: Ponds. See: http://www.lccdnet.org/wp-content/uploads/Ponds.pdf (accessed 18 July 2019).

Lee J. Y., Kim H. and Han M. (2016). Importance of maintenance in rainwater harvesting systems: a case study. *Water Science and Technology: Water Supply*, **16**(1), 97–103.

Leimgruber J., Krebs G., Camhy D. and Muschalla D. (2019). Model-based selection of cost-effective low impact development strategies to control water balance. *Sustainability*, **11**(8), 2440.

Leroy M. C., Portet-Koltalo F., Legras M., Lederf F., Moncond'Huy V., Polaert I. and Marcotte S. (2016). Performance of vegetated swales for improving road runoff quality in a moderate traffic urban area. *Science of the Total Environment*, **566**, 113–121.

Li Q., Wang F., Yu Y., Huang Z., Li M. and Guan Y. (2019). Comprehensive performance evaluation of LID practices for the sponge city construction: a case study in Guangxi, China. *Journal of Environmental Management*, **231**, 10–20.

Liu Y., Engel B. A., Flanagan D. C., Gitau M. W., McMillan S. K. and Chaubey I. (2017). A review on effectiveness of best management practices in improving hydrology and water quality: needs and opportunities. *Science of the Total Environment*, **601**, 580–593.

Locatelli L., Mark O., Mikkelsen P. S., Arnbjerg-Nielsen K., Wong T. and Binning P. J. (2015). Determining the extent of groundwater interference on the performance of infiltration trenches. *Journal of Hydrology*, **529**, 1360–1372.

Lucke T. and Nichols P. W. (2015). The pollution removal and stormwater reduction performance of street-side bioretention basins after ten years in operation. *Science of the Total Environment*, **536**, 784–792.

Ma J. S., Khan S., Li Y. X., Kim L. H., Ha S., Lau S. L., Kayhanian M. and Stenstrom M. K. (2002). First flush phenomena for highways: how it can be meaningfully defined.

Proceedings of the ninth International Conference on Urban Drainage, 8–13 September, Portland, Oregon.

Ma Y., He W., Zhao H., Zhao J., Wu X., Wu W., Li X. and Yin C. (2019). Influence of low impact development practices on urban diffuse pollutant transport process at catchment scale. *Journal of Cleaner Production*, **213**, 357–364.

Mani M., Bozorg-Haddad O. and Loáiciga H. A. (2019). A new framework for the optimal management of urban runoff with low-impact development stormwater control measures considering service-performance reduction. *Journal of Hydroinformatics*, **21**(5), 727–744.

Maniquiz M. C., Lee S. Y. and Kim L. H. (2010). Long-term monitoring of infiltration trench for nonpoint source pollution control. *Water, Air, and Soil Pollution*, **212**(1–4), 13–26.

Mentens J., Raes D. and Hermy M. (2006). Green roofs as a tool for solving the rainwater runoff problem in the urbanized 21st century?. *Landscape and Urban Planning*, **77**(3), 217–226.

Minnesota Stormwater Steering Committee (2013). *Minnesota Stormwater Manual*, Wiki version. Minnesota Pollution Control Agency, St. Paul, MN, US. See: https://stormwater.pca.state.mn.us/ (accessed 27 May 2021).

Monterusso M. A., Rowe D. B., Russell D. K. and Rugh C. L. (2004). Runoff water quantity and quality from green roof systems. *Acta Horticulturae*, **639**, 369–376.

Mooers E. W., Jamieson R. C., Hayward J. L., Drage J. and Lake C. B. (2018). Low-impact development effects on aquifer recharge using coupled surface and groundwater models. *Journal of Hydrologic Engineering*, **23**(9), 04018040.

Mtibaa S., Hotta N. and Irie M. (2018). Analysis of the efficacy and cost-effectiveness of best management practices for controlling sediment yield: a case study of the Joumine watershed, Tunisia. *Science of the Total Environment*, **616**, 1–16.

Newcomer M. E., Gurdak J. J., Sklar L. S. and Nanus L. (2014). Urban recharge beneath low impact development and effects of climate variability and change. *Water Resources Research*, **50**(2), 1716–1734.

Pazwash H. (2011). Urban Storm Water Management. CRC Press, Boca Raton, London & New York.

Pekarek K. A., Shelton D. P., Feehan K. A. and Holm B. A. (2011). Stormwater Management: Terminology. EC701. University of Nebraska-Lincoln, Extension Publication.

Pennsylvania Department of Environmental Protection (PaDEP) (2006). Pennsylvania Stormwater Best Management Practices Manual. PaDEP Report No. 363-0300-002. PaDEP, Harrisburg, PA.

Perez-Pedini C., Limbrunner J. F. and Vogel R. M. (2005). Optimal location of infiltration-based best management practices for storm water management. *Journal of Water Resources Planning and Management*, **131**(6), 441–448.

Petrucci G., Gromaire M. C., Shorshani M. F. and Chebbo G. (2014). Nonpoint source pollution of urban stormwater runoff: a methodology for source analysis. *Environmental Science and Pollution Research*, **21**(17), 10225–10242.

Pratt C. J., Newman A. P. and Bond P. C. (1999). Mineral oil bio-degradation within a permeable pavement: long term observations. *Water Science and Technology*, **39**(2), 103–109.

Qin H. P., Li Z. X. and Fu G. (2013). The effects of low impact development on urban flooding under different rainfall characteristics. *Journal of Environmental Management*, **129**, 577–585.

Rizwan A. M., Dennis L. Y. and Chunho L. I. U. (2008). A review on the generation, determination and mitigation of Urban Heat Island. *Journal of Environmental Sciences*, **20**(1), 120–128.

Sadeghi K. M., Loáiciga H. A. and Kharaghani S. (2018). Stormwater control measures for runoff and water quality management in urban landscapes. *Journal of the American Water Resources Association*, **54**(1), 124–133.

Scholz M. and Grabowiecki P. (2007). Review of permeable pavement systems. *Building and Environment*, **42**(11), 3830–3836.

Shafique M. and Kim R. (2015). Low impact development practices: a review of current research and recommendations for future directions. *Ecological Chemistry and Engineering S*, **22**(4), 543–563.

Sims A. W., Robinson C. E., Smart C. C., Voogt J. A., Hay G. J., Lundholm J. T, Powers B. and O'Carroll D. M. (2016). Retention performance of green roofs in three different climate regions. *Journal of Hydrology*, **542**, 115–124.

Sohn W., Kim J. H., Li M. H. and Brown R. (2019). The influence of climate on the effectiveness of low impact development: a systematic review. *Journal of Environmental Management*, **236**, 365–379.

Stagge J. H., Davis A. P., Jamil E. and Kim H. (2012). Performance of grass swales for improving water quality from highway runoff. *Water Research*, **46**(20), 6731–6742.

Stenstrom M. K. and Kayhanian M. (2005). First Flush Phenomenon Characterization (No. CTSW-RT-05-073.02.6). California Department of Transportation Division of Environmental Analysis, CA, USA.

Sun Y. W., Wei X. M. and Pomeroy C. A. (2011). Global analysis of sensitivity of bioretention cell design elements to hydrologic performance. *Water Science and Engineering*, **4**(3), 246–257.

Taguchi V. J., Weiss P. T., Gulliver J. S., Klein M. R., Hozalski R. M., Baker L. A., Finlay J. C., Keeler B. L. and Nieber J. L. (2020). It is not easy being green: recognizing unintended consequences of green stormwater infrastructure. *Water*, **12**(2), 522.

Teemusk A. and Mander Ü. (2011). The influence of green roofs on runoff water quality: a case study from Estonia. *Water Resources Management*, **25**(14), 3699.

The Groundwater Foundation (2018). National Groundwater Foundation: All about Rain Gardens. See: http://www.groundwater.org (accessed 1 July 2019).

United State Environmental Protection Agency (US EPA) (2018). Urban Runoff: Low Impact Development. See: https://www.epa.gov/nps/urban-runoff-low-impact-development (accessed 5 September 2019).

Vijayaraghavan K. and Badavane A. (2017). Preparation of growth substrate to improve runoff quality from green roofs: physico-chemical characterization, sorption and plant-support experiments. *Urban Water Journal*, **14**(8), 804–810.

Wang H. W., Zhai Y. J., Wei Y. Y. and Mao Y. F. (2019). Evaluation of the effects of low-impact development practices under different rainy types: case of Fuxing Island Park, Shanghai, China. *Environmental Science and Pollution Research*, **26**(7), 6706–6716.

Wardynski B. J. and Hunt W. F., III. (2012). Are bioretention cells being installed per design standards in North Carolina A field study. *Journal of Environmental Engineering*, **138**(12), 1210–1217.

Yang W., Brüggemann K., Seguya K. D., Ahmed E., Kaeseberg T., Dai Hua P., Zhang J. and Krebs P. (2020). Measuring performance of low impact development practices for the surface runoff management. *Environmental Science and Ecotechnology*, **1**, 100010.

Zahmatkesh Z., Burian S. J., Karamouz M., Tavakol-Davani H. and Goharian E. (2014). Low-impact development practices to mitigate climate change effects on urban stormwater runoff: case study of New York city. *Journal of Irrigation and Drainage Engineering*, **141**(1), 04014043.

Zhang D., Gersberg R. M., Ng W. J. and Tan S. K. (2017). Conventional and decentralized urban stormwater management: a comparison through case studies of Singapore and Berlin, Germany. *Urban Water Journal*, **14**(2), 113–124.

Zhang K., Chui T. F. M. and Yang Y. (2018). Simulating the hydrological performance of low impact development in shallow groundwater via a modified SWMM. *Journal of Hydrology*, **566**, 313–331.

Zhen S., Xun Y., Xu S. and Zhou Y. (2015). Analysis of Plans for Rainwater Reuse in a Residential Area in Jiangsu, China. Asia-Pacific Energy Equipment Engineering Research Conference 2015, Atlantis Press, Beijing.

Zhu Z., Chen Z., Chen X. and Yu G. (2019). An assessment of the hydrologic effectiveness of low impact development (LID) practices for managing runoff with different objectives. *Journal of Environmental Management*, **231**, 504–514.

Chapter 5

Land use planning and overseas production

Omid Bozorg-Haddad[1], Mohammad-Reza Rajabi[2] and Hugo A. Loáiciga[3]*

[1]Distinguished Professor, Department of Irrigation & Reclamation Engineering, Faculty of Agricultural Engineering & Technology, College of Agriculture & Natural Resources, University of Tehran, Karaj, Tehran, 31587-77871, Iran. E-mail: OBHaddad@ut.ac.ir

[2]M.Sc. Student, Department of Irrigation & Reclamation Engineering, Faculty of Agricultural Engineering & Technology, College of Agriculture & Natural Resources, University of Tehran, Karaj, Tehran, Iran. E-mail: mrzr504@gmail.com

[3]Professor, Department of Geography, University of California, Santa Barbara, CA 93016-4060, USA . E-mail: Hugo.Loaiciga@ucsb.edu

**Corresponding author.*

ABSTRACT

Land use planning is defined as the regulation of the relationship between space and human activities. Space occupied by human settlements, farmland and forests, parks, fallow land, rivers and lakes, and by transportation networks constitutes a network of areas dedicated to land uses such as agriculture, forestry, rangeland, industry, mining, recreation, and fishing. Land use planning is a comprehensive and long-term approach to planning human relations and their activities in space; it is the task of regulating and coordinating the strategies and general orientations of sectors, and it is the foundation of economies and human activities. The purpose of land use planning is to achieve an optimal distribution of economic and social activities. Land use selection and management are commonly done without regard to the carrying capacity of the land which, when exceeded, results in economic losses and in a reduction of environmental quality. The overall goal of land use planning is to achieve the optimal use of the land within the national interest framework.

Editor: Omid Bozorg-Haddad
doi: 10.2166/9781789062144_0087

Accordingly, formulating a successful plan for managing the development of a country requires proper attention to the roles of geography and land use. Creation of a suitable balance between the Earth and its use by human activities can be achieved by proper land management. The increasing importance of environmental issues and human alteration of natural environments calls for sustainable development and land use practices that conserve natural resources while benefiting society and the environment. Nowadays, due to many factors such as climate change, population growth, changing standards of living, poverty, access to education, and mismanagement of natural and water resources, there is a need to diversify food and agricultural production in a variety of ways. Overseas cultivation is one of them, and it has potential for improving agricultural production. The practice of overseas cultivation by a country means planting and harvesting of a variety of agricultural products in other countries to be sold in its domestic market or in foreign markets. There are five basic principles that underline the success of overseas cultivation involving any group of countries: (1) mutual respect for the governance of all countries; (2) making and abiding by agreements between the countries; (3) non-interference in the internal problems of each country and respect for the territorial integrity of all countries; (4) optimal use of human resources and capacities to support agricultural activities and trade; and (5) expanding security and military cooperation.

Keywords: Land use planning, Overseas cultivation, Overseas production.

5.1 INTRODUCTION

Attention was first drawn to land use planning as a form of centralized development program in France. General Charles de Gaulle, a leader of France in the post-World War II era, called land use planning a restructuring of the face of France. What de Gaulle said was not far off what the earliest presenters of the idea of land preparation – such as the then French Minister of Reconstruction and Urban Development Eugène Claudius-Petit – had proposed. They defined land use as the art or technique of ordering infrastructures and means of communication that serve human beings in the space of a country, taking into account natural, human and economic requirements (along with future human demand considerations and activities). Land use cannot be considered a new issue because land, as the center of human activity and habitation, has always been subject to some form of intervention, development, and conservation (Aerts *et al.*, 2002, 2003; Alier *et al.*, 1996; Barredo, 1996; Berger, 2002; Parker *et al.*, 2003; Suhaedi *et al.*, 2002; Ye *et al.*, 2018).

What characterizes innovation in this area is the purposeful and planned change of land use, and the methods of preparing and implementing such plans. France is one of the few countries in the 20th century to have had a land-use planning program and to have taken steps to prepare and implement such programs. The idea of planning land use in France came about in 1947 with the publication of a book, *Paris and the French Desert*, by Jean Francois Gravie. Even up until 1975 this approach to planning proved successful in France.

Today, poor water resources management, the uneven distribution of water throughout the world, and unsustainable resource use have caused some countries to face water scarcity. There are a number of methods that can be effective for mitigating the water scarcity problem: land-use is one of them (Ligtenberg *et al.*, 2001; Lu *et al.*, 2004; Makhdoum, 2005; Riveria & Maseda, 2006; Sharma & Lees, 2004; Weerakoon, 2002).

Global statistics and studies show that most developed countries in North America and Europe, and leading Asian countries and Middle Eastern oil exporting countries, are major buyers of agricultural land, whilst, on the other hand, mainly poor nations in Africa, South Asia, and in Europe sell or rent land. It should be noted that the most of the land leased or sold belongs to countries in Asia and Oceania (FAO, 2011). This is due to the large areas of arable land in these countries, which have resulted in the leasing or sale of these lands to other countries. Statistics show that Australia had leased or sold about 45 million hectares of its land by 2011. Indonesia, with 4.2 million hectares, is next in line in such land sales or rent. The smallest amount of land rented or sold is found in European countries. Among the European countries, Ukraine (the second largest country in Europe after Russia) had about one million hectares of land sold or leased by 2011. Ukraine is a country that has attracted foreign investment, such as from Australia and Indonesia, due to favorable weather conditions and rainfall, with no need for excessive irrigation, and with fertile soils. There is an average of about 1200 mm of rainfall throughout the year in the western part of the Ukraine. Among the countries of the American continents, South America's countries have the highest such land sales or rentals. By 2011, Argentina, Paraguay, the United States, and Brazil had 18, 7.8, 6.4, and 4.5 million hectares of such land, respectively, which all have suitable rainfall and soils.

The story is quite different in African countries, where the sale or lease of land is not necessarily due to high rainfall and suitable soil fertility but to the lack of necessary infrastructure in the host country. The low productivity in the agricultural sector in these countries attracts foreign investment. Statistics show that by the end of 2011 the Sudan, South Sudan, Tanzania, Ethiopia, and Guinea all had high sales or rental of land which had been coveted by the Persian Gulf states and countries such as India and China. The basic question that can be asked here is how and under what reasons is overseas cultivation justified? The answer to this question can be found in the food security of the citizens of a country. Access to adequate and proper food is a basic right that must be provided in any human settlement, and this is a basic concern involving policies about domestic food production and food imports.

Overseas cultivation must be practiced by countries due to their adverse climatic conditions, and scarcity of water and cropland – as by Saudi Arabia and the United Arab Emirates, for example. It is also worth noting that population growth, economic growth, and rising per capita income, in addition to environmental changes, may limit food production. In these cases, the need for more a diverse provision of agricultural products becomes factual, which justifies overseas cultivation in other countries. Countries such as South Korea, China, and the United Arab Emirates have been in this position for years (Diamond & Wright, 1988, 1989; Eastman *et al.*, 1995, 1998; Evans & Kelley, 2004; Li & Yeh, 2002; Ligtenberg *et al.*, 2004; Oliveira *et al.*, 2003).

Cegielska *et al.* (2018) studied land uses and land cover changes (LULCC) in Hungary and Poland. The study involved statistical data, including CORINE (Coordination of Information on the Environment) land cover, land registration, and statistical and GIS methods in northern Hungary and southern Poland. The results indicated a trend in each time series. Changes in land cover indicate a decline in agricultural areas and an increase in artificial areas (forests and semi-natural lands), and the authors argued that changes in Poland have been more severe than in Hungary.

Shrestha *et al.* (2018) evaluated the distinct and integrated effects of climate change and land use change on river flow and water nitrate-nitrogen content in the Songkhram River. Future river flow was modeled using a SWAT simulator, and future climate was projected with three regional climate models under two track concentrations. A Dynamic Land Use Model (Dyna-CLUE) was applied to generate two future land use scenarios (an economic scenario and a conservation scenario). Their results showed that climate change would decrease river flow and nitrate concentration under both scenarios, while land use change would increase river flow and decrease nitrate concentration under both scenarios.

Zhang *et al.* (2020) summarized case studies on the impacts of climate change and land use change on aquatic products focusing on a framework developed over the last decade. They discussed the future of water production. Accordingly, and as a result of this research, understanding the combined impact of climate change and land use change on water is essential for sustainable development of water resources.

Chowdhury *et al.* (2018) applied multi-purpose satellite data obtained from Landsat-2, Landsat-5 and Landsat-8 to analyze LULCC in the Halda River Basin, Bangladesh, over the past 40 years to assist decision makers, planners, and others to make the best land use management options in this basin. The catchment area was divided into five main categories of land use and land cover: agriculture, uncultivated soil, habitats, vegetation and water body. The LULCC results and the maps produced indicate a significant shift from vegetation and aquatic vegetation to agriculture, uncultivated soil, and habitat.

Deng *et al.* (2019) employed a spatio-temporal hybrid model (enhanced spatial and temporal adaptive reflectance fusion model (ESTARFM)) to integrate Landsat-2 information and MODIS data, and obtain Landsat-2 time series images. Land cover information was extracted using a subject-based classification method. The method presented in this study was applied to a case study in Changsha, China. The results indicated that the proposed method has good accuracy and efficiency. The accuracy and kappa coefficient were 94.38% and 0.88% respectively, and the accuracy of plant species was 85%.

Devátý *et al.* (2019) examined land use changes, spatial distribution of land units, and their impact on soil erosion risk in two areas with distinct geological, political, and historical development. These two study areas included one downstream agricultural catchment in central Bohemia in the Czech Republic, and the other an agricultural hillside and wine production complex in Northern Austria. Studies have established differences in land use based on the distinct political backgrounds of the two countries since World War II. The results showed that, although the spatial extent of agricultural land has not changed significantly, the internal organization of farmland can have a severe impact on soil erosion. Agricultural practices and land erosion are heavily influenced by the economy and political decision making.

5.2 DEFINITIONS AND TERMS

5.2.1 Land use

Land use is the spatial realization of the interactions between humans and space that meets the material and spiritual values of society based on cultural and scientific/technologic backgrounds.

5.2.2 The power of the environment

The power of the environment (both ecological and social-economic potential) is an estimate of the potential of land use for applications in agriculture, rangeland, forestry, parks (conservation and tourism), aquaculture, engineering, urban, industrial and rural development to provide agricultural, industrial, commercial and trade services and goods.

5.2.3 Ecological/natural resources

Ecological/natural resources include physical resources (such as climate, rocks, landforms and soils) and biological resources (such as vegetation, flora and fauna).

5.2.4 Land arrangement

Land arrangement is the regular or random distribution of a land parameter (such as the degree of slope). Land tends to be in balance with the forces of nature (essentially climatic forces and the Earth's gravity) only if the internal forces of the earth, such as volcanic, systemic (earthquake), and tectonic forces (orogenesis), work slowly or with a constant rate of change.

5.2.5 Resource identification methods

Resource Identification Methods (RIMs) include statistics and sampling, interpretation of aerial photographs, satellite and topographic maps, automatic interpretation of aerial photographs, satellite using computer applications (Artificial Intelligence) and geographic information systems.

5.2.6 Inventory

Inventory is the gathering of information for future use. Such a method involves quantifying or counting an area's characteristics. In other words, the inventory counts the information of a region in a classification scheme (such as farmland, river length, and forest volume) or over a period of time, providing quantitative statistics on factors such as annual growth, annual mortality, and seasonal rainfall.

5.2.7 Land shape unit

A unit of land that is selected on the basis of topographic or geological changes for land use purposes.

5.2.8 Overseas cultivation

The actions of a country to plant crops in other countries to produce agricultural (and sometimes industrial products) for importation after harvesting to achieve its food security and meet needs.

5.2.9 Agricultural ecology

The application of ecology to the use of agricultural land, taking into account natural conditions, social environments and technological conditions whose purpose is to extract the information needed to correctly select the target country and the type of crops for overseas cultivation.

5.3 BASICS AND LOGIC

The land and its resources are used to meet the material, social, cultural and spiritual needs of humans. In the process, humans change the earth to varying degrees. One example is the conversion of forests and meadows into agricultural areas to meet food demands from the world's growing population.

LULCC results from a complex interaction between social and environmental factors. As a result, changes in LULCC can lead to changes in environmental products and services for humans, livestock, agricultural production, and the type and extent of environmental services. In fact, the world is experiencing land use conversion pervasively. Mostly, the conversion transform forests into agricultural and pasture lands. Land cover depends on environmental variables and human activities to exploit the land. Physical factors such as elevation, climate, slope are among the many factors that govern LULCC, and mainly the deforestation process.

As long as many countries continue to degrade arable lands and their water resources there will be simultaneous decline of agricultural productivity, which may threaten the food security of these countries. Overseas cultivation is one of the options to improve security. Overseas cultivation can improve food security in water-scarce countries, and can generate economic benefits. Developed countries may engage in agricultural production in other countries' arable lands and export them to other countries in need of food supply, thus generating income while improving the food security of other countries.

5.4 IMPORTANCE AND NECESSITY

Analysis of the distribution of all water resources on Earth shows that 97% are composed of oceans and free waters, which are saline, whilst of the remaining 3% that are freshwater, 69% is glaciers and 30% are groundwaters and the remaining 1% constitute surface freshwater (Bozorg-Haddad, 2018). This amount of freshwater can meet human needs, but factors such as inadequate management of water resources, community development, population growth, and uneven temporal and spatial distribution of water resources on Earth put intense stress on water resources, causing severe water crises in many regions. Water scarcity occurs when the balance between resources and consumption is disrupted by either resource depletion or increased consumption, or both.

Reasons for resource depletion include changes in natural, hydrological (anthropogenic) or other human factors, including increased consumption due to agricultural development, industrial development, population growth, and changes in consumption patterns. The major consumers of water in the world are the agricultural, domestic, and industrial sectors, which account for 70%, 11% and 19% of the world's uses, respectively; the latter percentages vary through the world and vary by country. Water scarcity can be a limiting factor for development in some countries. Land use planning and land use may become central to resolving the water crisis in many parts of the world (Deng *et al.*, 2019).

Each country has unique conditions in terms of geographical range, climate, water, and soil resources, and therefore each region has its own capacities and vulnerabilities. Land use strategies determine the specific agricultural products appropriate to each area based on conditions prevailing in each area.

For example, one of the applications of land use planning is to improve food security, which is one of the key challenges of the present and future. Under present conditions, the world's societies cannot enjoy a high quality of life without necessary and comprehensive studies of land-based, long-term planning.

Therefore, as the world strives to achieve sustainable development, consideration must be given to land-use plans, which may consider economic, social, and environmental aspects, giving special attention to the spatial dimension of the development process. Research into land use planning, especially in rural agriculture, in various areas, including regional expertise in the selection of production sites and environmental protection, mandates that land use plans must be considered. Population growth, industrial growth, excessive depletion of natural resources, increased pollution, economic growth, and inadequate distribution of resources are factors contributing to environmental crises in many parts of the world. Furthermore, these crises may be especially acute in the agricultural sector. Ultimately, land use planning is a form of management, a strategy for reducing development gaps between regions of the world, between regions of a country, between cities and villages, and between different areas of a city.

Overseas cultivation is one option to improve food security in water-scarce countries. Countries that engage in overseas cultivation buy or rent others' arable lands for agricultural activities. Overseas cultivation is not confined to water-scarce countries, and nowadays countries with plentiful water resources also rent or buy arable lands in poor countries and sell the products produced there in their own domestic market or in other countries, thus generating economic benefits.

5.5 MATERIALS AND METHODS

5.5.1 Expert systems

Expert systems are computer programs that simulate the way experts reason. Such software identifies the logical patterns applied by an expert in decision making, and then creates patterns of reasoning that emulate human decision making. Expert systems are a specific type of artificial intelligence programs that make decisions in a specific field. These systems rely on large information databases that are built from human decision making about a particular topic. The process of building an expert system is also called knowledge engineering. It should be noted that creating an expert system requires an iterative process, i.e. the designer builds and tests part of the system and then improves the system, to be tested again until a desired performance level is attained. Expert systems require resources such as tools, languages, and shells. The tool is a developer environment that, besides the capabilities of the developer, provides other tasks such as managing files, displaying graphics and converting code to another language. The language used provides coding capabilities; for example, languages like Lisp, Prolog and C are used to create expert systems. Shell refers to an expert system excluding a knowledge database, in which case the user creates an expert system to access the knowledge.

Many expert systems have been developed to address spatial issues such as resource management, regional planning, land suitability analysis or land use allocation. Applications of expert systems have successfully focused on human expertise in the evaluation of natural resources. Land evaluation processes are stored in a knowledge base and used by an expert system. In this regard, GIS is used to process and display spatial data. A GIS is a powerful land evaluation system. Combining a GIS with economic analysis methods and environmental impact assessment yields a powerful tool for land use.

Another common use of expert systems is to choose the best locations for service facilities (following what is called location theory). The Islay Land Use Decision Support System (ILUDSS) is a knowledge-based location decision support system for strategic land use. It extends knowledge-based approaches to design and combines analytical, rule-based, and status-based modeling capabilities. The most important functions of ILUDSS are querying, formulating land use models to evaluate the potential of land use, and evaluating land use models through automated combination of databases, rules databases and different types of models.

5.5.2 Mathematical models

Mathematical models are often used in land use allocation related to multi-criteria evaluation methods, mathematical programming, and spatial simulation models.

5.5.2.1 Multi-criteria evaluation

Voogd (1983) presented several different applications of multi-criteria evaluation for regional planning. Similar methods were used to evaluate proposed land use management plans. An overview of texts that have proven useful in regional planning have the following traits:

(1) they are flexible;
(2) they match the characteristics of the political arena;
(3) they are able to integrate research strategies.

Land use planning methods should also have the following 'laws':

(1) be able to develop with awareness of the possible land use solutions;
(2) be able to manage changes quickly and effectively;
(3) be able to provide verifiable and interpretable results;
(4) be based on principles that are easy to describe;
(5) be cost effective in terms of manpower and finances;
(6) be able to create explicit values and rules.

But looking at land use preparation activities it can be concluded that these rules were not always followed. In particular, the second and sixth laws are often violated, which leads to the assumption that the land-use methods were deficient. This assumption motivated attention to the possibility of multi-criteria evaluation in urban and regional planning. One of the hallmarks of multi-criteria evaluation methods is that they start from a number of explicitly formulated criteria, and the criteria can show significant inter-differences in nature. Therefore, the criteria are not measured one dimensionally, but, rather, in multiple dimensions that represent diverse criteria in nature. Another important feature of multi-criteria evaluation method is the ability to explicitly pay attention to political priorities. This feature enables these approaches to appropriately combine different political views in the decision-making evaluations. However, land use requires special attention to the spatial components of the data, along with the analysis of multiple goals and criteria. The combination of the use of multi-criteria evaluation methods and GIS has been widely used to achieve land-suitability maps or to select locations for a particular activity.

The Idrisi software (see: https://clarklabs.org/terrset/idrisi-gis/) provides multi-criteria evaluation tools to construct electronic maps and an associated database of attributes. One of the important tools in this regard is the Multi-Criteria Evaluation (MCE) procedure. The process of carrying out the MCE procedure is performed in two main steps: (1) identifying and developing criteria; and (2) integration. Criteria are made up of two categories: constraints and factors. Constraints include those of the Boolean criteria that restrict the analysis to specific geographical areas and are provided as Boolean layers. Factors include fuzzy criteria that define the degree of utility for a region and represent it as continuous numbers. During step 2, the integration of layers of analysis, layers are merged in different ways. The most important ones presented in the MCE procedure are the Boolean approach, weighted linear combination, and sequential weighted average. The Boolean approach converts all criteria (constraints and factors) to Boolean values (zero and one), and the method of combining them is the Boolean equation (multiplication). The weighted linear combination method standardizes the criteria in a continuous numerical range and combines them based on weighted averages. Initially, the factors are aggregated based on the weight given to them. Thereafter, the calculated layer is multiplied by the constraint layers, and a fuzzy layer is obtained that represents the utility of the whole region. The sequential weighted average method works the same as the previous method, and another set of weights called sequential weights is used for the factors. Sequential weights allow for risk assessment and compensation in the evaluation process. Multi-criteria evaluation methods based on ideal point analysis have been heavily integrated into GIS.

Multi-criteria evaluation methods are categorized below.

5.5.2.2 *Adaptive planning*

Adaptive planning is a technique for multi-purpose conceptual analysis that has been modified to analyze discrete problems. This method identifies options that are close to ideal solutions. For example, adaptive planning was used to assess the quality of the red squirrel habitat. The latter study's results of multi-criteria models derived from adaptive planning were compared with real habitat use data according to the experts' perception of red squirrel habitat suitability. The results of the Berra (1993) study showed good agreement between habitat suitability classification using adaptive planning and the observed patterns of squirrel habitat use, and the predictive powers of the applied model were evaluated.

5.5.2.3 *The TOPSIS method*

The Technique for Order of Preference by Similarity to Ideal Solution (TOPSIS) ranks alternatives such that the best ranked one features the smallest distance to the best solution and the longest distance to worst solution. TOPSIS is applied to discrete situations where the number of alternatives is finite.

5.5.2.4 *Ideal point analysis, hierarchical optimization, similarity analysis*

Correlation Discriminant Analysis (CDA) is one of the widely used MCE methods based on pairwise comparison of options. This method measures the degree to which options (i.e., alternatives for action) and weightings of the criteria confirm or reject the superiority of the relationship between options. The differences between the weight of the criteria and the criteria scores are analyzed separately by similarity and non-similarity methods. It should be noted that Hierarchical Optimization (HO) models attempt to rank all criteria according to their relative priorities. This method performs optimization step by step.

5.5.2.5 Analytical hierarchy process

Another method of multi-criteria evaluation that is often integrated with GIS for land suitability analysis is the Analytical Hierarchy Process (AHP). The technique is based on pairwise comparisons and allows managers to examine different alternatives. This method enables the formulation of the problem in a hierarchical way, as well as the possibility of considering different quantitative and qualitative criteria in the problem. This process interferes with various options in decision making and enables the analysis of sensitivity to criteria and sub-criteria. Moreover, this method is based on pairwise comparisons that facilitate judgment and calculations. It also shows the degree of compatibility and incompatibility of the decision, which is one of the advantages of this technique in multi-criteria decision making. The AHP method can generate weights assigned to land suitability criteria or to fit maps to calculate a composite score.

5.5.2.6 Multi-objective land allocation

Multi-criteria evaluation methods have been used to generate multiple land use scenarios by selecting the optimal land use for each land unit. Among them, hierarchical optimization allocates the largest area to the highest priority, excludes it from other remaining uses, and repeats the process until the entire area is allocated. A suitable solution is the ideal point method that assigns to each location unit the user with the highest fitness when the hierarchy of goals is unknown.

Multi-Objective Land Allocation (MOLA) was developed to implement a concurrent solution to multi-objective allocation in the Idrisi software. Adaptive solutions to multi-objective problems are commonly provided by mathematical programming tools outside of GIS. Mathematical programming solutions (such as linear programming or correct programming) are applied successfully in problems with few options. However, raster GIS may involve large volumes of data that typically create costly computational burdens. As a result, the need for a solution to multi-purpose land allocation issues must resolve conflicting goals, so that large raster data can be managed as well as easily understood.

5.5.3 Mathematical programming

Mathematical programming, when used for land use purposes, searches for a set of land uses that optimize one or more objective functions while complying with a set of constraints. This section discusses a number of mathematical programming methods.

5.5.3.1 Linear programming

Linear programming is a mathematical method that has been used since late 1950 in many design and training programs. Although primarily used for business and management it has also been widely used in spatial analysis. Linear planning is not a spatial approach because it does not consider the spatial distribution of decision variables. However, if zoning is considered, this means linear planning may also be appropriate for spatial issues.

The main purpose of linear optimization planning is to maximize or minimize an objective function subject to a set of constraints. Functions and constraints must be formulated as linear equations. Each linear programming begins by defining decision variables as different problem options. For example, in land planning the decision variables are different types of land uses. The goals of land use planning scenarios should be clearly stated after defining the decision variables. The objective function is formulated by specifying the coefficients of the decision variables to achieve the problem goals. Problems in which there are two decision variables can be solved by graphical methods. A mathematical method is needed to solve more complex problems. There are several computer programs for solving linear programming problems.

5.5.3.2 Goal programming

Multi-objective problems can be solved using linear programming so that only one of the goals (the most important goal) is optimized and the other goals are taken into account in the constraints. One of the disadvantages of this approach is that the choice of the goal that is optimized is sometimes difficult to ascertain or is subjective. Multi-purpose planning provides a form of solving these problems, when the optimal solution to the linear programming problem is replaced by a set of solutions that are not necessarily optimal in the linear sense but are effective solutions nevertheless. Goal planning is one of the multi-purpose planning methods. A feature that distinguishes goal planning formulas is that one or more goals are directly involved in the goal function. Goal planning seeks a form of achievement that is as close to the ideal solution as possible. The purpose of this method is to minimize the sum of deviations between observations and predictions.

5.5.3.3 Interactive programming

Land use decisions often require interactive planning and information exchange between decision makers and systems. Interactive Multiple Goal Linear Programming (IMGLP) has been used for land preparation and agricultural development policy analysis.

5.5.3.4 Integer programming

Linear integer programming models allow results to be transformed into an optimal land use allocation map. User allocation optimization models are categorized into single-user allocation models and multi-user allocation models. Both categories of models allocate a set of one or more users across a set of one or more geographical areas to maximize a number of implicit or explicit utility functions. These models vary in the number of uses to be assigned, as well as in the way geographical areas are defined. The boundaries of the region, unlike in the multi-user model, are not known in advance, but are determined with a hybrid algorithm.

5.6 CASE STUDIES

5.6.1 LULCC assessment of the Halda watershed using remote sensing and GIS

The Halda is an economically and ecologically significant river which has undergone a number of changes as a result of various anthropological activities. This study shows that LULCC activities in the Halda Basin have changed significantly over the past 40 years. Currently, the main land uses in this watershed are agricultural. Vegetation and water levels decreased by 35.10% and 35.47%, respectively, while in agricultural and urban areas it increased (an average annual 860 and 428 ha, respectively). Changes in surrounding land cover have a significant effect on suspended solids, nitrate and nitrogen loads.

The Halda Basin study showed that the impact of agricultural activities is more severe than that of climate change. Therefore, increased agricultural activity and settlement in the Halda catchment area are major concerns. Several recommendations were been made for the proper management and protection of forests, water, and soil resources based on the results and conclusions of this study, as follows:

- because the catchment is large enough (175 691 ha) it must be divided into several sub-basins;
- forecasting population and development needs in this watershed are needed for the sound use and allocation of limited land resources among competing land uses. In addition, to ensure the best use of land resources the Ministry of Agriculture must identify areas suitable for cultivation and regulate the conversion of agricultural land to non-agricultural uses by laws.

5.6.2 Analysis of non-cultivated land use in Lithuania

Šalkauskienė *et al.* (2019) argued that human agricultural activities must be consistent with actual natural conditions in order for the environment to assimilate change. Accordingly, they conducted a systematic planning analysis of non-cultivated land use in Lithuania and developed a model to offer ways to use non-productive land. The results of statistical data demonstrated that the land use index of North-East Lithuanian hilly upland districts which did not have agricultural activities, was about 59% higher than Lithuania's average, which is about 16.59. The extensive land use index in Western Lithuanian uplands and plateau areas is about 67.9, whereas the index of non-cultivated lands is about 11.28 in the area, which is 32% lower than the Lithuanian average. This index is also about twice as high as the average of Lithuania, which is about 37.

The key contribution of Šalkauskienė *et al.* (2019)'s work was a method for land use planning of non-cultivated lands. The method involves inputs from experts who evaluate the land use criteria for agricultural options: crop production, animal husbandry, agricultural and environmental conditions, forestry, alternative activities. Expert surveys on land management, forest management, hydraulic engineering, and agriculture were carried out. A simple data processing system has been built into MS Excel for easy use of this method.

5.7 SUMMARY

Countries each have distinct features and conditions, in terms of geographical boundaries, climates, water and soil resources, and topography, so each region has its own capabilities and limitations. Land use strategies determine the specific products appropriate to each area based on the priorities, capabilities and feasibility of each area. Land use studies delineate the activities, capabilities, opportunities, and weaknesses of each region, which underpin sustainable development. Land use strategies contribute to the overall growth and integration of development plans that form the basis for sustainable development.

In summary, the objectives of land use planning are as follows:

(1) to provide optimal distribution of activities across spaces, attributes, and environmental features;
(2) to optimize the existence of a population in a space, based on its full potential;
(3) to prevent harmful spatial movement of populations;
(4) to apply the necessary actions required for the geographical stabilization of immigrant populations;
(5) to reduce environmental disputes, and encourage optimization, remediation and reconstruction of undesirable geographical locations.

Achieving the goals of land use planning for sustainable development requires optimal use of natural and human resources according to climate, culture, type of arrangement and placement of agricultural lands, pastures, hills, forests, and according to relations with other countries.

REFERENCES

Alier J. L., Cazorla A. and Martínez J. E. (1996). Optimization on Spatial Land Use Allocation: Methodology, Study Cases and Computer Package. Spanish Ministry of Agriculture, Madrid, Spain.

Aerts J. C. J. H. and Heuvelink G. B. M. (2002). Using simulated annealing for resource allocation. *International Journal of Geographical Information Science*, **16**(6), 571–587.

Aerts J. C. J. H., Eisinger E., Heuvelink G. B. M. and Stewart T. (2003). Using linear integer programming for multi-site land-use allocation. *Geographical Analysis*, **35**(2), 148–169.

Barredo J. I. (1996). Geographic Information Systems and Multicriteria Evaluation in Land Planning. Ra-ma Publications, Madrid, Spain.

Berger T. (2002). Multi-agent modelling applied to agroecological development, in agent-based models of land-use and land-cover change. In: Report and Review of an International Workshop, October 4–7, 2001, Irvine, California, D. C. Parker, T. Berger and S. Manson (eds), Indiana University, Bloomington, Indiana, pp. 50–56.

Bozorg-Haddad O. (2018). Water Resources Systems Optimization. Tehran University Publication No. 3, Tehran, Iran.

Cegielska K., Noszczyk T., Kukulska A., Szylar M., Hernik J., Dixon-Gough R., Jombach S., Valánszki I. and Filepné Kovács K. (2018). Land use and land cover changes in post-socialist countries: some observations from Hungary and Poland. *Land Use Policy*, **78**, 1–18.

Chowdhury M., Hasan M. E. and Abdullah-Al-Mamun M. M. (2018). Land use/land cover change assessment of Halda watershed using remote sensing and GIS. *The Egyptian Journal of Remote Sensing and Space Science*, **23**(1), doi: 10.1016/j.ejrs.2018.11.003

Devátý J., Dostál T., Hösl R., Krása J. and Strauss P. (2019). Effects of historical land use and land pattern changes on soil erosion – case studies from lower Austria and central Bohemia. *Land Use Policy*, **82**, 674–685.

Deng Z., Zhu X., He Q. and Tang L. (2019). Land use/land cover classification using time series landsat 8 images in a heavily urbanized area. *Advances in Space Research*, **63** (7), 2144–2154, doi: 10.1016/j.asr.2018.12.005

Diamond J. T. and Wright J. R. (1988). Design of an integrated spatial information system for multiobjective land-use planning. *Environment and Planning B: Planning and Design*, **15**(2), 205–214.

Diamond J. T. and Wright J. R. (1989). Efficient land allocation. *Journal of Urban Planning and Development*, **115**(2), 81–96.

Eastman J. R., Jin W., Kyem P. A. K. and Toledano J. (1995). Raster procedures for multi-criteria/multi-objective decisions. *Photogrammetric Engineering and Remote Sensing* **61**(5), 539–547.

Eastman J. R., Jiang H. and Toledano J. (1998). Multi-criteria and multi-objective decision making for land allocation using GIS. In: Multicriteria Analysis for Land-Use Management, E. Beinat and P. Nijkamp (eds), Kluwer Academic Publishers, Dordrecht, pp. 227–251.

Evans T. P. and Kelley H. (2004). Multi-scale analysis of a household level agent-based model of landcover change. *Journal of Environmental Management*, **72**, 57–72.

FAO (2011). The State of the World's Land and Water Resources for Food and Agriculture (SOLAW) – Managing Systems at Risk. Food and Agriculture Organization of the United Nations, Rome and Earthscan, London.

Li X. and Yeh A. G.-O. (2002). Urban simulation using principal components analysis and cellular automata for land-use planning. *Photogrammetric Engineering and Remote Sensing*, **68**(4), 341–352.

Ligtenberg A., Bregt A. K. and van Lammeren R. (2001). Multi-actor-based land use modeling: spatial planning using agents. *Landscape and Urban Planning*, **56**, 21–33.

Ligtenberg A., Wachowicz M., Bregt A. K., Beulens A. and Kettenis D. L. (2004). A design and application of a multi-agent system for simulation of multi-actor spatial planning. *Journal of Environmental Management*, **72**, 43–55.

Lu C. H., Van Ittersum M. K. and Rabbinge R. (2004). A scenario exploration of strategic land use options for the loess plateau in northern China. *Agricultural Systems*, **79**(2), 145–170.

Makhdoum M. (2005). Fundamentals of Land Use Planning. University of Tehran Publication, No. 1, Tehran, Iran.

Oliveira F., Patias N. M. and Sanquetta C. R. (2003). Goal programming in a planning problem. *Applied Mathematics and Computation*, **140**, 165–178.

Parker D. C., Manson S. M., Janssen M. A., Hoffmann M. J. and Deadman P. (2003). Multi-agent systems for the simulation of land-use and land-cover change: a review. *Annals of the Association of American Geographers*, **93**(2), 314–337.

Riveira I. S. and Maseda R. C. (2006). A review of rural land-use planning models. *Environment and Planning B: Planning and Design*, **33**(2), 165–183.

Šalkauskienė V., Gudritienė D. and Abalikštienė E. (2019). Analysis of the non-productive land use in Lithuania. *Land Use Policy*, **80**, 135–141, doi: 10.1016/j.landusepol.2018.10.010

Sharma S. K. and Lees B. G. (2004). A comparison of simulated annealing and GIS-based MOLA for solving the problem of multi-objective land use assessment and allocation. Proceedings of the 17th International Conference on Multiple Criteria Decision Analysis, Whistler, Canada, Open-File Report AP58.

Shrestha U. B., Sharma K. P., Devkota A., Siwakoti M. and Shrestha B. B. (2018). Potential impact of climate change on the distribution of six invasive alien plants in Nepal. *Ecological Indicators*, **95**, 99–107, doi: 10.1016/j.ecolind.2018.07.009

Suhaedi E., Metternicht G. and Lodwick G. (2002). Geographic information systems and multiple goal analysis for spatial land use modelling in Indonesia. 23rd Asian Conference on Remote Sensing, Asian Association on Remote Sensing (AARS), Kathmandu, Nepal.

Voogd H. (1983). Multicriteria Evaluation for Urban and Regional Planning. Pion, London.

Weerakoon K. (2002). Integration of GIS based suitability analysis and multicriteria evaluation for urban land use planning, contribution from the Analytic Hierarchy Process. 23rd Asian Conference on Remote Sensing, Asian Association on Remote Sensing (AARS), Kathmandu, Nepal.

Ye L., Huang X., Yang H., Chen Z., Zhong T. and Xie Z. (2018). Effects of dual ownerships and different land lease on industrial land use efficiency in Wuxi City, East China. *Habitat International*, **78**, 21–28, doi: 10.1016/j.habitatint.2018.04.011

Zhang Y., Tang C., Ye A., Zheng T., Nie X., Tu A., Zhu H., Zhang S. (2020). Impacts of climate and land-use change on blue and green water: A case study of the upper Ganjiang River Basin, China. *Water*, **12**(10), 2661. https://doi.org/10.3390/w12102661

Chapter 6

Water and sustainable development

Omid Bozorg-Haddad[1], Samira Komijani[2] and Erfan Goharian[3]*

[1]Distinguished Professor, Department of Irrigation & Reclamation Engineering, Faculty of Agriculture Engineering & Technology, College of Agriculture & Natural Resources, University of Tehran, Karaj, Alborz, Iran. E-mail: OBHaddad@ut.ac.ir.
[2]M.Sc. Student, Department Environmental Engineering, University of Tehran, Tehran, Iran. E-mail: Samira.komijani@ut.ac.ir
[3]Assistant Professor, Department of Civil and Environmental Engineering, University of South Carolina, Columbia, SC, USA. E-mail: goharian@cec.sc.edu
**Corresponding author.*

ABSTRACT

The never-ending exploitation of natural resources after the industrial revolution has caused irreversible damage to our environment and life. This vast resource deployment was expected to finally decelerate after the emergence of the Sustainable Development paradigm in the late 1980s. The concept of sustainable development aims to balance the tradeoff between supply and the needs of societies to also preserve the needs of future generations. Nonrenewable resources, especially water, have been at the heart of sustainable development goals. Sustainable Water Management (SWM) is an essential component of sustainable development. To achieve SWM, various water management challenges, in both developing and developed countries, should be addressed by attracting stakeholders to invest in water system efficiency and improvement projects, as well as by recycling and reusing water. To assess sustainable management policies and practices, evaluation of performance metrices – such as the Water Resources Sustainability Index, reliability, resiliency and vulnerability, and the Stability Index – is

Editor: Omid Bozorg-Haddad
doi: 10.2166/9781789062144_0103

imperative. A cost-efficient policy should also facilitate the attainment of sustainability in all socioeconomic, and environmental sectors.

Keywords: Society, Economy, Sustainability, Sustainable development.

6.1 INTRODUCTION

Water is the basis of human survival. It is the core of sustainable development and the key for socio-economic development, as production and development depend on it. Global water demand is predicted to increase significantly over the next decade. Currently, two-thirds of the world's population live in areas that experience water scarcity at least one month per year. About 500 million people live in areas with a water-consumption of two times the local renewable resources (UN-Water, 2017). This consistent increase in water demand does not match with the proportionate renewability of water resources. Groundwater resources are steadily decreasing and are vulnerable to depletion in these regions. The renewability of water resources is intrinsically linked with the continuous cycle of evaporation, precipitation, and runoff, which is driven by climatic forces. Due to the combined effects of climate change and human activities, the frequency and severity of droughts and floods have escalated (UN-Water, 2017). These events are often accompanied by a higher load of pollutants and higher concentrations of them in water. All these facts indicate an ever-increasing scarcity of water in the future, making it a crucial factor in development plans.

6.2 BASICS OF SUSTAINABILITY AND SUSTAINABLE DEVELOPMENT

6.2.1 The concept of sustainable development and its pillars

The concept and definition of sustainability and sustainable development might seem simple at first glance. However, the fact that there are over 200 different definitions of sustainable development is invites us to take a closer look at these definitions and their dissimilarities.

The most popular definition of sustainable development is that defined by the Brundtland Commission in 1987:

> 'Sustainable development is development that meets the needs of the present without compromising the ability of future generations to meet their own needs.'
>
> (World Commission on Environment and Development, 1987)

Another noteworthy definition is:

> 'In principle, such an optimal (sustainable growth) policy would seek to maintain an 'acceptable' rate of growth in per-capita real incomes without depleting the national capital asset stock or the natural environmental asset stock.'
>
> (Turner, 1988)

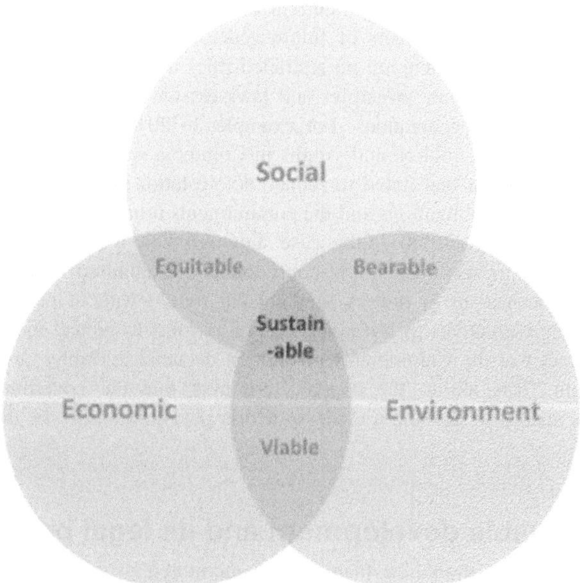

Figure 6.1 The three pillars of sustainability (Barbier, 1987).

However, the acceptable limits for development or how these limits should be defined are not clear from these definitions. How should 'needs' be defined and estimated, and how are those different from what is 'wanted'? Such questions highlight the potential for misinterpretation and, in turn, for conflicts.

Although the definitions are slightly different from each other, most of them highlight, directly or indirectly, the idea that the three independent pillars of sustainable development are environmental, economic, and social aspects (Elliot, 2006). To achieve sustainability, the equilibrium region between these three pillars needs to be found. This is often illustrated by a Venn diagram – as, for example, by the three interlocking circles presented by Barbier (1987); see Figure 6.1. The goal of sustainable development is to find the optimal and stable region which guarantees the achievement of objectives across all three systems (illustrated as the intersection of all three circles). Therefore, sustainable development should address present social, economic, and environmental needs while keeping an eye out for the needs and rights of future generations.

6.2.2 Sustainable development safeguarding future generations' rights

The definition of sustainable development emphasizes the 'right of future generations' to secure their interests. The General Conference of the United Nations Educational, Scientific and Cultural Organization (UNESCO) held in Paris 1997 passed a 'Declaration on the responsibilities of the present generations towards future generations'. This declaration officially recognized that the current generation has the great responsibility of ensuring that the needs and interests of future generations should be fully safeguarded (UNESCO, 1997). Although there are no restricted rules that can be enforced to preserve the rights of future generations, principles and laws exist which could send cases to the courts on behalf of future generations. For example, in 2018, the Colombian Supreme Court ruled in favor of 25 children and youths in Columbia who successfully argued that the Columbian government had failed to reduce deforestation in the Amazon, despite its national and international obligations and the commitments it had made at climate summits (Colombian Supreme Court, 2018). In this case, although there were no hard rules in place for the court to refer to specifically, the court found that children may use a *tutela* (a Columbian legal mechanism to protect fundamental rights) without permission of their parents or legal representatives to bring such claim to court to protect their rights. Later, the court found out that the Colombian Constitution declared a healthy environment as a fundamental right. This led to the court determining that the government had been ineffective in its actions to develop a series of action plans to combat the deforestation of the Amazon.

6.2.3 Sustainable development and its legal principles

Sustainable development should be planned and monitored over time, and thus can be considered as a dynamic system, where the sum of public policies causes profound social reform through time. Therefore, the Law of Sustainable Development cannot be conceived as a sum of static rules, but as a continuous process of decision-making at all hierarchical levels. The Law of Sustainable Development influences all other branches of law because the ideal sustainability condition strives for changes in public and private law. To realize this goal, we need to first meet the demand for environmental ethics and principles which are the origins of these laws and decisions. Some key principles that could be adopted are:

- *Principle of sovereignty:* Governments have a monopoly over their natural resources, but the exercise of this right should not cause damage to the environment of other countries or areas outside the jurisdiction of governments;
- Principle of commitment to cooperation, information, and assistance in emergencies: Governments and people are obliged to work together to achieve sustainability;
- *Principle of environmental protection:* The exploitation of natural resources shall be such that it would not threaten their sustainability (Li & Chen, 2019);
- *Precautionary principle:* Think forward, look backward and act. The precautionary principle (PP) goal is prediction and lessening potentially serious or irreversible risks under conditions of scientific uncertainty (Som *et al.*, 2009).

6.3 ROLE OF SUSTAINABLE WATER MANAGEMENT (SWM) IN SUSTAINABLE DEVELOPMENT

Increasing global concerns over water resources were given significant attention in 2015 by the World Economic Forum (World Economic Forum, 2015), after which these challenges were taken into careful consideration to develop new Sustainable Development Goals (SGDs). The UN recognized that water resources planning and management perspectives should contribute to the creation of the SDGs that deal with food, energy, and the environment. SDG6, in particular, is called the Water Goal.

6.3.1 Targets of the water goal (SDG6)

SDG6 seeks to ensure the availability and sustainable management of water and the provision of sanitation for all, with the following targets of the 'Water for Life' program (United Nations, 2015):

(1) achieve universal and equitable access to safe and affordable drinking water for all by 2030;
(2) achieve access to adequate and equitable sanitation and hygiene for all and to end open defecation by 2030, paying special attention to the needs of women and girls and those in vulnerable situations;
(3) improve water quality by reducing pollution, eliminating dumping and minimizing the release of hazardous chemicals and materials, reducing the proportion of untreated wastewater by half, and substantially increasing recycling and safe reuse globally by 2030;
(4) substantially increase water-use efficiency systems across all sectors, and ensure sustainable withdrawals and supply of freshwater to address water scarcity, and substantially reduce the number of people suffering from water scarcity by 2030;
(5) deploy integrated water resources management at all levels, including through trans-boundary cooperation as appropriate, by 2030;
(6) protect and restore water-related ecosystems, including mountains, forests, wetlands, rivers, aquifers, and lakes by 2020;
(7) by 2030, expand international cooperation and capacity-building support to developing countries in water and sanitation-related activities and programs, including water harvesting, desalination, water efficiency, wastewater treatment, recycling, and reuse technologies;
(8) support and strengthen the participation of local communities in improving water and sanitation management.

The inclusion of the Water Goal in the United Nations (2015) Development Agenda indicates the importance of water as a limiting factor, as opposed to a secondary factor for development and production.

6.3.2 Challenges to sustainable water management

Studies point to a future escalation of global water shortages, especially in urban areas. In one particular study, it is estimated that by 2050 about 993 million people will experience perennial water shortages, and 1–3.1 billion ($1–3.1 \times 10^9$) people will be experiencing seasonal water shortages around the world (McDonald *et al.*, 2011). Agricultural water use accounts for 70% of global water consumption, which can be as high as 90% in developing nations (Russo *et al.*, 2014). The majority of global water use can be divided into three main sectors of urban, agriculture, and ecosystem (NB: this may not constitute the majority of water use in a specific region or country; in those cases, a detailed study of SWM requirements needs to be completed for management and policy purposes). Some of the most important challenges and solutions for sustainable water management in each of the three sectors are shown in Tables 6.1 and 6.2.

Table 6.1 Challenges and solutions for sustainable water management in developing countries (Russo *et al.*, 2014).

Sector	Challenges	Solutions
Urban	• Intermittent operation • Lost or stolen water • Rapid urban growth • Political conflict	• Institutional improvements • Low-tech water capture and treatment • Graywater reuse • Cooperation, sharing riparian rights • Stakeholder engagement
Agriculture	• Irrigation infrastructure cost • Subsidies promoting irresponsible use • Low water use efficiency	• Optimize water productivity • Improve subsidy/pricing structure • Supplemental irrigation with rainwater harvesting
Ecosystem	• Economic development priorities	• Communication of ecosystem service value

According to Tables 6.1 and 6.2, urban water management in developing regions faces challenges such as inequitable delivery, especially under rapid urban population growth. Sustainable management plans should focus on continuously improving stakeholder involvement and on the development of water reuse in both developed and developing regions. Water reuse will reduce stress during drought periods, but its costs and uncertainties about its environmental impacts are still among major barriers for its implementation.

Table 6.2 Challenges and solutions for sustainable water management in developed country (Russo *et al.*, 2014).

Sector	Challenges	Solutions
Urban	• Demand Management • Cost of adopting new technology	• Conservation education • Water reuse and recycling • Stormwater management, green infrastructure
Agriculture	• Over-allocation of water resources • Cost of new technology vs. the low cost of water	• Improve water allocation, irrigation management • Affordable precision agricultural methods
Ecosystem	• Development priorities	• Identification and protection of most valuable resources • Definition of a degradation tolerance level

6.4 OBJECTIVES OF INTERDISCIPLINARY STUDIES

Today, interdisciplinary studies have an established place in leading scientific circles. Interdisciplinary research brings researchers from multiple fields and promotes collaboration among them to share different perspectives in order to reach a better understanding of the question under research. Conflicts between disciplines, differences in discourses and concepts, management and coordination of activities, boundaries between disciplines, and bias towards scholars' own particular specialties are among the problems that exist in many interdisciplinary studies (Carr *et al.*, 2018). Managing water resources is not an exception in this regard. For example, the largest US water management agencies, in addition to engineers and hydrologists, are composed of biologists, ecologists, economists, geologists, historians, lawyers, planners, policy makers, and sociologists. Inherent uncertainties associated with the status of future water resources and the community's needs and responses to changes has introduced various difficulties in studying and managing this resource (Vogel *et al.*, 2015; Walker *et al.*, 2015).

Water resources management is known as a complicated process because of the complexity not only of the physical components of water systems but also of the dynamic socio-economic interactions of these human–natural systems. This calls for the formation of multidisciplinary teams to tackle these issues. Questions such as 'how will water system decisions affect social tensions?' or 'how can we improve the status of water systems through social and cultural solutions?' should be addressed with the inclusion of researchers from different fields.

6.4.1 Hydro-social objectives

One of the main goals of engineering and management disciplines is to generate benefits for the communities they serve. Therefore, any decision that is taken regardless of the needs and interests of community members causes less profitable outcomes. The sustainability of communities depends directly and indirectly on the quantity, quality, reliability and affordability of essential decisions. Reducing or excessively increasing the amount of water, its pollution, or cost might cause uncertain community responses (Walker *et al.*, 2015). The need to examine the interactions between linked human-natural systems relies on better understanding of how a decision-making process affects or is affected by society by capturing public responses and understanding their mental models. Developing an analytical model might help with better prediction of how society reacts to perturbation in water resources and corresponding managemental decisions (Walker *et al.*, 2015). The effects of selective solutions and decisions made about aquatic systems can be mimicked with the help of simulation models and sociological theories. Still, it is difficult to understand how a community's activities will be formed under future scenarios, given the uncertainties that arise from socioeconomic and environmental impacts and changes of norms and values around the world through time.

6.4.2 Hydro-economic objectives

An interconnection and correlation exists between economic systems and water systems. Decisions made to a water system are required to take account of economic evaluations, while forming economic policies can be used as a regulatory tool for demand management. Therefore, economic evaluation is a key aspect of water resources management. In an economic evaluation, the best scenario can be chosen from different water management scenarios and used to optimize allocation among consumers (Simonovic, 2009). Still, economic evaluation cannot be the only criterion for decision-making. For example, in some places the environmental considerations are difficult to be valued. It may not be easy to determine the relationship between the economic system and the hydrological system. The relationship between the income of the agricultural and industrial sectors changes with the amount of allocated water, while this relationship also alters over time. Future commodity prices face uncertainties that make economic valuation a more difficult job. Also, some indirect effects of water allocation are simply not measurable, such as the rising price of real estate around a river due to the release of water into it. Interest rates are also used to evaluate the economic viability of projects, which is difficult to determine under inflation and fluctuations.

6.4.3 Hydro-environmental objectives

Conservation of natural ecosystems around wetlands, conservation of scarce species, conservation of aquifers in arid areas, protection of soil against salinity and pollution, and preservation of forests are among the environmental issues faced in water management. Environmental issues encountered in the management of water resources could mostly be solved by applying sustainability to the management approach but, in the long term, this can also improve the economic and social status of the region. For example, by preserving the environment and natural ecosystems, the local tourism industry has a chance to expand in the natural ecosystem, and the social health and vitality of the community will be increased (Cai *et al.*, 2002).

6.5 METHODOLOGY

6.5.1 Sustainability indicators

To better understand how a policy affects sustainability as a whole, it is vital to find parameters or indicators to measure the performance of the proposed policy for the system, and to compare it to other proposed policies. Relevant indices in environmental systems include an environmental index, environmental stress index, environmental sustainability index and multi-variate environmental index. As to SWM-specific indices, the Water Productivity Index (WPI), Water Sustainable Development Index (WSDI), Water Footprint (WF), Water availability index (WAI), Global Water Security index (GWS), Sustainable Water Management Index (SWaM), Human Sustainable Development Index (HSDI), Ecological Footprint (EF) and Social Development Index (SDI), can be named as the most commonly used ones.

In recent years, numerous studies have been conducted to assess the sustainability of water resources systems by developing sustainability indicators. Kundzewicz (1997) defined water resources indices for sustainable development, stating that sustainable development requires an integrated approach structure. This structure includes water resources and hydrological components as well as environmental, economic, demographic, social and cultural sub-systems. In 1997, Loucks developed the Stability Index (SI), first presented by Hashimoto , (1982), which shows that using performance metrics including reliability, resiliency, and vulnerability in conjunction can be used to quantify and monitor stability of water systems over time. Loucks (1997) shows that if a proposed policy makes a system more sustainable, the system will have high capacity to reduce vulnerabilities in the future. Additionally, it shows how multi-period economic efficiency goals can be defined and used in a multi-objective modeling framework to consider sustainability in the planning and management of water resources systems. Models can identify and estimate business exchanges between different goals for current and future generations.

Sophocleous (2000) showed that in order to ensure stability it is necessary to establish constraints based on the hydrological principles of mass balance. To create water use policies and planning horizons, Sophocleous used graph curves extracted from groundwater storage depletion levels to evaluate the amount of surface water needed. Raju (2000) presented an analysis of multi-criteria decision making for sustainable water resource planning in Spain. In this study economic factors including initial cost of irrigation system, maintenance cost, product profitability, European subsidies, environmental factors including water volume, post-irrigation water quality, water use efficiency, flood and drought resistance, and Social factors including the hiring of rural workers without cultivation and reconciliation planning, were used to rank the decision criteria. As a result, the concept of sustainability was introduced not only for planning problems by incorporating discussed criteria, but it was also used to compare with the traditional approach of surface irrigation. Cai *et al.* (2002) introduced a framework for sustainability analysis of water resources and management strategies, and tested its application for a coastal basin with regards to risk reduction in water supply, environmental protection, water allocation facilitation and economic efficiency. Sustainable water management in irrigation dominated river basins' attempts to ensure long-term, stable, and flexible water supply to meet crop water demands, as well as the growing municipal and industrial water demands, while mitigating negative environmental consequences. Fowler *et al.* (2003) presented the effects of climate change on the reliability, resiliency, and vulnerability of a water resource system. In this study, climate change was evaluated in terms of mean precipitation statistics and potential evapotranspiration. Fogg and LaBolle (2006) presented an incentive to integrate groundwater sustainability as a new idea in social sciences. They stated that sustainable water management is attainable in the vast majority of the world's river basins. However, this quest will become far less feasible if we wait until water supplies are further over-used. Ajami *et al.* (2008) introduced sustainable water resources management based on hydrological uncertainty and water resources management. In this study, the effects of hydrological uncertainties on the reliability, resiliency and vulnerability of a water resource system with different operating rules were investigated. Based on the presented results, it was concluded that a consistent operational approach, in which the decision maker adjusts the operating rules based on the forecast of the incoming climate and the current state of reservoir storage at any given time stage, can lead to efficiency. Hence, accurately accounting for various sources of uncertainty that influence our forecasted inflow provides more precise information to decision-makers and will help them operate a water resource system more sustainably. Staben (2010) described the sustainability of water reserves with their performance and application indices. They provided a comprehensive framework for measuring and reporting water resources liabilities, including environmental and sustainability issues, water conservation, social responsibility and long-term economic development. This kind of comprehensive framework is universally applicable in water and wastewater systems.

In order to meet the water demands of human beings or the environment, it is necessary to measure performance indicators to compare and evaluate them. This goal was presented by Sandoval-Solis *et al.* (2011) in the form of indicators of water resources sustainability that make it possible to evaluate and compare management policy options for water resources systems. Sandoval-solis *et al.* modified Loucks' SI using a new formulation of volumetric reliability, resiliency and vulnerability, and the ability of adding more criteria to the index. Goharian *et al.* (2017), in order to assess the performance of water systems, used the joint probability distribution of reliability and vulnerability by testing copula functions. They introduced the Water System Performance Index (WSPI) and employed it to present the performance of the water system during a historical period and then computed WSPI values for future climate projection. Results showed that the WSPI is a useful tool for both managers and stakeholders. Collet *et al.* (2015) presented the sustainability of water reserves and adaptation strategies based on pollution and climate change for a Mediterranean basin for urban, agricultural, and environmental needs. This study used local decision-makers to investigate possible future changes in the sub-basin's water allocation capacity using climate change, dam management and the use of water scenarios. Studies have shown that participatory management is an effective approach to involve people and other water stakeholders in deciding on water management programs. Brown *et al.* (2015) presented an analysis of water resources systems in terms of the most challenging water issues such as water scarcity, climate change, food and energy, social and economic incentives. Findings have shown that water resource analysis today suffers from low scientific visibility and a difficulty of generalization of localized findings within a scientific framework of sustainable water management.

McEwen *et al.* (2017) introduced a result-oriented approach from interdisciplinary research in the field of 'sustainable flood memory' based on flood risk management. This project aimed to create social learning opportunities to gain the public's adaptive capacity for resilience. In this research, they explored key factors in the concept of resilience, community, and flood risk through interviewing floodplain residents in the UK. The paper outlines research methods, explores key findings and critically evaluates sustainable flood memory for the strategy, process and practice of developing flood resilience.

6.5.2 Reliability, resiliency and vulnerability

A reliability indicator shows the probability of successful allocation of water to the consumer (Hashimoto *et al.*, 1982). The reliability of a system (α) can be calculated according to Equation (6.1):

$$\alpha = Prob[X_t \in S] \qquad \forall t \tag{6.1}$$

where the system reliability (α) is the probability of the allocated water (X_t) meeting the desired outputs (S) in each time step (t).

The reliability of a system can also be expressed using Equation (6.2) (Goharian *et al.*, 2016, 2018):

$$Rel = \frac{\sum_{t=1}^{T} Z_t}{T} = 1 - \left(\frac{n_f}{T}\right) \tag{6.2}$$

where Z_t is a binary variable. When water demands have been fully met (or when the water shortage is zero), Z_t is equal to 1; and when there is water shortage in each time step (t), Z_t is equal to 0. T is the total number of time steps during the simulation period, and n_f is the number of failure periods out of the total period, T.

Resiliency is the ability of a system to change the conditions of that system. In fact, this parameter for water supply systems is defined as the probability of a system returning to its desired state after failure. Because climatic conditions and water resources systems are often uneven, resiliency is also considered as a statistical parameter for examining system flexibility for changing management conditions and different policies (Chambers *et al.*, 2019; Maiolo & Pantusa, 2019; Nabinejad *et al.*, 2017; Sarzaeim *et al.*, 2017).

Moy *et al.* (1986) defined the maximum number of consecutive periods in which a system is in failure mode before returning to its desired state as the resiliency or flexibility of a water resources system. Resiliency is defined by Equation (6.3) (Ashofteh *et al.*, 2015):

$$\text{Res}^i = \frac{N_{t=1}^{T}(D_{t+1} \leq R_{t+1} | D_t > R_t)}{N_{t=1}^{T}(D_t > R_t)} \times 100 \qquad \forall t = 1, 2, \ldots, T \tag{6.3}$$

where N is the number of time steps during the study period in which the condition shown in parentheses holds true. $N_{t=1}^{T}(D_{t+1} \leq R_{t+1} | D_t > R_t)$ is the number of successful time periods in terms of providing a desired demand after a failure state, and $N_{t=1}^{T}(D_t > R_t)$ is the number of time periods in which water shortage occurred in system.

Vulnerability shows the magnitude of damage that a system receives during failure (McMahon *et al.*, 2006). Vulnerability can be defined as: (1) average failure (McMahon *et al.*, 2006); (2) average maximum deficiencies during successive periods of system failure (Hashimoto *et al.*, 1982). The vulnerability criterion can be quantified by estimating the ratio of the sum of the total deficiencies to the number of time-steps in which the deficiency occurred, divided into the total time-steps measured annually or in the length of the period (Chambers *et al.*, 2019; Maiolo & Pantusa, 2019; Nabinejad *et al.*, 2017; Sarzaeim *et al.*, 2017).

According to various references (Chambers *et al.*, 2019; Maiolo & Pantusa, 2019; Nabinejad *et al.*, 2017; Sarzaeim *et al.*, 2017), vulnerability can be defined as the maximum deficiency during the studied periods where the failure occurred. The ratio of water supply (supply) to water demand (demand) for each time step (t), called (RIA^t), is presented in Equation (6.4), whilst vulnerability (VUL) can be calculated by Equation (6.5).

$$RIA^t = \frac{\text{Supply}^t}{\text{Demand}^t} \tag{6.4}$$

$$VUL = \max(1 - RIA^t) \tag{6.5}$$

As mentioned above, there are different methods to calculate the vulnerability of a water resource. Goharian *et al.* (2018) presented a new method for calculation of vulnerability of a water supply system (Equation (6.6)):

$$\text{Vulnerability} = \text{Exp} \times W_e + S \times W_s + PS \times W_{ps} \tag{6.6}$$

In this method, high vulnerability means that a system has been exposed to greater severity of failure (S), exposure of a system to a new climate condition (Exp), or potential severity (PS). W_e, W_s, and W_{ps} are weights of exposure, severity, and weight of potential severity, respectively (Goharian *et al.*, 2018).

Equation (6.6) is based on a top-down approach. Goharian *et al.* (2017) insert two new factors (sensitivity and adaptive capacity) into Equation (6.6), and proposed a further new method, according to Equation (6.7) (Goharian *et al.*, 2017):

$$\text{Vulnerability (vul)} = f\left(\text{exposure, sensitivity}, \frac{1}{\text{adaptive capacity, severity, potential severity}}\right) \tag{6.7}$$

6.5.3 Sustainability index

A sustainability index consists of system performance criteria to facilitate comparison and decision-making between different management options and water resources plans (Lane *et al.*, 2015). The sustainability index Equation (6.8) is defined according to performance criteria (Sandoval-Solis *et al.*, 2011):

$$SI^j = \{\text{Rel}^j \times \text{Res}^j \times (1 - \text{Vul}^j)\}^{1/3} \tag{6.8}$$

where Rel^i, Res^i, and Vul^i are reliability, resiliency, and vulnerability of the *j*th water user, respectively. According Equation (6.8), the main components of a sustainability index estimation are the system performance measures. The value of SI can change from zero to one. In other words, higher values of SI mean the better performance of the system in terms of sustainability.

As discussed above, sustainability can be achieved by considering the three main pillars: the economic, social and environmental systems. This requires the expansion of a sustainability index to cover not only environmental system performance indicators, but the other pillars as well. Figure 6.2 shows a summary of common sustainability

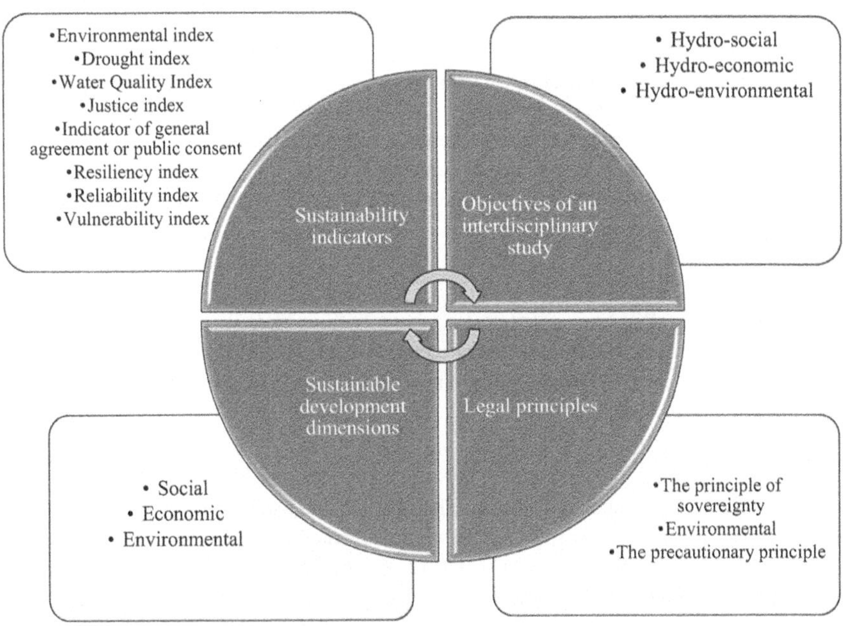

Figure 6.2 Water and sustainable development.

indicators, dimensions, legal principles, and objectives for sustainable development of water systems.

6.6 SUMMARY AND DISCUSSION

Sustainability of water management in urban, agricultural and environmental contexts is essential for sustainable development. There are many ways to evaluate sustainable management practices. Integrating socio-economic characteristics of environmental systems should be quantified in order to represent and implement developments. Urban water management in developing regions faces challenges such as unequal distribution of water due to rapid population growth. Sustainable management plans should focus on issues such as attracting shareholders to invest in developing areas and revitalizing and reusing water in developed areas. Water reuse reduces the stresses caused by drought periods. However, there are still obstacles, including the cost of using these technologies and the risks involved in developing and developed countries alike. Improving water productivity in the agricultural sector is beneficial to all other sectors through the use of new irrigation technologies and methods, as it reduces competition between agricultural, urban and environmental sectors. Achieving water sustainability only in the environmental field could be inconsistent with other development plans, and occur as a result of a restricted and uncomprehensive approach. However, long-term economic development and social goals of societies are linked to both environmental health and environmental fracture, which is why developed countries emphasize socio-economic development. Sustainable water management can provide sustainability in all socioeconomic and environmental sectors in areas with diverse geographies.

Sustainability indices (SIs) such as reliability, resiliency, and vulnerability are indices that determine the probability of successful water allocation, probability of a system returning to its desired state after a failure, and the magnitude of damage that system receives during failures, respectively. Various types of studies in the field of sustainable water resources management have been carried out and have unveiled the advantages of sustainable development management of these human-natural systems. Furthermore, sustainability in the field of water management can be applied to different types of water bodies; such as sustainable groundwater management, sustainable river basins, coastal basins, and to other kinds of water bodies. Current research has introduced a wide range of applications of sustainable water resources research, not only making systems more sustainable but also reducing vulnerabilities in the future.

There are different ways to apply sustainability in water resources management. One practical way to model sustainable water management is a multi-criteria decision-making process that provides higher efficiency in water allocation. Another way is the risk reduction approach in water supply, environmental protection, and economic efficiency, which ensures long-term and flexible water supply in agricultural, municipal, and industrial sectors. In addition to reliability, resiliency and vulnerability, the Water Poverty Index (WPI) summarizes the information required to evaluate the sustainability of water systems and can be used as a singular measure to assist decision-makers in determining priorities and coming up with proper development plans.

Some of the most influential factors in the field of sustainable environmental management are climate change and hydrological uncertainty that will cause dramatic water losses in the environment in the future. Results have shown that adjusting operational rules based on climate projections can lead to efficiency (Goharian *et al.*, 2017). Participatory management is also an effective approach in sustainable water resources management, especially in integrated human-natural systems, because it involves people and other water stakeholders in deciding on their own water management programs.

REFERENCES

Ajami N. K., Hornberger G. M. and Sunding D. L. (2008). Sustainable water resource management under hydrological uncertainty. *Water Resources Research*, **44**, W11406, doi: 10.1029/2007WR006736

Ashofteh P. S., Haddad O. B., Akbari-Alashti H. and Marino M. A. (2015). Determination of irrigation allocation policy under climate change by genetic programming. *Journal of Irrigation and Drainage Engineering*, **141**(4), 04014059.

Barbier E. B. (1987). The concept of sustainable economic development. *Environmental Conservation*, **14**(2), 101–110, https://doi.org/10.1017/S0376892900011449 (accessed 27 May 2021).

Brown C. M., Lund J. R., Cai X., Reed P. M., Zagona E. A., Ostfeld A., Hall J., Characklis G. W., Yu W., Brekke L. and Brekke L. (2015). The future of water resources systems analysis: toward a scientific framework for sustainable water management. *Water Resources Research*, **51**(8), 6110–6124.

Cai X., McKinney D. C. and Lasdon L. S. (2002). A framework for sustainability analysis in water resources management and application to the Syr Darya Basin. *Water Resources Research*, **38**(6), 1085, doi: 10.1029/2001WR000214

Carr G., Loucks D. P. and Blöschl G. (2018). Gaining insight into interdisciplinary research and education programmes: a framework for evaluation. *Research Policy*, **47** (1), 35–48.

Chambers K. F., Baker A. C., McKay S. K. and Morgan H. (2019). Retrospective Sustainability and Resilience: Complementary Concepts for Managing Systems (Report No. ERDC/TN EMRRP-SR-86). US Army Corps of Engineers, Engineer Research and Development Center (ERDC), Vicksburg, United States.

Collet L., Ruelland D., Estupina V. B., Dezetter A. and Servat E. (2015). Water supply sustainability and adaptation strategies under anthropogenic and climatic changes of a meso-scale Mediterranean catchment. *Science of the Total Environment*, **536**, 589–602.

Colombian Supreme Court. (2018). Climate Change Case. *Supreme Court of Justice of Columbia, April 5, 2018 (STC 4360-2018).*

Elliot J. A. (2006). An Introduction to Sustainable Development, 3rd edn. Taylor & Francis e-Library (available online), London.

Fogg G. E. and LaBolle E. M., (2006). Motivation of synthesis, with an example on groundwater quality sustainability. *Water Resources Research*, **42**, WO3S05, doi: 10. 1029/2005WR004372

Fowler H. J., Kilsby C. G. and O'Connell P. E., (2003). Modeling the impacts of climatic change and variability on the reliability, resilience, and vulnerability of a water resource system. *Water Resources Research*, **39**(8), 1222, doi: 10. 1029/2002WR001778

Goharian E., Burian S. J., Lillywhite J. and Hile R. (2017). Vulnerability assessment to support integrated water resources management of metropolitan water supply systems. *Journal of Water Resources Planning and Management*, **143**(3).

Goharian E., Burian S. J., Bardsley T. and Strong C. (2016). Incorporating potential severity into vulnerability assessment of water supply systems under climate change conditions. *Journal of Water Resources Planning and Management*, **142**(2).

Goharian E., Zahmatkesh Z. and Sandoval-Solis S. (2018). Uncertainty propagation of hydrologic modeling in water supply system performance: Application of Markov Chain Monte Carlo method. *Journal of Hydrologic Engineering*, **23**(5).

Hashimoto T., Stedinger J. R. and Loucks D. P. (1982). Reliability, resiliency, and vulnerability criteria for water resource system performance evaluation. *Water Resources Research*, **18**(1), 14–20.

Kundzewicz Z. W. (1997). Water resources for sustainable development. *Hydrological Sciences Journal*, **42**(4), 467–480, doi: 10.1080/02626669709492047

Lane B. A., Sandoval-Solis S. and Porse E. C. (2015). Environmental flows in a human-dominated system: integrated water management strategies for the Rio Grande/Bravo Basin. *River Research and Applications*, **31**(9), 1053–1065.

Li C. and Chen J. (2019). The ethical turn of the rule of law in the global environment under the influence of the concept of a community of shared future for mankind. *Beijing Law Review*, **10**, 98.

Loucks D. P. (1997). Quantifying trends in system sustainability. *Hydrological Sciences Journal des Sciences Hydrologiques*, **42**(4).

Maiolo M. and Pantusa D. (2019). Sustainable water management index, SWaM_Index. *Cogent Engineering*, **6**(1), 1603817.

McDonald R. I., Green P., Balk D., Fekete B. M., Revenga C., Todd M. and Montgomery M. (2011). Urban growth, climate change, and freshwater availability. *Proceedings of the National Academy of Sciences*, **108**(15), 6312–6317.

McEwen L., Garde Hansen J., Holmes A., Jones O. and Krause F. (2017). Sustainable flood memories, lay knowledges and the development of community resilience to future flood risk. *Transactions of the Institute of British Geographers*, **42**(1), 14–28.

McMahon T. A., Adeloye A. J. and Zhou S. L. (2006). Understanding performance measures of reservoirs. *Journal of Hydrology*, **324**(1–4), 359–382.

Moy W. S., Cohon J. L. and ReVelle C. S. (1986). A programming model for analysis of the reliability, resilience, and vulnerability of a water supply reservoir. *Water Resources Research*, **22**(4), 489–498.

Nabinejad S., Mousavi S. J. and Kim J. H. (2017). Sustainable basin-scale water allocation with hydrologic state-dependent multi-reservoir operation rules. *Water Resources Management*, **31**(11), 3507–3526.

Raju K. S., Duckstein L. and Arondel C. (2000). Multicriterion analysis for sustainable water resources planning: a case study in Spain. *Water Resources Management*, **14**, 435–456.

Russo T., Alfredo K. A. and Fisher J. (2014). Sustainable water management in urban, agricultural, and natural systems. *Water*, **6**(12), 3934–3956.

Sarzaeim P., Bozorg-Haddad O., Fallah-Mehdipour E. and Loáiciga H. A. (2017). Climate change outlook for water resources management in a semiarid river basin: the effect of the environmental water demand. *Environmental Earth Sciences*, **76**(14), 498.

Sandoval-Solis S., McKinney D. C. and Loucks D. P. (2011). Sustainability index for water resources planning and management. *Journal of Water Resources Planning and Management*, **137**(5), 381–390.

Simonovic S. P. (2009). Managing Water Resources: Methods and Tools for a System Approach. UNESCO, London.

Som C., Hilty L. M. and Köhler A. R. (2009). The precautionary principle as a framework for a sustainable information society. *Journal of Business Ethics*, **85**(Suppl. 3), 493–505, https://doi.org/10.1007/s10551-009-0214-x (accessed 27 May 2021).

Sophocleous M. (2000). From safe yield to sustainable development of water resources the Kansas experience. *Journal of Hydrology*, **235**, 27–43.

Turner R. K. (1988). Sustainable Development Management. Belhaven, London.

UNESCO. (1997). Declaration on the Responsibilities of the Present Generations Towards Future Generations. The General Conference of the United Nations Educational, Scientific and Cultural Organization (UNESCO), Paris.

United Nations. (2015). Water and the Sustainable Development Goals (SDGs). See: https://www.un.org/waterforlifedecade/waterandsustainabledevelopment2015/open_working_group_sdg.shtml (accessed April 2021).

UN-Water. (2017). Wastewater: The Untapped Resource. The United Nations World Water Development Report 2017. UNESCO, Paris.

Vogel R. M., Lall U., Cai X., Rajagopalan B., Weiskel P. K., Hooper R. P. and Matalas N. C. (2015). Hydrology: the interdisciplinary science of water. *Water Resources Research*, **51**(6), 4409–4430.

Walker W. E., Loucks D. P. and Carr G. (2015). Social responses to water management decisions. *Environmental Processes*, **2**(3), 485–509.

World Commission on Environment and Development. (1987). Our Common Future. World Commission on Environment and Development, Geneva.

World Economic Forum. (2015). Global Risks 2015. World Economic Forum, Geneva.

Chapter 7

Comprehensive integrated water management

Alireza Rezaee[1], Omid Bozorg-Haddad[2], Ronny Brendtsson[3] and Vijay P. Singh[4]*

[1]M.Sc. Student, Department of Irrigation & Reclamation Engineering, Faculty of Agriculture Engineering & Technology, College of Agriculture & Natural Resources, University of Tehran, Karaj, Alborz, Iran. E-mail: AliRezaRezaee75@ut.ac.ir
[2]Distinguished Professor, Department of Irrigation & Reclamation Engineering, Faculty of Agricultural Engineering & Technology, College of Agriculture & Natural Resources, University of Tehran, Karaj, Tehran, 31587-77871, Iran. E-mail: OBHaddad@ut.ac.ir
[3]Professor, Division of Water Resources Engineering & Centre for Middle Eastern Studies, Lund University, Lund, Sweden. E-mail: ronny.berndtsson@tvrl.lth.se
[4]Distinguished Professor, Regents Professor, Caroline & William N. Lehrer Distinguished Chair in Water Engineering, Department of Biological and Agricultural Engineering & Zachry Department of Civil Engineering, Texas A & M University, 321 Scoates Hall, 2117 TAMU, College Station, Texas 77843-2117, USA. E-mail: VSingh@tamu.edu
**Corresponding author.*

ABSTRACT

Improper utilization of water resources has the potential to result in reduced availability of high-quality water and adverse effects on societal development. In fact, what appears to be a serious gap in comprehensive water resources studies is the lack of a coherent approach that can link different social, economic and environmental parts within the framework of the integrated water management paradigm to extract strategies and operational plans. Comprehensive water resources management (CWRM) is a process that intends to develop and manage water, land and other resources in a way that maximizes the social and

 Water Resources: Future Perspectives, Challenges, Concepts and Necessities
Editor: Omid Bozorg-Haddad
doi: 10.2166/9781789062144_0119

economic well-being of human societies, without compromising the integrity and sustainability of vital ecosystems and future benefits. This chapter discusses the definitions of integrated and comprehensive water resources management describing the steps of using integrated management in practical examples.

Keywords: Comprehensive water resources management (CWRM), integrated water resources management (IWRM).

7.1 INTRODUCTION

Water has always been vital. Population growth, agricultural development in arid and semi-arid regions of the world, industrial expansion, inadequate distribution of fresh water in time and at various locations, and increasing constraints and problems of water quality in many countries have made water supply a major challenge of this century. Improper utilization of water resources in many parts of the world has resulted in reduced levels of current water resources, and other negative societal and environmental effects. What can be considered as a serious gap in the comprehensive studies of water resources is the lack of a coherent methodology that can link different social, economic, and environmental aspects within the framework of an integrated water management paradigm and derive strategies from their interaction. Comprehensive water resources management (CWRM) is a process that leads to the development and management of water, land, and other resources in a way that maximizes the social and economic well-being of societies without compromising the integrity and sustainability of vital ecosystems and future benefits.

Integrated water resources management (IWRM), while offsetting some of the shortages, increases the availability of existing water resources, minimizing the negative impacts of separate resource use and efficient water management. IWRM requires increased integration and cooperation to ensure fair, effective and sustainable management of scarce water resources in both national and international sectors.

Sustainable development will be difficult to achieve without considering water development, which is a key element. We need to invest in water resources infrastructure and improve existing water resources management to achieve sustainable development goals and overcome the specific challenges of the water sector. In this regard, an IWRM approach enables optimization of water sector participation in achieving sustainable development. The idea of using an IWRM strategy is not just to create one or more plans for water resources development and management, but to develop a dynamic framework for better planning and decision making for improved water management. Based on the principles of economic efficiency, ecological safety and social equity, IWRM includes an overall approach to water management and the integration of water, land, and other resources. IWRM came about when water professionals realized that water problems can only be solved through a multi-sector, multi-regional integrated approach that is achieved through stakeholder collaboration.

The IWRM paradigm was developed to address local to global water management issues as a process that promotes the coordinated development and management of water, land and related resources in order to maximize economic and social welfare in a fair manner without regard to how sustainability of critical ecosystems is defined. CWRM is a suitable solution to water resource problems because of the complexity and multifaceted nature of water problems, the diversity of stakeholders, and interaction between nature and humanity. Population growth and increasing water demands have adverse effects on the environment and ecosystems. Often, traditional water resources management practices are not responsive to the idea of the sustainability of land and ecosystems, and to sustainable development in general. Therefore, important water resource decisions and policies should be taken using a holistic approach, taking into account the different political, social, economic, hydrological, and environmental interactions on a more comprehensive spatial and temporal scale.The role of IWRM depends on the general level of development of each country. Underdeveloped countries, developing countries, and developed countries have different ways of implementing IWRM and its benefits are often different. Underdeveloped countries see IWRM as a factor in addressing issues, such as poverty, hunger, health, and environmental sustainability; and developing countries see IWRM as an approach to sustainably manage water resources with economic growth. Therefore, IWRM has different dimensions and contexts in different parts of the world.

7.2 THEORETICAL DEFINITIONS AND APPROACHES TO IWRM

In order to solve a problem in a system formed by a set of interdependent individuals, an integrated approach is necessary. Indeed, a holistic viewpoint considering the regular interaction of all components in a system is the proper way to face a problem.

IWRM was presented in four principles of the Dublin 5th EPRA meeting (1997). At that meeting, the principles of freshwater resource use, the use of CWRM, the coordination of economic and social development, and natural resource conservation were all defined. Through participatory actions and by raising public awareness, the role of women in water resources management, the economic value of water, and water management as an economic commodity were all emphasized (Snellen & Schrevel, 2004).

The most important of the different views of IWRM can be set out in the following five perspectives:

(1) World Water viewpoint: IWRM is a philosophy that states that water must be considered in a comprehensive way, including its natural state, whilst balancing different agricultural, industrial, and environmental needs.
(2) Global Water Partnership, Technical Advisory Committee: IWRM is used to build confidence in water performance to advance the economic and social development goals of countries, so as not to jeopardize the sustainability of their vital ecosystems or their future productive power and the fulfilment of water needs.
(3) United Nations Development Program viewpoint: This program was created on the CapNet internet site (for capacity building in water management) and defines IWRM as a fundamental process for the sustainable development, allocation, and monitoring of water resources in social, economic, and environmental contexts.
(4) World Bank viewpoint: The integrated approach to water resources ensures that social, economic, environmental and other aspects of water resources management and development are taken into account.
(5) United States Agency for International Development viewpoint: IWRM defines the convergence and collaboration of governors, communities and stakeholders to choose an option for how to use freshwater and coastal resources.

7.3 BASICS AND LOGIC

7.3.1 The IWRM cycle

Planning with the IWRM approach includes steps that are performed in a sequential cycle (Zare, 2011). Planning is a logical sequence of steps that are implemented and supported by online management support activities, and it will have the most impact when viewed as a continuous cycle. The planning steps in the IWRM approach include startups, stakeholder planning, strategic vision building, status assessment, water management strategies, IWRM implementation, and evaluation.

7.3.2 IWRM actors

Actors, such as the Global Water Partnership, have significantly contributed to the global implementation of IWRM. The GWP pioneered the concept and replicated the Dublin concept (Petit & Baron, 2009), adopted at the International Water and Environment Summit in 1992; which led to the formulation of recommendations contained in Chapter 18 of the Convention on Freshwater Resources of the 21st UN Convention on Environment and Development (UNCED). IWRM has received support from many other international organizations, such as the International Water Management Institute (IWMI), the Food and Agriculture Organization (FAO), and World Bank (WB), and other regional bodies. Another global endorsement of the IWRM approach is the Johannesburg Executive Program, held by governments at the World Summit on Sustainable Development (Bizikova *et al.*, 2013).

7.3.3 Integration of water resources management

Recent water management research has depicted many problems in contemporary water management. To improve this situation, a rigorous framework is needed that ensures the integration of all departments, which requires a change in existing oversight structures in departments and requires co-operation between them. IWRM is a framework that requires significant changes in laws, regulations, institutions, policies, and actors (Rogers & Hall, 2003) leading to a change in governance over its traditional approach. Therefore, the concept of IWRM has evolved from non-integrated and fragmented to an integrated approach (Molle, 2008).

This paradigm created an organizational change of water management from government to regional authorities, from a goal to multi-purpose projects, from canal to watershed management. In addition, this has led to changes in global water policies and institutions. Countries have updated their water laws using IWRM to ensure environmental and socio-economic requirements.

7.3.4 Dimensions of CWRM

CWRM is not an end in itself, but rather an approach and tool to increase the benefits and productivity of limited water, as well as financial and human resources. CWRM is a good solution to water resource problems, because it addresses the complexity and multifaceted nature of those problems, the diversity of stakeholders, and the interaction between nature and humans. Although CWRM is recognized as an effective and efficient way to manage water resources, it cannot be viewed as a simple universal idea. With all the work and agreements on CWRM being made globally, various definitions and understandings of the concept of comprehensiveness come into play. It is worth noting that focusing solely on water does not guarantee comprehensive management of water resources; instead, water, land, and the environment must all be considered in parallel.

In the context of water resources management, 'geographical community' refers to the use of hydrological boundaries (river basin and sub-basins) to define units rather than political or administrative boundaries. Geographically, water resources management should be undertaken at least on a 'basin' scale. In other words, the river basin should be managed as a 'hydrological unit' with a single identity. In deciding the development and management of natural resources, this geographical community includes interactions between land and water users in the aquatic and terrestrial ecosystems, and in human systems, as well as upstream–downstream effects within and between the main basin and sub-basins.

Similarly, 'integrity of water use' refers to the competition for water between agriculture, hydroelectricity, urban water supply and health, industry, and environment. By contrast, economic, social, and environmental integrity refers not only to the costs and economic benefits of water management decisions but also to the costs and benefits of social and environmental considerations, and must be taken into account in any decision making.

7.3.5 Implementation of CWRM

Human life is under increasing pressure. In many parts of the world, natural resources have been declining and these problems can no longer be resolved by mere engineering solutions. An integrated approach is needed to coordinate development across sectors, and individuals as well as institutions must contribute to improve conditions. The concept of comprehensive water management has emerged to guide the process that leads to the development and management of more sustainable water resources. These are the ultimate goals of CWRM:

(1) to increase economic efficiency to understand the economic and social values of water for people;
(2) to achieve equality between different regions and peoples;
(3) to achieve environmental sustainability, especially in critical ecosystems.

These goals require balancing water consumption at a subsistence level and protecting this resource to sustain its functions and properties. Excessive exploitation and mismanagement can lead to unstable conditions that have severe effects on humans and the environment. This requires that all stakeholders agree on the problems they are involved in and are eager to work together to solve them. Another principle of comprehensive water management emphasizes that the development and management of water resources must be based on a participatory approach and involve all consumers, planners, and policymakers at all levels. The role of women should also be taken into consideration, as women always play a key role in foresight, management, and conservation of water resources at the home level.

The final principle explains that water should not be considered a free resource. Consumer water supply (household, industrial, agricultural) requires investment and operating costs, and consumers must contribute to these costs. This also gives them an incentive to increase their water use efficiency.

Applying the above principles, of course, is time-consuming. Sometimes comprehensive management of water resources is regarded as a medicine to treat all kinds of pains. But this is not true and there is no one universal solution. It is an ongoing learning process and a collaborative approach that brings about gradual change and improvement for a system and its management. However, experiences gained from other comprehensive water management projects can teach us some of the (successful) practices of implementing comprehensive water management:

- there must be a functional organizational structure for communication, exchange and decision making that somehow involves all stakeholders. To facilitate this process, a strong technical secretariat and working groups are also needed;
- there must be an agreement on the final goals, and on any other goals that are relevant to the macro-development goals and processes of national development planning;
- sufficient human and financial resources must be devoted to implementation so that stakeholders can benefit from CWRM;
- it is important to use monitoring and evaluation mechanisms that critique and improve the process.

7.3.6 Principles of CWRM

The widespread concept of comprehensive water management focuses not only on the provision and development of water resources but also on a systematic approach to the allocation and monitoring of water use in the context of social, economic and environmental objectives. The basics of this management are not only based on principles of engineering but also include social foundations, public participation, environmental, and economic perspectives. There are a range of basic aspects to water-related policy-making, including setting priorities, project definition, reviewing impacts, return on capital, centralization and decentralization, environmental protection, and social and technological development. Water-related laws also include rights and ceding of rights, exploitation of water structures, water quality, stakeholder engagement, responsibility and conflict of interest management. To implement these principles, the World Water Partnership Organization recommended the Fourth Principles of the Dublin International Conference on Water and Environment (1992), as follows:

- Principle 1: Freshwater is a limited and vulnerable source and is essential for life, development and environmental health.
- Principle 2: Water resources management and development should be based on a collaborative approach among consumers, planners and policymakers at all levels.
- Principle 3: Women play a key role in the provision, management and conservation of water.
- Principle 4: Water has economic value in all its uses and must be recognized as an economic commodity as a fundamental social necessity.

According to the four Principles of the Dublin Conference, it is necessary for officials and the public to recognise that water is a cross-cutting issue, and that sustainable development focuses on decentralizing water management, fostering the involvement of all stakeholders and decision makers, even at the lowest level. Therefore, this type of managerial attitude requires a line of cooperation by all stakeholders who may also have a conflict of interest.

On the issue of water governance, this partnership should also be created, especially in countries where one organization/department has responsibility for agricultural water, another is responsible for drinking water, and protection and quality control of water resources is the responsibility of another body. However, there is poor legal communication between these different organizations. Therefore, part of the comprehensive planning and management of water resources requires new legislation and definitions of structures and responsibilities for water related organizations in a country, which may require planning and regulating of feedback functions.

One of the main components of land planning may be the comprehensive management of water resources, because of the importance of determining land use, locating growth areas, urban and industrial development, agricultural areas and cultivation patterns, preserving ecosystems and observing health and so on. The most important constraint of planning is sustainable access to freshwater and its supply and demand. Therefore, the following points should be considered in this regard:

- investigation and evaluation of existing water resources, both surface and ground water, taking into account the qualitative aspects of water;
- temporal and spatial distribution of precipitation as the main input to the water balance of the region;
- investigation of limiting conditions of water resources, such as droughts and floods;
- investigation of past water use, and of present and forecasted future needs;
- survey and evaluation of non-conventional water resources;
- investigation of water resources exchanges, and quantitative and qualitative systematic processes and the possibility of recycling of wastewater;
- description of current management, entitlements, intra- and inter-organizational agreements, stakeholder conflicts of interest, and examination of the ability of current management to sustainably develop water resources; and, finally
- analysis of how people and stakeholders are involved in the current situation, and of the potential for increased cooperation and participation in IWRM.

Accordingly, basins are the most rational unit for comprehensive planning and management of water resources that can be taken into consideration in land preparation.

7.3.7 Stakeholder analysis

In its basic definition, stakeholders included groups of both product producers and end-users of the products, but in fact these stakeholders are individuals or organizations affected by some form of organization activity and able to enhance or hinder the success of the organization. In water resources management, stakeholders are divided into three groups:

- primary stakeholders – stakeholders who are most dependent on the source and are more likely to actively participate in management (these individuals can definitely also be categorized as primary stakeholders);
- secondary stakeholders – extremely loud voices, including local government departments and people who live close to the source but are not heavily dependent on it; and
- tertiary (or third party) stakeholders – national and international government agencies.

There is also a group called key stakeholders that includes individuals, groups or organizations heavily influenced by a program. They may be experts in the subject.

In the first stage of stakeholder analysis, key stakeholders are identified. At this stage, 'potential stakeholders' are identified by looking specifically at 'vulnerable groups' and 'stakeholder relationships'. Secondly, stakeholder evaluation is undertaken, in which the benefits and potential impacts of integrated watershed management on stakeholders and vice versa are assessed, where stakeholder expectations should be identified, together with the potential benefits of this management system for stakeholders, as well as potential conflicts of interest under the new management approach. The degree of influence and importance of the stakeholders is also assessed by their status and power (political, economic, and social), degree of organization, strategic resource control, and informal influence (personal relationships). When identifying who the main stakeholders are, and who benefits most from the management of these areas in an ecosystem, both the determination of ecosystem boundaries and of the people, groups and institutions that may in the medium to long term be able to protect, manage and make decisions within these boundaries is made. Given that conservationists have a long-term perspective, the practicality of coordination and management must be reinforced from the bottom through stakeholder associations, rather than being ordered from above. Last, the final stage is formulated, based on the results of previous stages of stakeholder engagement strategy.

7.3.8 Obstacles to implementing IWRM

Experience from some countries suggests that this type of management also encounters obstacles and difficulties in implementation, including:

(1) its orientation and the poor participation of different institutions in its operation. The main reason for this is the inefficiency of education, training and specialized systems, which make the various departments insufficiently familiar with the benefits of this management style;
(2) a separate and multicast infrastructure. In other words, the existing institutional capacities of each organization are individually active, so the capabilities are used separately and are not used for a single purpose;
(3) inefficient and outdated legislation;
(4) unrealistic data and statistics in the field of water resources, as some institutions may manipulate statistics to create a positive image of their performance; and
(5) a culture of over-consuming water, causing an over-demand.

7.4 THE IMPORTANCE OF IWRM

Today, with the misuse of surface and groundwater resources, water scarcity has become a critical issue and competition for water resources has become more intense. In the past, increasing demand for water was met with the development of new water resources, but in present circumstances, the economic, social, and environmental costs mean that unlimited development of water resources cannot continue.

No country can maintain its economic, social and political stability without sustainable use of resources. Water is one of the most important environmental and life-saving resources, and so the current water crisis is a serious issue for mankind. Many factors put pressure on water resources, most of which are socio-economic drivers, including economic growth in many developed or developing countries. Water crises in the past have not had as perceptible an effect on the global population, but today, with increasing global population and global warming, over-utilization of environmental resources and the human need for more food, as well as a significant decrease in rainfall, the effects are widely felt. Water plays a vital role in the social, economic, and environmental dimensions of sustainable development, and this crisis has implications for all three dimensions.

Nearly 6.3 billion (6.3×10^9) people live in basins with scarce water which damages the ecosystem and livelihoods. Studies show that water scarcity leads to a decrease of almost 6.5% of GDP in countries. Water-related natural disasters are the most devastating social and economic disasters, with floods and droughts threatening nearly 4.2 billion people worldwide and causing 6.7 trillion (6.7×10^{12}) dollars in damage.

Water is a precious resource but if managed properly and recycled it plays a vital role in social, economic and environmental aspects, even in the face of a water deficit. Years of top-down planning, sometimes with and sometimes without regard to the broader planning area, precludes a comprehensive understanding of dynamic relationships. The relationships between the various departments involved in water resources management have become complex. Improper utilization of water resources causes loss of resources, drought and ultimately inequality of living conditions in settlement areas. Inequality of access to water resources between settlements, and any inequality in social, economic, and environmental opportunities, is compounded by water's association with other dimensions of development in drier habitats.

These inequalities cause instability and drive regions to a circle of disadvantage, in which regional imbalance drives migration, so that centralization and population density all increase within more developed cities and settlements, disrupting the proportional distribution of the population and putting further pressure on urban areas, with all the additional problems which that involves.

The growth of urbanization and the lack of a proper relationship between urban and rural populations, more developed cities and other settlements, and concentration in those areas all exacerbates instability and environmental problems, on top of the water crisis. These problems all add to the importance and necessity of having IWRM as a solution.

7.5 EXAMPLES

7.5.1 Shadegan Wetland

In this example, a comprehensive program developed for the Shadegan Wetland is presented. Shadegan Wetland in Iran covers an area of 5377 km^2 with a large ecological system consisting of a very high carbonate sedimentary layer, with a relatively high yield.

In these studies, the management of the entire range of internationally protected ecological wetlands, including coastal areas and the Khore Musa region, covering more than 527 000 hectares, have been considered. It was the third Wetland to receive a conservation plan in Iran, and the purpose of the project is to enhance wetland ecosystem management through the development of a comprehensive management plan using an ecological management approach to reduce major threats to biodiversity of international importance.

Over the past decade, inflows to the wetland have been steadily declining due to the development of irrigation in the upstream areas, with increasing volumes of agricultural drainage, and urban and industrial wastewater continuously damaging the wetland. These problems, exacerbated by the recent prolonged droughts, are already having detrimental effects on the globally important biodiversity of the site and on the enormous benefits that the wetlands provide to local communities.

The Shadegan Wetland catchment area is 24,310 km^2 and consists of three main sub-basins: the Maroon, Allah, and Jarrahi. In each of the Maroon and Allah rivers, a reservoir controls the flow of the rivers for irrigation development in the plains of Behbahan, Ramhormoz, Jaizan, Khalafabad, and Shadegan.

Storage dams control flood currents, which are the main source of wetland water supply. The water quality of the wetland varies from freshwater in the northern part which directly receives river water flows, to saline water in the southernmost Khoriyat, including Khormosi. Between these two areas, a mixed water quality is produced, where freshwater and seawater combine to form an area of brackish water. Wetland regimes have been subject to major changes in quantity, quality, and seasonality over the past several decades due to the implementation of large-scale irrigation development plans upstream of the river basin. These development plans not only extract and consume large amounts of water from the river system, but also drain lots of water containing various pollutants into the wetland.

7.5.2 Kazakhstan

Another country that has used IWRM over the past decade is Kazakhstan in central Asia. There, the problem has been the development of the agricultural sector and rising water demand. The region's concerns have been the lack of supply and inefficient use of water resources. In addition to these problems, other challenges such as worn-out technical structures and a discussion of water pollution in the country have raised concerns about water management in general. National water management emphasized that a holistic approach should be adopted and that stakeholder participation and environmental sustainability should be ensured. The program began in 2004 and provided stakeholders with the necessary information during meetings. After studies, the program began with a number of urban and rural households. The steps identified and implemented for this project were:

(1) description of the current status of water resources management;
(2) presentation of an overview of the status of water resources and outline of future decisions;
(3) development of a transition plan from the current to the desired situation.

After targeting, measures to implement IWRM began in the country with actions like:

(1) creating groups to coordinate activities and partnerships of government units;
(2) conducting meetings with stakeholders to get feedback;
(3) preparation and completion of National IWRM.

Although not actually part of the implementation of the plan, relevant experts were also able to achieve goals by conducting meetings with key committee members and removing negative views about the plan.

7.6 SUMMARY

Various definitions and perspectives on IWRM were presented at the beginning of this chapter, and it was emphasised that IWRM is a process not a product, and a tool not a design. A summary of the differences between the IWRM approach and the more traditional approach are outlined in Table 7.1.

IWRM goals vary across countries and scales based on the importance of economic, environmental and social impacts. Chile, for example, emphasizes the importance of economic productivity, while South Africa and the Netherlands tend to place greater emphasis on social and environmental goals, respectively. However, it is thought that there is always a balance between these goals and that a fully integrated approach to water resources management can be achieved to achieve win-win outcomes that will accomplish more than one goal. For example, using an appropriate plan for pricing water resources can promote economic productivity, generate environmental benefits from reduced water demand, and generate social benefits if invested to expand service provision.

Whilst executive differences can make a definition of IWRM difficult, it can broadly be described by a number of fundamental tendencies. First, there is a tendency to move away from command and control tools that focus on water supply management, like large-scale water infrastructure, along with demand management using economic tools. Second, under IWRM, emphasis is on stakeholder engagement and community involvement in decision making.

Some of the benefits of broader collaboration include combining specialized knowledge, encouraging innovative solutions to problems due to a greater variety of perspectives, encouraging collaboration and reducing the risk of water resources conflicts, and developing solutions that are fully open, inclusive, and democratic, thus resulting in broader support and lasting results.

This chapter presents a definition of comprehensive IWRM as an approach to sustainable development. In this regard, while emphasizing the necessity of integrated management of physical resources, emphasis is also placed on the use of policy and organizational tools along with planning for integrated management of water resources. The following is important in this regard:

- issues affecting IWRM can be identified by changing contexts in the policy, organizational and management sectors. The effectiveness of IWRM depends on the specific contexts of these three sectors;
- starting an integrated management plan is a priority to address existing development needs. Addressing the various challenges presented in this chapter will further ensure the success of IWRM;

Table 7.1 differences between the traditional approach and the IWRM approach.

Traditional Approach	IWRM Approach
Partial benefits	General benefits
Limited institutions are involved	Many institutions are involved
Decisions by department	Collective decisions
Only issues of interest are resolved	Various issues are resolved

- the emphasis is on reinforcing economic analyses and integrating them with environmental, social, and water resources as a way of creating a common language among different groups;
- approaches to the study of consequences and effects require a transformation of physical and technical studies. Comprehensiveness is the first condition for such studies and requires all three groups of physical hydrological, social, and environmental studies.

REFERENCES

Bizikova L., Roy D., Swanson D., Venema H. D. and McCandless M. (2013). The Water-Energy Food Security Nexus: Towards a Practical Planning and Decision-Support Framework for Landscape Investment and Risk Management. International Institute for Sustainable Development, Winnipeg, Manitoba.

Molle F. (2008). Nirvana concepts, narratives and policy models: Insights from the water sector. *Water Alternatives*, **1**(1), 131–156.

Petit O. and Baron C. (2009). Integrated water resources management: from general principles to its implementation by the state. The case of Burkina Faso. *Natural Resources Forum*, **33**(1), 49–59.

Rogers P. and Hall A. W. (2003). Effective Water Governance. Global Water Partnership, Stockholm.

Snellen W. B. and Schrevel A. (2004). IWRM: for sustainable use of water 50 years of international experience with the concept of integrated water management (Background document to the FAO/Netherlands Conference on Water for Food and Ecosystems). Ministry of Agriculture, Nature and Food Quality, The Netherlands.

Zare F. (2011). Integrated physico-socio-economic water resources system modeling in a watershed scale (Case study: Gorganrood Watershed – Iran). Master's thesis, Faculty of Agriculture, Tarbiat Modares University, Tehran, Iran.

Chapter 8

Virtual water

Mohammad Delpasand[1], Omid Bozorg-Haddad[2] and Erfan Goharian[3]*

[1]M.Sc. Student, Dept. of Irrigation & Reclamation Engineering, Faculty of Agricultural Engineering & Technology, College of Agriculture & Natural Resources, University of Tehran, 3158777871 Karaj, Iran. E-mail: Delpasand.m@ut.ac.ir.
[2]Distinguished Professor, Department of Irrigation & Reclamation Engineering, Faculty of Agricultural Engineering & Technology, College of Agriculture & Natural Resources, University of Tehran, Karaj, Tehran, 31587-77871, Iran. E-mail: OBHaddad@ut.ac.ir.
[3]Assistant Professor, Civil and Environmental Engineering, College of Engineering and Computing, University of South Carolina, 300 Main St. Room C206 Columbia, SC 29208 (803) 777-4625, E-mail: Goharian@cec.sc.edu.
**Corresponding author.*

ABSTRACT

In many regions of the world, water scarcity has created numerous problems for the supply of domestic water use, agricultural production, and for the human life cycle in general. Perhaps in the near future water will be synonymous with gold. Over the past two decades, the virtual water concept has become known as a mitigating water scarcity crisis solution which can balance the unequal distribution of world water resources. Virtual water is the amount of water that a product consumes in a production process from start to end. Developers of the virtual water idea believe that by exporting and importing commodities, large volumes of water can be traded which is referred to as virtual water trading. In the theory of virtual water trade, in order to reduce the pressure on water resources, it is recommended that water-scarce countries import products instead of producing them from internal water sources and allocate these water resources for other lucrative business activities. Therefore,

Editor: Omid Bozorg-Haddad
doi: 10.2166/9781789062144_0131

in this chapter, the virtual water concept and its specifications and evaluations are fully explained.

Keywords: Virtual water, Virtual water trade, Agricultural virtual water, Industrial virtual water.

8.1 INTRODUCTION

Tony Allan (Allan, 1993) was the first to define the term 'virtual water' as the amount of water consumed to produce a product from a range of goods, farm products or services. Since then, many studies have been conducted using this concept, which used the concept of virtual water expressed by Allan in their research, and measured the amount of water needed to produce a 2-megabyte electronic chip, finding that 2 m^3 of water were needed, equivalent to that needed to produce 5 kg of red meat or about 2–5 kg of wheat.

Hoekstra and Hung (2003) quantified the flow of virtual water trade between nations in the context of the international crop trade between 1995 and1999. There calculations showed that 13% of the water consumed for the production of goods in the world is not used for domestic consumption but is exported in virtual form. It should be noted that this percentage is global and conditions vary across countries. The results also show that during the same period, the US, Canada, Thailand, Argentina and India had the largest amount of virtual water exports while Sri Lanka, Japan, The Netherlands, Korea and China had the largest volume of virtual water imports.

Jiang *et al.* (2015), using a multi-regional input–output model in China, found that some provinces in China, despite their very poor water resources, are still among the first exporters of virtual water, and that some provinces are heavily dependent on imported products (imported virtual water). For the future, they proposed a pricing system based on the market and in the end, in order to avoid the pressure of certain provinces, new technologies to increase water use efficiency.

Sun *et al.* (2016) examined China to assess the amount of water exported and imported through grain and cereal trade. They surveyed the water pressure index by assessing the different regions of China, as well as the volume of exports and imports. Finally, they recommended that China should focus on water saving policies as well as boosting water use efficiency in order to prevent future food crises.

Katyainia and Baruaa (2016) examined the virtual water trade in India between 1996 and 2014 using the input–output analysis method. They chose India because it is the world's largest consumer of water. Despite the severe water shortage, agriculture makes up the largest portion of water use, and their goal was to identify the gaps in virtual water research in India. By analyzing virtual water in food grains and then linking virtual water to water shortages in different provinces, they found that water supply was increasing on an inter-provincial scale while water storage in some national subsectors was decreasing.

Zhang *et al.* (2016) examined China's relationship with the oceanic states between 2001 and 2013. The study was about virtual water flow from plant production, which eventually divided the countries of the Trans-Pacific Partnership into four categories: countries which import a huge volume of virtual water but do not export virtual water to China, and vice versa; countries with bilateral relations with China; and those with equal relations with China.

Tamea *et al.* (2016) assessed different countries in terms of local food crises and the virtual water perspective. They determined the vulnerability of countries by specifying the amount of food they exported and imported, and eventually the food produced. For example, due to high imports and low production in 2009, Iran was exposed to moderate to high vulnerabilities. In this research, they did not look at issues such as water shortages in countries and consumption management or other aspects of this issue.

Thompson and Jeffords (2016) used the concept of virtual water to obtain an indicator of pollution and eventually suggested that increased freedom in virtual water trade and the introduction of products with high water levels may reduce pollution. They used BOD in their assessments.

Marston and Konar (2016) examined the central valley area of California for virtual water research and investigated how severe drought can affect agricultural products, water footprints, and virtual water. They found that during three years of drought, a 3% increase in virtual water and a 12% reduction in crop yields occurred. These results are useful for water management in future droughts.

Tuninetti *et al.* (2016) examined the empirical relationships between countries engaged in global trade and divided relationships between countries into two active and inactive categories. Active connection is defined as when the amount of water transported between the two countries is more than 1000 m^3/year. In assessing these connections, the geographical distance, population, and use of fertilizers were the main factors.

Qu *et al.* (2017) examined the risk of virtual water shortages in the world trade system between 1995 and 2009 and identified countries that were endangered by their water exports. They showed a snapshot of the vulnerability in the global economy but did not consider how the environmental or local factors could distort production.

Chouchane *et al.* (2017) conducted their research using a dynamic analysis of virtual water trade and Tunisia as the case study area. They considered different regression models for different plants and eventually recommended that the governments and decision makers in Tunisia should imports plants with high water consumption instead of producing them.

Bekchanov *et al.* (2017) modeled a virtual water-based economy in a number of input–output models and simulation and optimization models.

Angelis *et al.* (2017) examined the role of virtual water in military disputes and concluded that bilateral or multilateral communication on virtual water could reduce the probability of war between countries.

Lamastra *et al.* (2017) conducted their research on virtual water trade between Italy and China. They concluded that Italy has imported more water in the field of agriculture than China, rather than exporting it, and China has met the demand of Italy by putting pressure on its water resources.

Chen *et al.* (2017) monitored the input and output of water in China between 2000 and 2012 using the input–output method to balance the virtual water trade in China, and monitored and categorized different countries in terms of export, imports, and their quantities.

Brindha (2017) assessed the virtual water trade of plants and animals in India between 1982 and 1986, calculating virtual water for each plant and animal that India exported or imported. The products in which there are more exports of virtual water were also evaluated. Finally, Brindha determined which countries import or export virtual water to India.

Sartori *et al.* (2017) used network combinations and gravity models to determine how network topology and virtual water quantity properties can change over time under different socioeconomic and climatic scenarios. They used these models to make predictions for the years 2020 and 2050. The advantage of using this method was the ability to reproduce the observed characteristics of the virtual water trade. They concluded that some countries will be strained due to the high import and water scarcity risk of the countries giving them virtual water.

Liu *et al.* (2017) used a noncompetitive input–output method to investigate water allocation. Their goal was to calculate the water flow and study strategic policies in the management of water resources. They chose Urumqi as their case study and ultimately provided a strong and early diagnosis of virtual water flow. They concluded that general water management depends on two improvements: (1) the expansion of virtual water trade with rich countries to increase virtual water resources; and (2) the development of a construction mechanism and infrastructure to increase the physical resources of water.

Zhi *et al.* (2018) developed a semi-quantitative model for analyzing, accepting, and developing economic and business patterns in the water trade. In order to achieve mutual benefits in virtual water trade, they proposed improvement in water use efficiency.

Cheng Fu *et al.* (2018) used China as their case study area to assess the virtual water trade and quantify the amount of virtual water generated by intra-regional and inter-regional trade.

Miglietta *et al.* (2018) evaluated five wine factories in Italy in order to find the highest water use efficiency. They monitored these factories between 2011 and 2015 and concluded that to achieve sustainable development in the future, the best way is to use water more efficiently, which will cost less and bring more value to the product.

Han *et al.* (2018) used the input–output method to provide systematic accounting of virtual water and conducted a comprehensive analysis of transmission patterns. They conducted their research in 182 countries and eventually classified and ranked them in terms of efficiency, pressures, etc.

Tan and Bi (2018) researched virtual water trade in China and developed an integrated framework for integration of inter-regional inputs–outputs and social network analysis, which not only measures the scale and structure of water trade but also specifies the topology of the transmission network. The purpose of their research was to estimate the pattern of water trade with regard to provincial water trade, which includes both exports and imports. Finally, by comparing virtual water accounting and topology information of the region, they determined the interrelationships of provinces from the virtual water trade perspective.

Qian *et al.* (2018), using a combination of input–output analysis and structural analysis, examined Yunnan, China, to assess the potential for socioeconomic development in relation to water sustainability. The data used spanned 2002 to 2012 and was completed in two ways as follows: integrated and one-dimensional. Analysis showed that Yunnan, despite its abundant water resources, made little contribution to reducing the pressure of water scarcity through virtual water trade. While technological advances have reduced water consumption in the world, in the time frame considered, Yunnan's water footprint has increased. As a result, the study suggests they optimize industry and increase efficiency and use of new technologies. If Yunnan wants to achieve economic development, it must increase its exports using its rich water resources.

Liao *et al.* (2018) reviewed the virtual water transmitted between the provinces of China between 2002 and 2010 through the production of electricity in hydroelectric plants and thermal power plants through burning coal. They found that virtual water is used to generate electricity from the northeast; Center and south of China which is transmitted to the east and north of China. By simulating four different scenarios, they also carried out a

2030 forecast. These scenarios were as follows: (1) stay in the current state; (2) use fuel resources without carbon; (3) use low carbon fuel sources; and finally (4) replace water-cooling methods with air cooling methods.

Ridoutt *et al.* (2018) examined the agricultural sector in Australia using an input–output method for water shortage in the Australian agricultural sector and compared this deficit to conventional water consumption, but did not consider food and export needs. In the livestock sector, they concluded that, to reduce their water use, instead of reducing their production they could change the livestock's diet.

Wahba *et al.* (2018) examined the impact of virtual water trade and domestic use on available water sources using the input–output method. They chose Egypt as a case study and assessed import, export, and amounts, and compared household water consumption with rural households in order to assess domestic consumption.

Abdelkader *et al.* (2018) designed a model that makes two simulations: (1) a simulation of agricultural, industrial and household uses; and (2) a simulation of the export and import of virtual water on an international scale. They used a system of dynamic models for this assessment and Egypt as a case study.

Bae and Dall'erba (2018) examined the Arizona region in relation to virtual water exports using the input–output method. Their goal was to save 19% more water, which they examined in three different scenarios: (1) increased irrigation efficiency; (2) increased water prices; and (3) reduced exports. Method 1 was determined to be the most advantageous approach.

Wang *et al.* (2019) examined the flow of virtual water from grain and cereal trade in China. They found that this type of production in China has reduced water use efficiency. They offered a linear optimization model with the goal of minimizing costs, finding that the present situation in China is inappropriate for cereal exports and that the virtual water from this trade is 1179.24 billion (1179.24×10^9) m^3 a year.

Virtual water trade as a comprehensive policy for countries requires a deep understanding of its impact not only on the interdependencies and interactions of international trade, but also on a proper understanding of local, social, environmental, economic and cultural conditions. In addition, having a sustainable virtual water trade requires fair trade rules and assured access to local, national and regional food security for all peoples (Chapagain & Hoekstra 2003*a*, *b*). However, countries' water managers need to be aware of the advantages and disadvantages of virtual water theory in order to use this tool to better protect their countries' water resources.

8.2 DEFINITIONS AND TERMS

8.2.1 Virtual water

The term 'virtual water' was first introduced by Tony Allan in 1993 and was defined as the sum of water used to produce a certain amount of product (whether commodities, agricultural products or even services) (Allan, 1993). The amount of virtual water is equal to the sum of the total water consumed at different stages of the production chain, from start to finish. For example, an average of 1300 liters of water is needed to produce 1 kg of wheat. The virtual attribute is used in the definition to mean that the bulk of the water consumed during the production process is not physically present in the finished product or product. A few years later, Hoekstra (2003) provided a more complete definition of virtual water. Virtual water is the sum of the water needed to produce a certain amount of product, depending on climatic conditions, location, production time and efficiency. According to the latest definition, climatic conditions, place and time of production, management and planning, people's culture and habits are influenced by the amount of virtual water present, and it will certainly vary by product in different regions.

8.2.2 Virtual water trade

Developers of the idea of virtual water believe that by exporting and importing goods and products, large volumes of water are transferred – which is referred to as virtual water trading. The average volume of virtual water exchanged in the world is estimated to be more than 1900 billion m^3/year. In the theory of virtual water trading, in order to reduce the pressure on water resources, it is recommended that low-water countries import products instead of producing products from domestic water sources and allocate domestic water resources to profitable business activities.

8.2.3 Water productivity

Water productivity is the amount of product produced per unit volume of water (for example, how many potatoes are produced per m^3 of water) and its unit is usually kg/m^3, whereas virtual water represents the amount of water used to produce a given amount of product and its unit is l/kg (or m^3/kg). In other words, in productivity, the emphasis is on the amount of water produced, whilst in virtual water, the emphasis is on the amount of water consumed during production. Therefore, with increasing water productivity, the amount of virtual water in the product will decrease and vice versa.

8.2.4 Virtual water flow

Virtual water flow is defined as a volume of virtual water transfer through goods, products and services from one country to another in the form of trade.

8.2.5 Virtual water import

Virtual water associated with goods or services imported into a country or region is called virtual water import (VWI). Its value is also equal to the amount of water consumed (in the exporting countries) for the production of those goods and services. Importantly, this type of water has the potential to add to available water resources.

8.2.6 Virtual water export

Virtual water which is exported with services or products is called virtual water export (VWE). Its value is equal to the amount of water consumed (in exporting countries) for its production and services. From the export point of view, this type of water is one of the factors in the availability of water resources.

8.2.7 Virtual water level

The difference between import and export of virtual water over a significant period of time is the virtual water level. Positive virtual water level occurs when virtual water imports from other countries exceed virtual water exports. Negative virtual water level occurs when virtual water exports exceed imports.

8.2.8 Water neutral

Water neutral is the maximum amount of reasonable and feasible reduction in water footprint, plus compensation for the adverse effects of human waste. In other words, a person can reduce their water footprint as far as possible and compensate for the negative effects of their residual water footprint. The purpose is to reduce the negative effects of waste, reduce waste production, recycle it and manage gray water. The water neutral concept aims to encourage people and companies which consume and contaminate water to develop a water mitigation strategy to reduce water consumption and pollution, and also to offset the negative impacts of water pollution (by investing in environmentally friendly projects).

8.3 IMPORTANCE AND NECESSITY

A tangible decline in water resources is one of the major challenges faced in many countries. Water resource surveys show that, by 2025, approximately 3.5 billion people will live in countries with water stress (about 6.5 times the population living in similar circumstances in 2000). Of these, at least 1.1 billion people will be without access to safe drinking water and about 4.2 billion people will be without access to appropriate sanitation systems.

Agriculture is responsible for the largest share of water consumption in the world, and more than 70% of fresh water is consumed in agriculture globally. Therefore, the limitations of fresh water resources for food supply in agriculture require that both soil and water resources be managed simultaneously. Water consumption management is the most important integrated management factor for water resources in the agricultural sector.

Diet has a huge impact on the total volume of water consumed by the population living in a geographic area. For example, if the entire population of the planet's diet were changed to that of the typical American, the amount of water needed to produce enough food to feed the world's population would be 75% greater than it is currently. Water is an economic commodity (World Meteorological Organization, 1992) and, with this in mind, the importance of an economic view of the water trade and an in-depth look at the concept of virtual water is justified.

When countries do not have reliable surface water resources (due to rainfall because of arid climates or poor management of water resources etc.) and at the same time have a significant decline in aquifers (meaning their groundwater resources are also facing serious threats), whilst facing high population densities and rising water demand, they can expect to face serious crises and conflicts. In other words, given resource constraints and demand growth, it seems clear that resource management and water use should be more important than they were in the past. Therefore, identifying resource allocation priorities and the adoption of rational policies by the decision-making systems of individual countries can be considered as the most fundamental solutions to the water crisis. In this regard, one of the issues that needs to be considered by authorities and planners is virtual water management.

Allan (1997) reconsidered his earlier pessimism about the outlook for future water warfare. He saw trade of agricultural products as a solution to resolving the unequal distribution of water resources. The main reason for this is that end-product trade is much cheaper than physical water trade. Water-scarce countries can therefore spend their precious water resources on more valuable activities by trading products or, in other words, by trading virtual water. The concept was initially limited to the Middle East and North Africa, areas where there was a high likelihood of war due to water disputes in the

Jordan and Euphrates basins. In fact, the purpose of the concept of virtual water was to examine the impact of free trade on the severity of tensions in the political economy of water. From an economic standpoint, two effects can be seen: (1) smoothing inequality; and (2) providing efficiency in trade. Therefore, the concept in practice is both profitable and contributes to better distribution. According to the theory of virtual water, if a country's water resources are scarce, the country's virtual water imports should increase in order to decrease water scarcity.

Nowadays, the concept of virtual water has become a multidimensional issue. There is ample evidence which shows that water system use changes without regard to other factors – such as land use, partial planning, soil management, climate change, population development, economic production and consumption, public health, environmental management, business policies, national security and development co-operatives. Virtual water import is a method to reduce pressure on domestic water resources, which can be attractive to water-scarce countries. Examining the theories expressed about virtual water, it can be said that virtual water should be considered as part of any sustainable agriculture implementation and for sustainable use of water resources.

8.4 BASICS AND LOGIC

Annually, large volumes of water are used for purposes such as drinking, cooking, washing and for other human activities that can be regarded as direct uses for water. However, much larger volumes of water are spent on agricultural production and industrial products and services. The concept of virtual water is a relatively new approach to the consumption of indirect water which is used in the production of products or services for human beings. Originally, it took nearly ten years after Allan first debated the virtual water concept for it to come to public attention. The first international conference on the subject was held in December 2002 in Delft, The Netherlands. At the Third World Water conference in Japan in March 2003, a special session was devoted to the issue of virtual water. It is worth noting that the amount of virtual water for each product depends on the production conditions of the product. For example, to produce 1 kg of grain in dry and favorable weather conditions, between 1 and 2 m^3 of water are needed. To produce the same amount of grain in adverse weather conditions (high temperatures and with high evapotranspiration) between 3 to 5 m^3 of water are needed. The virtual water required for livestock is much more than that of agricultural products. Of course, the concept of virtual water is not only for use in the context of agricultural products and in recent years several studies of virtual water in the field of industrial production have also been conducted.

The global trade in goods creates an international flow of virtual water – the so-called virtual water trade. The inter-country virtual water trade was estimated at 1340 billion m^3 in 2000, of which 60% was for agricultural products, 14% for fish and seafood, 13% for livestock and 13% for the meat trade. While the development of the virtual water trade will result in sustainable development of water resources, social and cultural problems that could challenge the food security some of the importing countries make the virtual water trade an important and complex issue.

8.5 MATERIALS AND METHODS

Accurate calculation of virtual water for agricultural products, industrial goods, and services can vary from being simple to very complex depending on the type and time of production.

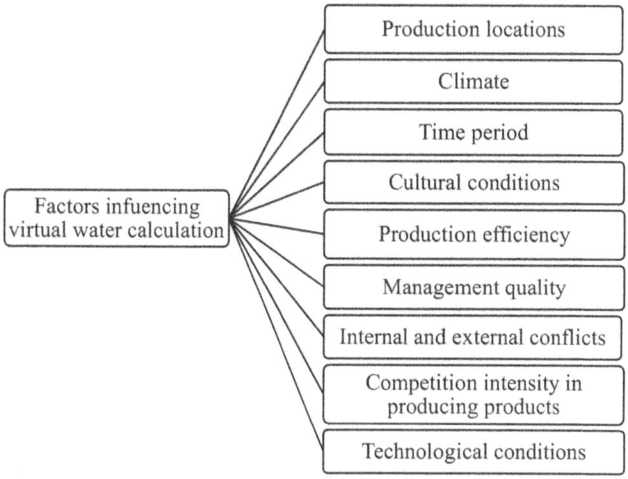

Figure 8.1 Factors influencing virtual water calculation.

If the final product is obtained by combining several smaller components, then accurate calculation of virtual water will be much more difficult. For such composite products, the virtual water for each component of the product should first be calculated separately. Virtual water is a multidimensional concept and the final amount depends on many factors (efficiency, climate, etc.). There are many variables involved and with a small change in any of them the final volume of virtual water is also subject to change. Although this has led to the complexity of virtual water computing, it is advantageous in that the amount of virtual water can be controlled by the number of controllable factors of any materials.

In this section, the basics of virtual water computing are presented. It should be noted that the methods presented can be varied according to different perspectives and are only for initial acquaintance with virtual water calculation methods.

8.5.1 Virtual water calculation at a national scale

The virtual water for a country is the sum of the amount of water consumed directly or indirectly by its residents for drinking, sanitation, production of products, service provision, etc., over a specified period (e.g. one year).

8.5.2 Agricultural virtual water

8.5.2.1 Water used in crop production

The amount of water consumed in agriculture for crop production can be calculated from Equation (8.1):

$$\mathrm{AMU} = \sum_{c=1}^{n_c} \mathrm{CWU}[c] \tag{8.1}$$

where AMU = the total amount of water consumed for agricultural production in m^3/year, and CWU[c] = the volume of water consumed by plant c, which is the sum of the water consumed for the growth of a plant over a given time period, for example in m^3/year, which can be calculated in Equation (8.2) as follows:

$$\mathrm{CWU}[c] = \mathrm{CWR}[c] \times \frac{P_y}{P_a} \tag{8.2}$$

where, CWR[c] = the water requirement of plant c in m^3, P_y = the amount of crop produced per year in tons/year and P_a = the production rate per unit area in tons/hectare.

8.5.2.2 Crop virtual water

The virtual water of a crop is calculated from the ratio of the total water used to produce that crop to the total product produced, which is usually measured in m^3/ton, and its equation is as follows:

$$\mathrm{VWC}[c] = \frac{\mathrm{CWU}[c]}{T_p[c]} \tag{8.3}$$

where VWC[c] = virtual water of plant c in liters/kg or m^3/ton and T_p = the product yield in tons or kg.

8.5.3 Livestock and animal virtual water

The amount of virtual water for a completely grown animal (which is ready to sell at market) is the total volume of water used to grow that animal including water for drinking, sanitation, feed, and the like. It depends on various conditions such as environmental status and farm system (traditional or modern), type of feed used, farm management and climatic conditions of the animal's growth environment. In summary, the following three factors can be considered in calculating the amount of virtual water in animals: (1) feed (forage and other nutritional supplements); (2) drinking water; and (3) services. With this description, the amount of virtual water of an animal is calculated from Equation (8.4):

$$\mathrm{VWC}[a] = \mathrm{VWC}_f\,[a] + \mathrm{VWC}_d[a] + \mathrm{VWC}_s[a] \tag{8.4}$$

where VWC[a] = total amount of drinking water consumed by animal [a] in liters/kg, $\mathrm{VWC}_f[a]$ = the amount of virtual water used in the preparation of animal [a]'s feed in liters/kg, $\mathrm{VWC}_d[a]$ = the amount of virtual water which animal [a] receives through drinking, equivalent to the total volume of water given to the animal from birth to delivery to the slaughterhouse in liters/kg, and $\mathrm{VWC}_s[a]$ = the water used to keep animal [a], including grooming, health, etc., expressed in liters/kg.

The amount of available virtual water in animal or poultry feed consists of two main parts. The first is the amount of actual water used to mix the foods and the second is the virtual water that is contained in the various feed components. The following equation illustrates the method of calculating virtual water in animal feed:

$$\mathrm{VWC}_f[a] = \frac{\int_b^{sl} \left\{ q_{\mathrm{mixing}}[a] + \sum_{c=1}^{n} \mathrm{VWC}[c] \times F[a,c] \right\} \mathrm{d}t}{W[a]} \tag{8.5}$$

where q_{mixing} = the volume of water needed to combine animal feed in m^3/day (for example, bran in animal diets should be mixed with water.), $F[a, c]$ = the amount of crop feed [c] (of the desired type) consumed by animal [a] in tons/day, and $W[a]$ = animal [a]'s weight at the time of delivery to the slaughterhouse in tons; b represents the time of birth and sl represents the time of slaughter of the target animal.

The amount of virtual water an animal receives through drinking equals the total volume of water given to the animal from birth to delivery to the slaughterhouse and can be obtained from Equation (8.6):

$$\mathrm{VWC}_d[a] = \frac{\int_b^{sl} q_d[a]\mathrm{d}t}{W[a]} \tag{8.6}$$

where q_d= daily water needed for drinking, in m^3/day.

Additionally, the amount of virtual water used to maintain an animal, including the amount of water used to clean the animal and other services needed to protect the environment around it, is also calculated from the time period of the animal's growth and development which can be obtained from the Equation (8.7):

$$\mathrm{VWC}_s[a] = \frac{\int_b^{sl} q_s[a]\mathrm{d}t}{W[a]} \tag{8.7}$$

where $q_s[a]$ = daily water used to provide maintenance services for animal $[a]$ in m^3/day.

8.5.4 Virtual water in processed crops and livestock

Virtual water of a processed agricultural or animal product depends on the amount of virtual water for the primary plant or live animal. It should be noted that the virtual water for the primary plant or live animal is transferred to the secondary product during the processing stages. It is generally assumed for ease of computation that the product produced is derived from only one primary plant or animal species. For example, commodity (A) imported from a country of origin has been obtained only from plant (B) or animal (C) and must also have been used solely for its production or development in its domestic water resources.

The level of production can be defined for ease of systematic analysis of virtual water. Accordingly, products derived directly from the primary plant or live animal are called primary products. For example, milk and carcasses of slaughtered animals (including meat, bones, and skin) are considered as primary products of cattle. As an example, in a paddy field, rice is the primary crop.

Some primary products become secondary products during the processing phase. For example, cheese and butter are called secondary products derived from primary production, using the same milk, or flour which is a secondary product from wheat (which is the primary product). In the same sense, sausages and hamburgers as secondary products of beef carcasses.

8.5.5 Virtual water flow in agricultural product trade

Virtual water flow in agricultural commodity trading can be calculated by multiplying the weight of the commercial commodities by an amount of virtual water. The virtual water for an agricultural product or an animal depends on its location and production. Therefore, it is easier to assume that all the products are manufactured in the same exporting country. The virtual flow of agricultural products from an exporting country to an importing country can be calculated as follows, and the unit is usually m^3/year:

$$\text{VWF}[e,i,p] = \text{PT}[e,i,p] \times \text{VWC}[e,p] \tag{8.8}$$

where, VWF = virtual water flow in m^3/year, pt = weight of commercial product in tons/year, e = exporting country, i = importing country and p = desired product.

8.5.6 Virtual water flow in industrial product trade

The calculation of virtual water flow for industrial products is similar to that of agricultural products, but the manufacturing methods and standards of production of industrial products differ in different countries. Therefore, given the varied production patterns and differences in standards, the same amount of water should not be used in the production of a particular commodity and as a result their amount of virtual water will vary from country to country.

The total volume of water allocated to the industrial sector is about 716 billion m^3/year, which is about 10% of the total water consumed in the world. The average value added to virtual water in the industrial sector for each country can be calculated as follows:

$$\text{VWC}[e] = \frac{\text{IWW}[e]}{\text{GDP}_i[e]} \tag{8.9}$$

where IWW = the amount of water allocated to the country's industries in m^3/year and GDP = the value added to the industry sector in the currency of each country per year.

Thus, global average virtual water content of industrial products (VWC_g) can be derived from Equation (8.10):

$$\text{VWC}_g = \frac{\sum_{e=1}^{n} \text{IWW}[e]}{\sum_{e=1}^{n} \text{GDP}_i[e]} \tag{8.10}$$

The total volume of virtual water exported from a country is obtained by multiplying the amount of virtual water by the export value of industrial products (EVIP):

$$\text{VWE}[e] = \text{VWC}[e] \times \text{EVIP}[e] \tag{8.11}$$

where VWE = virtual water exports through industrial production in m^3, VWC = virtual water for industrial production in m^3 per currency of each country, and EVIP = export value of industrial production in the currency of each country.

Virtual water imported through imported goods and products can also be calculated as follows:

$$\text{VWI}[e] = \text{VWC}_g \times \text{IVIP}[e] \tag{8.12}$$

where VWI = virtual water imported in m^3 and IVIP = import value of industrial production in monetary terms.

8.5.7 Virtual water balance of a country

The balance of the virtual water flow to the lands can be obtained by examining the amount of exports and imports of the virtual water. The difference between the amount of virtual water exported and imported to a country is its net virtual water. This calculation will help determine the positive and negative water balance and also specifies which countries are generally exporters and which are importers Investigating the total imported and exported virtual water is of great importance for understanding the balance of a country's virtual water flow over a period of time. A positive balance indicates the entry of virtual water into that country and a negative balance indicates the export of virtual water from that country.

8.5.8 Water scarcity, dependence on imported water and water self-sufficiency

Water scarcity (WS) in a region or country is derived from the ratio of the water footprint of that country or region to the amount of water available (WA):

$$WS = \frac{WFP}{WA} \times 100 \tag{8.13}$$

where, WS = water scarcity (expressed as a percentage), WFP = water footprint in m^3/year, WA = water availability in m^3/year.

If national food supply, services, households, and water industries require more than a nation's available water, national water shortages can rise significantly. The criterion for renewable water resources in computation can be used to facilitate understanding and measurement. Dependence on imported water is also defined as the ratio of external water footprint to the total water footprint of a country:

$$WD = \frac{EWFP}{WFP} \times 100 \tag{8.14}$$

where, WD = dependence on imported water (as a percentage), EWFP = external water footprint in m^3/year and WFP = total water footprint of the country in m^3/year.

However, there is another term called water self-sufficiency (WSS) at the national level, which is the ratio of internal water footprint to the total water footprint of a country, and is derived from Equation (8.15):

$$WSS = \frac{IWFP}{WFP} \times 100 \tag{8.15}$$

where IWFP = internal water footprint in m^3/year, and WFP = total water footprint of a country in m^3/year.

One hundred percent water self-sufficiency occurs when all water needs in a country are supplied from domestic water sources, and water self-sufficiency comes close to zero when all the needs of a country's services, industries, etc. are heavily dependent on virtual water imports.

8.6 CASE STUDIES

8.6.1 The virtual water content of major grain crops and virtual water flows between regions in China

In this case study, the benefits of saving virtual water flows in relation to the transfer of three crops between regions is evaluated. It is clear that agriculture, as a major consumer of water throughout the world, consumes significant amounts of water which puts pressure on the demands of providing for other sectors. The study, conducted by Sun *et al.* (2012), was undertaken because virtual water and its trade theory provided a potential solution for improving water resource management in the agricultural sector and could solve water crises in low-water areas. They evaluated the quantities of blue and green virtual water for wheat, corn and rice at a regional scale in China. Food security issues in China have attracted a great deal of attention worldwide, and uneven distribution of water and agricultural resources have contributed to this. Croplands are mainly located in the northern areas and water resources are mainly in the southern areas of the country. Northern arable lands make up about 60% of China's total arable, while northern arable lands make up only about 19% of China's total water resources. In contrast, arable land in the south accounts for about 40% of China's total land area. However, water resources in these areas account for almost 81% of China's total water resources. Due to favorable climatic conditions and abundant water resources, southern areas have traditionally been superior in food production. Therefore, the southern regions were used as the main grain export areas until 1990. However, due to economic, social, and agricultural changes, cereal production in the southern areas has declined over the past 20 years and these areas have become importers of cereals, which has had a significant impact on redistribution of agricultural land and water resources in China. Cereal transportation from north to south has caused the transfer of water resources from north to south. This transition raises a number of issues to consider:

(1) regional disputes over green and blue water for the production of primary cereal products;
(2) the increase in the amount of virtual water flows associated with the grain transfer process; and
(3) the water-saving benefits associated with cereal transportation between regions.

In order to investigate the problems mentioned above, the authors selected wheat, maize and rice, being the main crops in China and accounting for more than 86% of total national cereal output. They first analyzed the virtual water for wheat, maize, and rice at a regional scale in China and then assessed the virtual water flows and the water saving volume of the three crops between regions. The study sites selected were 31 independent and non-independent provinces of China selected based on geographical location, climatic conditions, and agricultural production conditions.

As a result of this research, various calculations were made, including the national average of the virtual mean water content of wheat per ton (50.98% green water, 49.02% blue water), maize per ton (76.27% green water, 23.73% blue water) and rice per ton (61.9% green water, 38.1% blue water). With the regional transfer of wheat, maize and rice, the flows of virtual water reached 59.91% green water and 40.09% blue water. The authors concluded that in order to ensure food security in China, the government should improve water efficiency during the grain production process. Meanwhile, under the conditions of economic feasibility and availability of groundwater resources, China must equip its southern cereal crop area to fully exploit green water resources and reduce the pressure on water resources.

8.6.2 Embodied water for urban economy: A three-scale input–output analysis for Beijing

Han *et al.* (2015) presented a three-scale input–output analysis plan to examine the characteristics of urban water consumption in Beijing. In their study, virtual water for the urban economy is defined as direct and indirect water, which is supported by extensive domestic and foreign trade and can be divided into nine categories in relation to three sources (local outlet, domestic imports, foreign imports) and three destinations (local final demand, domestic exports and foreign exports). According to statistics, in Beijing in 2010, only 3.53 billion m^3 of local water extraction was attributed to the urban economy, while total virtual water demand was estimated at 13.61 billion m^3. Also, an additional 10.08 billion m^3 of indirect water consumption was obtained through cross-border trade. In general, the water that is unintentionally provided by internal and global trade has an important role to play in meeting water demand, which has important implications for decision making to reduce urban water scarcity.

8.7 SUMMARY

In this chapter, after introducing virtual water as the sum of water consumed to produce a certain amount of product (whether commodity, agricultural product or service), virtual water trade and other related terms have been introduced. The reason for the importance of this issue was then given. The calculation method of virtual water in various sectors such as agriculture and livestock as well as calculations related to virtual water balance in trade was provided. There is always a question of whether researchers and managers are choosing the most appropriate way to use available water resources. Is crop production remaining the same? Are the environmental effects of untreated water being monitored? How and to what extent can virtual water offer a new policy to the international community? One of the main focuses in virtual water analysis and water footprints is to draw attention to the type of water, location, and time of production. Understanding the amount of virtual water used in various commodities and foods should guide macroeconomic management policies in countries. Water loss in various forms through the waste of goods and food is also an issue that can be further investigated. The loss of agricultural products around the world causes a significant portion of agricultural output to be extracted from the production cycle, so that, on the basis of available statistics, on average 35% of agricultural products in the world are wasted at various stages. Agricultural wastes occur in three stages: before harvest, at harvest and after harvest. Most of the waste is created at harvest and after (until consumption). If the amount of virtual water produced by agricultural waste in countries is estimated, it will show how much water is wasted in this way.

Thus, one of the benefits of paying attention to virtual water is being aware of the huge volume of wasted water through the waste of products and food. In this regard, the following comments and suggestions are made, inspired by virtual water theory:

- Informing the public through the media has been identified as one of the ongoing tasks of water resources management and should be taken seriously into account in planning.
- The amount of virtual water in goods as well as the place of production should be labeled. Dissemination of this information will have a positive psychological impact on a population's water saving.
- Many goods and products consume huge volumes of water during their production process. Disposal of used goods ignores much of this virtual water. Collecting a large number of goods, including paper, plastics, metals (iron, copper, etc.), and recycling them is a very effective way of reducing water consumption.

- Genetic characteristics of plants, planting times, climate and other factors all cause the virtual water for similar crops to vary in different regions. Therefore, if the relative advantage of each crop is identified in terms of the minimum amount of virtual water in each region and the cropping pattern program is adjusted to the relative advantage, then the average virtual water for crops produced in the country will be reduced. In addition, it is possible to map the relative advantage of agricultural products to countries using the data obtained about the amount of virtual water needed.
- Finding inexpensive technologies to reduce evaporation from the soil and transpiration from the plant will help to reduce the virtual water needed for crops.
- More efficient use of green water resources, with the aim of reducing the pressure on water resources, can be applied as a strategy for food security.

In the end, although virtual water has a lot of potential for demonstrating the intrinsic value of water as a scarce element, and can be a guide for people, particularly water resource policymakers, critics still believe that the concept of virtual water does not provide policy makers with much guidance on water efficiency and productivity. While this concept is simple and looks appealing, it has some defects as an efficient tool for analyzing a country's economic and public policy. Thus, the following criticisms have been levelled at virtual water:

- A virtual water strategy on its own will not ensure that water resources are used in a way that will lead to greater profits.
- Virtual water cannot provide researchers with information about water supply and demand, as it cannot by itself determine whether water resources are being used to the extent permitted or not.
- The dangers of food dependence are among the concerns of critics of the virtual water trade who claim that food imports could become a tool for abuse of power and political pressure. Over-reliance on global food markets could have a major impact on the world's food security.

REFERENCES

Abdelkader A., Elshorbagy A., Tunninetti M., Liao F., Ridoulfi L., Fahmi H. and Hoekstra A. Y. (2018). National water, food, and trade modeling framework: the case of Egypt. *Science of the Total Environment*, **639**, 485–496.

Allan J. A. (1993). Fortunately, there are substitutes for water otherwise our hydro-political futures would be impossible. Proceedings of the Conference on Priorities for Water Resources Allocation and Management, Southampton, July 1992, Overseas Development Administration, London, UK. pp. 13–26.

Allan J. A. (1997). "Virtual water": A long term solution for water short Middle Eastern economies. Presented at the 1997 British Association Festival of Science, Water and Development Session, University of Leeds, London.

Angelis E., Metulini R., Bove V. and Riccaboni M. (2017). Virtual water trade and bilateral conflicts. *Advances in Water Resources*, **110**, 549–561.

ArabiYazdi A., Alizade A. and Mohamadian F. (2009), Investigation of water ecological footprints in Iranian agriculture, *Water and Soil Journal*, **23**(4), 15–11.

Bae J. and Dall'erba S. (2018). Crop production, export of virtual water and water-saving strategies in arizona. *Ecological Economics*, **146**, 148–156.

Bagheri A. (2005). Introduction to the principles of hydrology in the integrated management of water resources. Water Rights Education Workshop, University of Tarbiat Modares.

Baghestani A. A. and Mehrabi H. (1386). The concept of virtual water and its application in determining the pattern of agricultural trade in Iran. National Irrigation Seminar and Reduction of Evaporation, Kerman, *No 9.*

Banooei A., Pashazanoos P. and Bahrami J. (2013). Political analysis importance role in measuring the importance of Iranian economy sectors, *Journal of Commerce*, **67**, 81–100.

Bekchanov M., Sood A., Pinto A. and Jeuland M. (2017). Systematic review of water economy modeling applications. *Water Resources and Planning Management*, **143**(8).

Brindha K. (2017). International virtual water flows from agricultural and livestock products of India. *Journal of Cleaner Production*, **161**, 922–930.

Brown A. and Matlock M. D. (2011) A review of water scarcity indices and methodologies. The Sustainability Consortium, University of Arkansas.

Chapagain A. K. and Hoekstra A. Y. (2003a). Virtual water flows between nations in relation to trade in livestock and livestock products. *Value of Water Research Report Series No. 13*. UNESCO-IHE Institute for Water Education, Delft, The Netherlands.

Chapagain A. K. and Hoekstra A. Y. (2003b). The water needed to have the Dutch drink coffee. *Value of Water Research Report Series No. 14*. UNESCO-IHE Institute for Water Education, Delft, The Netherlands.

Chen W., Wu S., Lei Y. and Li S. (2017). Virtual water export and import in China's foreign trade: A quantification using input–output tables of China from 2000 to 2012. *Resources, Conservation and Recycling*, **132**, 278–290.

Chouchane H., Krol M. S. and Hoekstra A. Y. (2017). Virtual water trade patterns in relation to environmental and socioeconomic factors: A case study for Tunisia. *Journal of Environmental Management*, **613**, 287–297.

Dietzenbacher E. and Velázquez E. (2007). Analysing andalusian virtual water trade in an input–output framework. *Regional studies*, **41**(2), 185–196.

Ehsani M., Khaledi H and Barghi Y. (2008). *Introduction to Virtual Water. National Drainage Irrigation Committee*, No 1, Tehran, Iran.

Food and Agriculture Organization (FAO) (2013). Technical conversion factors for agricultural commodities. FAO, Rome, Italy, Vol. **916**.

Fu Y. C., Zhao J., Wang C., Peng W., Wang Q. and Zhang C. (2018). The virtual water flow of crops between intraregional and interregional in mainland China. *Agricultural Water Management*, **208**, 204–213.

Han M. Y., Chen G. Q. and Li Y. L. (2018). Global water transfers embodied in international trade: tracking imbalanced and inefficient flows. *Journal of Cleaner Production,* **184**, 50–64.

Han M. Y, Chen G. Q., Mustafa M. T., Hayat T., Ling Shao J. S. and Xia X. H. (2015). Embodied water for urban economy: A three-scale input–output analysis for Beijing *Ecological Modelling*, *Ecomod*-7551 (No. 7).

Hoekstra A. Y. (2003). Virtual water trade: A quantification of virtual water flows between nations in relation to international crop trade. Proceedings of the International Expert Meeting on Virtual Water Trade 12, Delft, pp. 25–47.

Hoekstra A. Y. and Hung P. Q. (2003). Virtual water trade: a quantification of virtual water flows between nations in relation to international crop trade. Proceedings of the International Expert Meeting on Virtual Water Trade 12, Delft. pp. 25–47.

Jiang Y., Cai W., Du P., Pan W. and Wang C. (2015). Virtual water in interprovincial trade with implications for China's water policy. *Journal of Cleaner Production*, **87**, 655–665.

Katyainia S. and Baruaa A. (2016). Assessment of inter-state virtual water flows embedded in agriculture to mitigate water scarcity in India. *Water Resources Research*, **53**(8), 7382–7400.

Lamastra L., Miglietta P. P., Toma P., Leo F. D. and Massari S. (2017). Virtual water trade of agri-food products: evidence from Italian–Chinese relations. *Science of the Total Environment*, **599**, 474–482.

Liao X., Zhao X., Hall J. W. and Guan D. (2018). Categorising virtual water transfers through China's electric power sector. *Applied Energy*, **226**, 252–260. https://doi.org/10.1016/j.apenergy.2018.05.132.

Liu Y., Yang D. and Lü G. (2017). Water scarcity and generalized water resources management in urumqi: contributions from noncompetitive water input–output model. *Water Resources Planning and Management*, **8**(143).

Marston L. and Konar M. (2016). Drought impacts to water footprints and virtual water transfers of the central valley of California. *Water Resources Research*, **53**(7), 5756–5773.

Miglietta P. P., Morrone D. and Lamastra L. (2018). Water footprint and economic water productivity of Italian wines with appellation of origin: managing sustainability through an integrated approach. *Science of the Total Environment*, **633**, 1280–1286.

Montazer A. (2009). Developing a virtual water assessment model for irrigation networks using the analytical dynamic hierarchy process. *Water and Soil Agriculture Sciences and Industries*, **23**(4)

Qian Y., Dong H., Geng Y., Zhong S., Tian X., Yu Y., Chen Y. and Moss D. A. (2018). Water footprint characteristic of less developed water-rich regions: case of yunnan, China. *Water Research*, **141**, 208–216.

Qu S., Liang S., Konar M., Zhu Z., Chiu A. S. F., Jia X. and Xu M. (2017). Virtual water scarcity risk to the global trade system. *Environment Science and Technology*, **226**, 252–260.

Ramirez J. and Rogers P. (2004), Virtual water flows & trade liberalization. *Water Science & Technology*, **49**(7), 25–32.

Renault D. (2003). Value of virtual water in food: Principles and virtues. Proceedings of the Expert meeting on Virtual Water, Delft, December 2002, Food and Agriculture Organization of the United Nations, Delft, the Netherlands.

Ridoutt B. G., Hadjikakou M., Nolan M. and Bryan B. A. (2018). From water use to water scarcity footprinting in environmentally extended input–output analysis. *Environmental Science and Technology*, **52**(12), 6761–6770.

Sartori M., Schiavo S., Fracasso A. and Riccaboni M. (2017). Modeling the future evolution of the virtual water trade network: A combination of network and gravity models. *Advances in Water Resources*, **110**, 538–548.

Singh N. K., Gupta R and Gupta B. M. (2016). "A scientometric assessment of global publication output on rfid with reference to India during 2006–15." *DESIDOC Journal of Library & Information Technology*, **36**(4).

Sun K. S., Wu P. T., Wang Y. B. and Zhao X. N. (2012). The virtual water content of major grain crops and virtual water flows between regions in China. *Journal of the Science of Food and Agriculture*, **93**(6),1427–1437.

Sun S., Wang Y., Engel B. A. and Wu P. (2016). Effects of virtual water flow on regional water resources stress: A case study of grain in China. *Science of the Total Environment*, **550**, 871–879.

Tamea S., Liao F. and Ridolfi L. (2016). Global effects of local food-production crises: a virtual water perspective. *Scientific Reports*, **6**, 18803

Tan F. and Bi J. (2018). An inquiry into water transfer network of the Yangtze river economic belt in China. *Journal of Cleaner Production*, **176**, 288–297.

Thompson A. and Jeffords C. (2016). Virtual water and an EKC for water pollution. *Water Resources Management*, **31**(3), 1061–1066.

Tuninetti M., Tamea S., Liao F. and Ridoulfi L. (2016). To trade or not to trade: link prediction in the virtual water network. *Advances in Water Resources*, **110**, 528–537.

Wahba S. M., Scott K. and Steinberger J. K. (2018). Analyzing Egypt's water footprint based on trade balance and expenditure inequality. *Journal of Cleaner Production*, **198**, 1526–1535.

Wang Z., Zhang L., Zhang Q., Wei Y. M., Wang J. W., Ding X. and Mi Z. (2019). Optimization of virtual water flow via grain trade within China. *Ecological Indicators*, **97**, 25–34.

Welch A. and Cai H. (2011). Enter the dragon: the internationalization of China's higher education system. In: China's higher education reform and internationalization, J. Ryan, (ed.), Routledge, London, pp. 9–33.

Wichelns D. (2010). Virtual water: A helpful perspective, but not a sufficient policy criterion. *Water Resources Management*, **24**(10), 2203–2219.

World Meteorological Organization (1992). International Conference on Water and the Environment: Development Issues for the 21st Century, World Meteorological Organization, January 26–31, World Meteorological Organization, Dublin, Ireland.

Zehnder A. J. B., Yang H. and Schertenleib R. (2002). Water issues: the need for action at different levels. *Aquatic Sciences*, **65**(1), 1–20.

Zhang Y., Zhang J., Wang C., Cao J., Liu Z. and Wang L. (2016). China and trans-pacific partnership agreement countries: estimation of the virtual water trade of agricultural products. *Journal of Cleaner Production*, **140**, 1493–1503.

Zhao J., Fu Q. and Yu B. (2011). Virtual water measure and analyses of structure change for regional agricultural products in Sanjiang plain. International Conference on Electrical and Control Engineering, Yichang, China, pp. 3757–3760.

Zhi Y., Hamilton P. B., Wang X., Zhang Z. and Liang L. (2018). Game theory analysis of the virtual water strategy. *Water Resources Management*, **32**(14), 4747–4761.

Zimmer D. and Renault D. (2003). Virtual water in food production and global trade: review of methodological issues and preliminary results. In: 'virtual, water trade: proceedings of the international expert meeting on virtual water trade', *Value of Water Research Report Series*, **12**(1), 1–19.

Zygmunt J. (2007). Hidden waters. A waterwise briefing. Waterwise, London, UK.

Chapter 9

Water footprint

Mohammad Delpasand[1], Omid Bozorg-Haddad[2] and Erfan Goharian[3]*

[1]M.Sc. Student, Department of Irrigation & Reclamation Engineering, Faculty of Agricultural Engineering & Technology, College of Agriculture & Natural Resources, University of Tehran, 3158777871 Karaj, Iran. E-mail: Delpasand.m@ut.ac.ir.
[2]Distinguished Professor, Department of Irrigation & Reclamation Engineering, Faculty of Agricultural Engineering & Technology, College of Agriculture & Natural Resources, University of Tehran, Karaj, Tehran, 31587-77871, Iran. E-mail: OBHaddad@ut.ac.ir
[3]Assistant Professor, Civil and Environmental Engineering, College of Engineering and Computing, University of South Carolina, 300 Main St. Room C206 Columbia, SC 29208, USA. E-mail: Goharian@cec.sc.edu
**Corresponding author.*

ABSTRACT

Uneven water distribution in the world is the main reason today that some countries face problems due to water scarcity. Human activities consume and pollute large amounts of water. Globally, agriculture is the largest water user by volume. However, the water used by industrial and household sectors is still significant. Water consumption and pollution are caused by specific activities such as irrigation, bathing, washing, cleaning, cooling and by various other processes. Little attention has been paid to how much water use and pollution ultimately result from such activities, and how much water is consumed by communities, compared to the attention paid to the structure of the public economy that supplies consumer goods and services. Overall, to mitigate water scarcity problems, there are several approaches that can be made, such as inter-basin water transfer, increasing efficiency of water consumption and also using new concepts such as virtual water and the

Editor: Omid Bozorg-Haddad
doi: 10.2166/9781789062144_0151

water footprint. The footprint of a product is the amount of fresh water used to produce it, measured across the complete supply chain. Water footprint is a multidimensional indicator that shows the amount of water consumed by the source as well as the amount and types of contamination.

Keywords: Water footprint, blue water footprint, green water footprint, gray water footprint, white water footprint, water footprint trade.

9.1 INTRODUCTION

Human activities consume and pollute large amounts of water. On a global scale, the largest volume of water is used in agricultural production, but a significant amount is consumed and contaminated in the industrial and household sectors. Water consumption and pollution are caused by specific activities such as irrigation, bathing, washing, cleaning, cooling and various processes. Less attention has been paid to how much total water consumption and pollution is ultimately caused, and how much water communities consume compared to the attention paid to the structure of the general economy that supplies consumer goods and services.

Although there is little known about how continuous water production and consumption cycles contribute to increasing pollution, in recent decades several scientific and practical ideas have been put forward in the field of water management concerning water consumption and pollution, such as modification of consumption patterns, and development of the concept of water footprints, etc. (Delpasand *et al.*, 2020; Hoekstra & Chapagain 2007*a*, *b*).

Hoekstra and Chapagain (2008) showed that visualizing the use of hidden water in production could help to better understand the general index of freshwater use, quantify the effects of consumption and trade on water use, which could form the basis for better management of freshwater resources on Earth. In addition to regional markets, with the growth of international trade in water supplies, fresh water is increasingly becoming a global resource. With global markets for the production of water products such as crops and animals, natural fibers, bio energies and other products, consumption of this important resource has increased in various places in the world. In other words, today, all production stages do not happen in one place and in many instances there can be long distances between them and various stages of production. An example of this is flax which, as it moves from the farm to final production line, goes through many different stages with different impacts on water resources, and these stages of production often occur in different locations, and end-use can be located elsewhere. Flax is not grown in Malaysia, for example, but Malaysia imports raw flax from China, India and Pakistan for use in the textile industry, and exports linen to Europe. As a result, the effects of end-use consumption of flax on the world's water resources can only be observed by looking at the supply and traceability cycles (sources of production). The discovery of the hidden relationship between water use and consumption can form the basis for formulating water governance because new drivers can be identified for change. Since end-users, retailers, the food industry, and merchants have traditionally had no place in product development as a researcher or manager of water, these new entrants will provide new insights as potential change agents. Given their impact, not only can they be direct users, but also indirect users as well.

The water footprint concept was introduced by Hoekstra (2003) but other researchers had also studied the ecological footprint of water. Bicknell (1998) measured the ecological footprint of water before introducing the concept of a water footprint in the context of the input–output pattern. In this study, the ecological footprint of New Zealand was calculated using land use information and the combination of energy efficiency coefficients used in the framework of an 80-part input–output table. Ferng (2001) cited some of the shortcomings of the method proposed by Bicknell (1998) and then used a composite incremental land coefficient to estimate the ecological footprint associated with productive activities, rather than a cumulative incremental coefficient of land ecological footprint. Lenzen (2001) modified Bicknell's method and proposed a new method for calculating the Australian ecological footprint based on actual land use in an input–output framework. Two years later, Lenzen and Murray (2003) studied how to analyze input data to calculate national and regional ecological footprint to analyze footprint calculations in production layers and quantify the relationship between socioeconomic factors (such as household spending) and ecological footprint measurement expansion.

After the introduction of the water footprint concept in 2003, many studies have been made, each one studying various dimensions. Below, several of these recent studies are described in two categories: (1) water footprint; and (2) a combination of water footprint and virtual water trade.

9.1.1 Water footprint (WF)

Feng *et al.* (2015) examined the water footprint in Zhangye, China, by combining two methods: input–output analysis and structural analysis. In Zhangye in 2001–2011, unlike in many other cities that were facing drought due to the rapid urbanization process and ever increasing water footprints, there was a slight decrease in the water footprint because of technological issues. Nevertheless, due to the large export volume, the city had the potential to undergo water crisis, so the authors suggested that: (1) minimum water for irrigation be used; and (2) plants with high economic value be cultivated.

Chini *et al.* (2016) investigated the use of direct and indirect urban water in the United States, which introduced indirect water consumed indirectly through energy and food, and calculated this type of water in terms of the energy–water–food nexus. They calculated the average daily water footprint in the United States per person to be 1700 liters per day and eventually came to three conclusions:

(1) in the urban environment, indirect water flow plays a key role;
(2) reducing both direct and indirect energy consumption should be considered; and
(3) assessing the total water footprint can help formulate urban water cycle policy and management.

For the first time in Iran, using the Agwat program, Ababai and Ramezani-Etedali (2016) examined the water footprint of wheat, maize and barley plants in 2006–2012. In this regard, they categorized water footprint into blue, green, gray and white water. They introduced the concept of white water, as a water which penetrates more than 50 meters. As a result, they suggested that water resources management and environmental protection be emphasized in Iran to reduce water footprint in the four aforementioned categories.

Hoekstra (2017) wrote a review article on the evolution of water footprint and water resource management, and the reasons why using this concept is necessary today. As an example, he pointed out that because of the deterioration of renewable water resources, instead of solving water problems at a local scale, they should be solved at an international scale.

Wang *et al.* (2017*a*, *b*) aimed at optimizing the price of water and redistributing stored water in the agricultural sector between subsectors. They introduced a system called agricultural water rights trade to help regional irrigation administrations and high consumers of water to maximize their benefits in a saving water game. The authors of this paper used a combination of the above system with a system of compensation for water exports and introduced a proposed system as an effective solution for increasing water allocation efficiency, which in agriculture leads to increased water resource efficiency and consequently increased agricultural production.

Ti Ngo *et al.* (2017) studied the amount of water used in the agricultural sector in Vietnam from a virtual water perspective and tracked the water footprints of four main plants cultivated there (rice, coffee, sugar cane and corn) for domestic consumption and exports. They conducted this study on blue, green, and gray water, using the Cropwat program for rice and the evapotranspiration formulas for the rest of the products. To calculate the amount of gray water, they also used the concept of leaching.

Luo *et al.* (2017) considered the water footprint in Cheng Ming in China in the years 2011–2014 and considered four scenarios for predicting water resources in 2040, eventually realizing that production of products with less water consumption will save 11.5% of current water use.

Sun *et al.* (2017) studied 30 provinces in China to investigate water footprint inequality and discussed the role of virtual water transfer in determining water footprint from an inequality perspective. They concluded that the greatest inequalities are in agricultural, industrial and domestic use, respectively. Also, the water footprint in different provinces is independent of physical available water.

Madugundu *et al.* (2018) examined the water footprints of carrot and wheat crops in 2015–2016. They used two ways to capture data, one of which was the use of weather station information, and the other to use Landsat-8 satellite imagery. The Landsat Satellite encountered shooting problems on rainy days, but is generally a reliable source for calculating water footprints.

Fan *et al.* (2019) investigated the water footprint in domestic use using the input–output method in China. This study was conducted by comparing urban and rural water consumption. They concluded that the rapid growth of urbanization has increased the amount of virtual water consumption in cities.

Cui *et al.* (2018) studied the catchment area of the Heihe River and provided an assessment system for a virtual water strategy from the perspective of water, environment and economics. By using entropy weights and fuzzy models, regions were classified into five groups of suitability for virtual water strategy: (1) excellent; (2) good; (3) moderate; (4) fairly good; (5) weak. These groups were categorized as follows: (1) economic investment; (2) technology; (3) economic development; and, finally (4) social acceptance. In the study area, more than half of the regions were found in the excellent group and the rest, with the exception of one region that was ranked weak, were scored good and moderate.

Megdal (2018), in a review article, examined two independent efforts in the United States and the rest of the world to further pay attention to the importance of rational governance and the management of groundwater resources. During these two efforts, some of the principles were as follows: (1) understanding of groundwater and the importance of these waters as critical, limited, valuable and vulnerable resources; (2) drainage of groundwater in aquifers; (3) the uniqueness of irrigation systems and the need for them to be studied; (4) the need for integrated groundwater resources management; (5) artificial feeding of aquifers should significantly increase globally; (6) effective groundwater management requires collaboration, continuous stakeholder participation, and community participation.

Salmoral and Yan (2018), using the Life Cycle Inventory (LCI) method, investigated the potential of understanding the complex and hidden connection of life cycle to the virtual water cycle. They investigated virtual water upstream and energy of food consumption in the Tamar basin and, by making a distinction between domestic consumption and imports, identified important imports and identified food, vegetables and animal feeds among them. Finally, the share of households, agriculture, and other sectors for blue and green water, and the energy consumption in the basin, were all calculated.

Ye *et al.* (2018) presented a multi-objective optimization model to optimize the physical and virtual allocation of water to different consumers in Beijing, China. In the first stage, the simulation of water allocation was traditionally composed of groundwater, surface water, etc., and the percentage of agricultural and industrial consumption was calculated. By adding virtual water to the observation model by importing five selected agricultural products, water consumption dropped from 54% to 45%, and industry saw a 4% drop in water consumption as imports increased.

By investigating the water use of Spain in 2014–2015, Blas *et al.* (2018) assessed the water footprint in food and its wastes in domestic use in three water footprint categories (blue, green and gray). The focus of this paper was on the implications of water and food consumption in Spain in order to identify possible policy recommendations and, as a result, they found that in the event of waste disposal, only a 4% saving in water supplies would be observed.

9.1.2 Combination of water footprint and virtual water trade

Cazcarro *et al.* (2016) studied the gray water footprint of agricultural products at regional and international scales in 17 regions of Spain. In this study, they investigated point source and non-point source pollution by a multi-regional input–output method. They also determined vulnerable areas. Finally, they determined what types of pollutions entered which regions. They linked the input–output method and GIS for the first time.

Wang *et al.* (2017*a*, *b*) monitored China to evaluate and find a pattern for Chemical Oxygen Demand (COD) footprint. They discovered, agriculture, paper-producing and food-processing, were the key sectors for discharge reduction from the supply side perspective.

Shtull-Trauring and Bernstein (2017) assessed the water footprint of plants and the export and import of virtual water in Israel. They assessed the total amount of water used for agriculture as well as the amount of water displaced by exports and imports.

Cai *et al.* (2017), considered China as a case study and examined the quantity and quality of virtual water. They compared the quantity and quantity of blue water with that of gray water footprint. The results showed that the gray water transferred through virtual water was eight and a half times the blue water content of the virtual water.

Li *et al.* (2017) examined the export of the gray water footprint using the input–output method in Beijing, China, and found that the export of gray water in that region was higher than that of non-polluting water exports. Because of its industrial nature, the region produces a lot of gray water that needs to be addressed in the direction of sustainable development.

Tian *et al.* (2018) examined virtual water and water footprint in China between 1995 and 2009 using a multi-regional input–output method with respect to blue, green and gray waters, and concluded that green water was the largest water exported and imported in China; also, China was a virtual exporter of water during these years. Further, they suggested that, in order to achieve sustainable development, technology development and better political decision-making should be highly considered.

Bazrafshan *et al.* (2019) examined the production of saffron in Iran from the perspective of virtual water and water tracks in 2008–2014. Iran ranks first in the world in production and export of saffron. The average amount of virtual water of saffron is 4,659 m^3/kg, the total amount of which will be equal to 1,541 million $m^3/year$, of which 1,354.16 million m^3 will be exported. The amount of gray and white water used in the production of saffron is about 46%, which indicates the high use of fertilizer and low irrigation efficiency.

9.2 DEFINITIONS AND TERMS

9.2.1 Water footprint

The term 'water footprint' (WF) is actually inspired by corresponding and similar concepts in the environmental, industrial and other related sciences. In general, WF is used as a quantitative measure to indicate the amount of human consumption of natural resources, and is in fact a freshwater use index, in which not only direct consumption of water by a consumer or producer is conceptualized, but indirect consumption is also considered. Water footprint can be an indicator of freshwater resource allocation. Although the standard per capita water consumption in different countries around the world varies from 90 to 200 liters per day, the reality is that water consumption exceeds the standards set by each country. Each person consumes water directly and indirectly throughout the day. In addition to direct consumption of water, which includes drinking, bathing, cooking, washing, sanitation, etc., they also indirectly consume large quantities of water in food and services, and also in their production process chains. In fact, in an indirect way, the water which is consumed in the production processes of goods is transmitted to humans. The volume of water that humans receive virtually is far greater than their direct consumption of water, and varies according to their cultural habits, diet, and income.

According to the above description, the water footprint can be defined as the sum of the total water consumed directly and indirectly by an individual, company, village, city or country, at a daily, weekly, monthly or yearly interval. The footprint of a product is the volume of fresh water used to produce a product that is measured across the complete supply chain. Water footprint is also a multidimensional indicator that shows the amount of water consumed by a source, and the amount of contamination as well as type.

9.2.2 Blue water

The water that is found in rivers, behind dams, and in lakes, surface water and groundwater is called blue water. Virtual blue water, therefore, is the volume of surface water or groundwater consumed directly or indirectly during the process of producing a product or service.

9.2.2.1 Blue water footprint

This concept refers to the consumption of surface and groundwater resources during the supply cycle of a product. The term consumption refers to the use of water from a surface or groundwater body at a watershed. On the other hand, losses occur when evaporated water returns to another watershed or sea level or is used in the process of production (Hoekstra, 2008*a*, *b*).

9.2.3 Green water

Water stored as moisture in the soil is called green water. Although green water is one of the main sources of water supply for plants, especially in rainfed lands, in recent years, less attention has been paid by experts and planners to this source and as such, it has remained neglected.

9.2.3.1 Green water footprint

This concept refers to the direct and indirect consumption of green water resources. Also, it should be noted that green water refers to rainwater until it is runoff water.

9.2.4 Gray water

Water that is contaminated in the production cycle of a product is called gray water. Some industrial products in the production cycle require a certain volume of water for dilution of chemicals, washing or other purposes. Such waters, which become chemical and industrial effluents after consumption, and are also commonly considered to be environmental hazards, are called gray water. In other words, wastewater can be classified as gray water (Li *et al.*, 2016).

9.2.4.1 Gray water footprint

The idea of a gray water footprint has been put forward to express the contamination of water with a so-called contaminated volume, which is measured in terms of gray water consumption expressed in terms of volume. The amount of gray water consumed in the agricultural sector or in the industrial sector as effluent is called virtual gray water. Gray water footprint refers to contamination and is considered to be the volume of fresh water needed to match the load of the pollutants created, according to natural background concentrations and in accordance with environmental water quality standards.

9.2.5 White water

The concept of white water was introduced by Ababai and Ramezani-Etedali (2016). According to their study, white water is considered to be water that penetrates more than 50 m in the ground.

9.2.6 Direct and indirect water footprint

The general recommendation is that both direct and indirect water footprints should be taken into account. Although there has been a traditional focus on direct water footprints for consumers and companies, indirect water footprints are usually higher. Thus, by only considering direct water footprints, consumers neglect the fact that most of their water footprint is related to the products they use and not the water they consume at home. For most traders, the supply chain footprint is much larger than the water footprint of their operations. Ignoring the supply chain component may lead to investments in upgrading operational water, while investments in supply chain development could be more cost-effective. But ultimately, for the purposes of a study, it should be decided whether to consider direct, indirect or both water footprints in any analysis.

9.2.7 Trade water footprint

Trade water footprint is the sum of fresh water consumed directly and indirectly to produce and enhance a product, in which two main components are involved (Hoekstra, 2008c):

(1) the operational (direct) water footprint of a business, which is the volume of fresh water consumed or contaminated as a result of business operations; and
(2) the water supply chain (indirect) footprint of a business, which is the volume of fresh water consumed or contaminated for all the goods and services.

It is also worth noting that terms such as company water footprint or organization water footprint can also be used interchangeably with trade water footprint.

9.3 IMPORTANCE AND NECESSITY

Rapid population growth, socioeconomic development of communities, excessive extraction of fresh water and increasing levels of pollution are some of the factors that make some countries more vulnerable to water shortages and crises. Agriculture is a major consumer of water and accounts for about 80% of the consumption of surface and groundwater resources, and reducing water consumption is one of the most important economic strategies to reduce water scarcity. In fact, water harvesting and consumption by communities is not just about household consumption, but most of the water consumed by human beings is the water consumed through food that is passed through several chains for its production and then distribution. For instance, 140 liters of water is consumed to produce one cup of coffee, and 2600 liters of water is used to make a hamburger.

As discussed above, the water we consume is definitely more than the amount directly visible to us. This water embedded in commodities is called virtual water. Allan (1993) first introduced the concept of virtual water to determine how much water is consumed to produce a product. The amount of virtual water in food production is significant, and knowing the amount of virtual water in a person's diet, or used by wider society, can help manage water consumption. In order to identify the water consumed in the production cycles of different resources, it was necessary to come up with a more complete concept of virtual water, and so the concept of water footprint was developed, which is a more comprehensive concept, yet closely related to virtual water.

A water footprint is a multidimensional indicator for the consumption of fresh water that is not limited to direct consumption of water by the consumer or producer, and also includes indirect consumption of water. But, in addition to traditional water consumption measurements, water footprints can also be considered as a comprehensive indicator to identify available fresh water resources, water supply sources and water use management strategies, since it shows the volume of water consumed and the type of water used. The type of water used, in this case, shows how the water was supplied (whether surface water, groundwater or residual moisture from the soil) as well as how much of the initial and desired water quality was lost in exchange for the intended crop. It should be noted that water footprint indicates the amount of water resources allocated to individuals or specific activities in different areas. In other words, this concept can be used to formulate strategies to allocate water resources to a region or basin.

Since the introduction of the water footprint concept, many studies have been conducted to evaluate this indicator in different areas. At first, most studies were conducted to estimate all three water footprint components (blue, green and gray water footprints) on a global scale, but given the increasing importance of smaller scale water resources management, evaluating this concept at local scales has become particularly important.

9.4 BASICS AND LOGIC

In recent decades, several scientific and practical ideas have been put forward in the field of water management about water consumption and pollution during production and supply lifecycles. One of these ideas was the development of the concept of water footprints.

Water footprint evaluation refers to a wide range of activities such as:

(1) finding the quantity and location of water in a process or production;
(2) assessing the environmental, social and economic sustainability.

In short, the purpose of evaluating water footprints is to analyze how human activities relate to specific products and the related water scarcity and pollution issues, and how activities and products can be sustainable from the perspective of water consumption. It is proper to note that how the water footprint is evaluated depends on the researchers' interests and is essential for solving a problem. One researcher may be interested in the water footprint of a given process in a complete production chain or the water footprint of a final product, while another may be interested in the water footprint of a consumer or a group of consumers, or the water footprint of a producer or a comprehensive economic sector. Also, another researcher can consider a geographical area and examine the total water footprint for a specific area such as city, province, country, watershed or river basin. Accordingly, the total water footprint is the sum of the water footprints of a number of separate steps (processes) in one area. Water footprint evaluation is an analysis tool that can be useful in understanding water activities and products from the water scarcity perspective as well as for pollution and its associated effects. Therefore, water footprint evaluation is a tool that can enhance research insights. It does not force researchers to do research in specific fields but rather helps them understand what they can research.

9.5 MATERIALS AND METHODS

9.5.1 Blue water footprint

As mentioned in Section 9.2, the blue water footprint is an indicator of the consumption of blue water (surface or groundwater freshwater). It depends on one of the following four components:

(1) water evaporation;
(2) water (virtual water) used in production;

(3) water that does not return to the same watershed – for example, water that returns to another basin or the sea;
(4) water that does not return during the same period – for example, that is used during a drought period and returned during a wet season.

The first component, evaporation, is one of the most important components. Total production related evaporation, including water coming from a reservoir (e.g., from artificial water tanks), transfer (e.g., from open channels), processes (e.g., evaporation of warm water that is not collected again) and collection and disposal (e.g., from drainage channels and from sewage treatment plants) should be calculated. Usually, the water footprint of a process is expressed in units of volume of water per unit of time, and when divided by the amount of output produced by the process (units of production per unit time), it can be expressed as the volume of water per unit of production.

Water consumption does not mean the disappearance of water resources, because the water remains in the hydrologic cycle balance and goes to another point of the cycle. Water is a renewable resource, but that does not mean it is infinite. In a given period, the amount of water that feeds underground water reservoirs and the flow of a river are often limited. Underground rivers and aquifers are usually used for irrigation, industrial, or household use, but over a given period, most of the available water cannot be consumed. Blue water footprint measures the amount of water consumed over a given period. In this method, the amount of blue water consumed by humans is determined. The rest of the groundwater and surface water, which is not consumed by humans, is released into the ecosystems that depend on groundwater and surface water to survive.

The blue water footprint is calculated as follows in one stage of the process (Hoekstra *et al.*, 2011):

$$WF_{proc,blue} = NRW + BW + EWV \tag{9.1}$$

where $WF_{proc,blue}$ is the blue water footprint in a process stage in terms of volume of water per unit of time, NRW is the water flow that is not returned in terms of volume of water per unit of time, BW is the water in the product in terms of volume of water per unit of time, and EWV is evaporation of blue water in terms of volume of water per unit of time.

The not returned water flow component (NRW) refers to a portion of the return flow that is not available for reuse within the same catchment area during the same harvest period because it returns to another catchment, is discharged into the sea or returns at a later time. It is possible to distinguish between different types of water resources in assessing water footprint of a process. The main divisions are in the form of surface water, current (renewable) groundwater, and fossil groundwater. Now, after getting acquainted with this division, it is necessary to state the order of these divisions.

The divisions are arranged as follows: (1) blue water footprint of surface water resources; (2) blue water footprints of renewable groundwater; (3) the blue water footprint of fossil groundwater. In practice, it is very difficult to distinguish between these because the data is often insufficient, so no distinction can be made. If available data is sufficient, a distinction can be made based on type of water resource.

When determining source-based blue water footprints, a researcher may be interested in clearly distinguishing the intended consumption of precipitation water. Rainwater intake is a special case because whether the incoming rainwater is blue or green is disputed. Basically, receiving rainwater refers to collecting rain, otherwise it is included in the equations as runoff. As the consumption of rainfall water reduces runoff, it is recommended to count rainwater as blue water. There are different types of techniques to consume rainwater for drinking, for livestock or for irrigating crops and orchards. It should be noted that the consumption of locally collected runoff through the receiving of rainwater through rooftops or other hard surfaces, or directing water to small pools, can be classified as blue water footprint as well. Conversely, when a researcher talks about measuring the increase in soil water storage capacity or about green roofs for rainwater storage, the consumption of this water for crop production is considered as green water footprint.

9.5.1.1 Water recycling and reuse

Often, the terms 'recycling' and 'reusing' of water can be used interchangeably. Here, the term water recycling is particularly used for reusing water within a basin for the same purpose, whilst the term water reuse is used for reusing water elsewhere, and possibly for another purpose. In the recirculation mode, a sub-distinction can be made between recirculation of wastewater (treatment for reuse) and recirculation of evaporated water (taking into account evaporation for reuse).

9.5.2 Green water footprint

Green water refers to rain that does not make it into runoff and does not feed groundwater, but is stored in soil or temporarily stays on soil surfaces or on vegetation. In the end, this part of precipitation evaporates or is used by plants. Green water can produce crop growth, but plants cannot absorb all green water because it usually evaporates from the soil and is not suitable for crop growth at all times of the year.

In fact, green water footprint is the volume of rainfall consumed in the production process, which is specifically related to agricultural (crop) and forestry (timber) production and here the sum of rainfall evapotranspiration (from fields and vegetation) plus associated water refers to the product or harvested wood. The green water footprint of a process step can be given as (Hoekstra *et al.*, 2011):

$$WF_{proc,green} = GW + GWE \tag{9.2}$$

where $WF_{proc,green}$ is the green water footprint in a volume of water per unit of time, GW is the green water consumption of a product by volume of water per unit of time, and GWE is evaporation of green water by volume of water per unit of time.

The distinction between green and blue water footprints is important because the hydrological, environmental, social, and economic opportunity costs of using surface and groundwater for production differ markedly from the effects and costs of rainwater. Green water consumption in agriculture can be measured and estimated using empirical formulas or a crop model based on estimation of evapotranspiration based on input data of climate, soil and crop characteristics.

9.5.3 Gray water footprint

The gray water footprint of a process stage is an indicator of the degree of freshwater contamination that can be associated with that process stage and which is defined as the amount of fresh water needed to match pollutant loads based on natural state concentrations and environmental water quality standards. The concept of gray water footprint can be defined based on the amount of water pollution, which is the volume of water needed to dilute harmful water pollutants:

$$WF_{proc,gray} = \frac{L}{c_{max} - c_{nat}} \tag{9.3}$$

where $WF_{proc,gray}$ is the gray water footprint in a volume of water per unit of time, L is the contamination load in terms of unit volume per time per mass, c_{max} is the maximum acceptable concentration of contamination in terms of unit volume per time per mass, and c_{nat} is the normal concentration of contamination in the body of water received in terms of unit volume per time per mass.

The natural concentration of contamination in the receiving body is the concentration that the body receives in the absence of human activity in the catchment area. For naturally occurring human materials, c_{nat} is equal to zero, or can be assumed to be zero when natural concentrations are accurately known. But this leads to underestimation of the gray water footprint, since in fact c_{nat} is not equal to zero. It may be asked why natural concentration is considered as a reference rather than the actual concentration that the body receives. The answer is that the gray water footprint is an indicator of the reception capacity assigned, and the reception capacity of a receiving water body depends on the difference between the maximum permissible and natural concentration of an element. If the maximum permissible concentration is compared with the actual concentration of an element, the residual acceptance capacity is observed which varies markedly with time. This is a function of the actual level of contamination at a given time. Gray water footprints are calculated using ambient water quality standards for the receiving freshwater body or, in other words, at the maximum standard permissible concentrations. On the other hand, environmental water quality standards are a special group of water quality standards. There are also other types of standards such as drinking water quality standards, irrigation quality standards and emission standards (for wastewater) that must be monitored. For a particular element, ambient water quality standards may vary from one water body to another. In addition, natural concentrations may also vary from place to place. As a result, certain pollution loads can lead to one gray water footprint in one location and another gray water footprint in another location. This is possible because the volume of water needed to match a given contamination time varies widely between the maximum permissible concentration and the natural concentration. Water quality standards are often contained in national or state laws or are regulated by national laws or local agreements for a watershed or body of water (similar to the European Water Framework Directive (European Union, 2000). But they do not exist for all elements and for all locations. First and foremost, of course, it is important to determine what standards of water quality and natural concentrations should be used in the preparation of water footprint calculations. It should be noted that both ambient water quality standards and natural water concentrations in surface and groundwater vary. Groundwater thresholds are determined based on drinking water needs. However, the maximum permissible concentrations in surface water are largely determined by environmental considerations. On this basis, gray water footprint calculations can be performed separately for surface and underground water resources systems. But the problem is that available groundwater volume is usually smaller than the available surface water, but the standard water quality difference and the natural concentration of materials for the most critical water body can be better used for groundwater pollution loads (either groundwater or surface water). When accurately identifying incoming loads to groundwater and surface water systems in the gray water footprint component, they are coded as follows: gray groundwater footprint and gray surface water footprint. Typically, gray water footprint greater than zero does not indicates a violation of water quality standards, but shows that only a part of the contamination capacity is already in use. As long as the calculated gray water footprint is smaller than the existing river water or groundwater flow, there is still enough water to dilute the pollutants to below the standard limit. When the calculated gray water flow is exactly equal to the ambient water flow, the obtained concentration is just below the standard. On the other hand, when wastewater is heavily loaded with chemicals, the gray water footprint calculated from the existing river or groundwater flow may increase, with contamination advancing to the receiving water body's receiving capacity. The fact that the gray water footprint can be larger than the available water flow indicates that the gray water footprint does not indicate the volume of contaminated water (because it cannot pollute more than the available water volume). Therefore, gray water footprint is an indicator of the severity of water pollution that is defined as the amount of fresh water needed to accept the load of existing pollution.

The method of calculating the gray water footprint is also called the critical load method. In both cases, the basic diagnosis is that the absorption capacity of the wastewater in a water body is limited by the difference between the maximum and natural concentrations. Critical load refers to the situation where space has been completely used to absorb wastewater. At critical times, dilution should be carried out to an extent such that chemical concentrations are below the permitted levels.

9.5.3.1 Point source pollutions

At point sources of water pollution, or when chemicals are released into wastewater or directly into a surface water body, the amount of pollution load can be estimated by measuring the volume of the wastewater and the concentration of chemicals in that wastewater.

$$\mathrm{WF}_{\mathrm{proc,grey}} = \frac{L}{c_{\max} - c_{\mathrm{nat}}} = \frac{E_{\mathrm{ffl}} \times C_{\mathrm{effl}} - \mathrm{Abstr} \times C_{\mathrm{act}}}{c_{\max} - c_{\mathrm{nat}}} \tag{9.4}$$

where E_{ffl} is the wastewater discharge in terms of volume of wastewater per unit of time, C_{effl} is pollutant concentration in wastewater in terms of mass per unit of volume, Abstr is the discharge of water harvested in volume per unit of time, and C_{act} is the actual concentration of water consumed by mass per unit of time.

9.5.4 Calculating the green, blue and gray water footprints of a crop or tree

Many products contain agricultural or forestry materials. Agricultural products are used for such things as foods, textiles, fuel, soap, cosmetics and the like. Forest products are mainly timber and shrubs for use in the paper and fuel industries. Since the agricultural and forestry sectors are major consumers of water, agriculture or forestry crops are often subject to significant water footprints. For all these crops, it is appropriate to take a special look at the water footprint of the product or tree growth process. This section discusses the details of water footprint evaluation of a product and of tree growth. This method can be used for both annual and perennial crops. In this section, the term 'product' is used in a wide sense, including for the growth of trees for wood. The total water footprint of a product or tree growth process (WF_{proc}) consists of a combination of blue, green, and gray components:

$$WF_{proc} = WF_{proc,green} + WF_{proc,blue} + WF_{proc,grey} \tag{9.5}$$

where WF_{proc} is the total water footprint of the process of growing crops or trees by volume of water per mass, $WF_{proc,green}$ is the green water footprint of a process by volume of water per mass, $WF_{proc,blue}$ is the blue water footprint of a process by volume of water per mass, and $WF_{proc,grey}$ is the gray water footprint of a process by volume of water per mass.

All process water footprints in this section are expressed in terms of a unit of production, i.e. volume of water per mass. Process water footprints in agriculture or forests are usually expressed in terms of m^3/ton which is equivalent to the unit of liter per kilograms (L/kg). The green component in the water footprint of the process of growing a crop or tree is calculated as follows:

$$WF_{proc,green} = \frac{CWU_{green}}{Y} \tag{9.6}$$

where CWU_{green} is the green component in water consumption of crops by volume of water per area, and Y is the crop yield in terms of water mass per area.

Likewise, the blue component in the water footprint of the process of growing a crop or tree is calculated as follows:

$$WF_{proc,blue} = \frac{CWU_{blue}}{Y} \tag{9.7}$$

where CWU_{blue} is the blue component in water consumption of crops by volume of water per area.

For perennial crops, the average annual efficiency over the lifetime of a crop should be considered. In this method, for the calculation of a perennial crop, the yield is low or zero in the first year of planting, reaches its maximum after a few years and often decreases at the end of the life cycle. Also, for water consumption of the crop, the average annual water use of the crop during its lifetime is required.

The gray component in the water footprint of the growth of a crop or tree is calculated as follows:

$$WF_{proc,grey} = \frac{(\alpha * AR)/(c_{max} - c_{nat})}{Y} \tag{9.8}$$

where α is the leaching–runoff coefficient, and AR is the application of chemicals in an area by mass per unit area.

The green and blue components in crop water consumption with total daily evapotranspiration during the growing period are fully calculated:

$$CWU_{green} = \sum\nolimits_{d=1}^{lgp} ET_{green} \tag{9.9}$$

$$CWU_{blue} = 10 \times \sum\nolimits_{d=1}^{lgp} ET_{Blue} \tag{9.10}$$

where ET_{green} is the daily evapotranspiration of green water during the growth period in units of length per unit of time, ET_{Blue} is the daily evapotranspiration of blue water during the growth period in units of length per unit of time, and lgp is the length of growing period in days.

Since the growth period of different varieties of crops is significant, this factor is very important in calculating crop water consumption. For permanent (perennial) crops and evapotranspiration production, the whole duration of the year must be considered. In addition, to calculate evapotranspiration discrepancies in a product or tree life, the annual average evapotranspiration of the product or tree should be taken into account. For example, suppose a permanent product has a 20-year life cycle, and produces a product from year six. In this case, the water consumption of the crop over the whole 20 years is divided into the total crop yield of 15 years. Green water consumption of the crop represents the sum of evaporated precipitation from the farm during the growing season. The product's water use represents the sum of water evaporated from irrigation. Evapotranspiration (ET) of a farm can be measured or estimated using a model based on empirical formulas. Measuring evapotranspiration is expensive and unusual. Usually, evapotranspiration is estimated indirectly using a model that uses weather data, soil properties, and crop characteristics as inputs. There are various ways to model ET and product growth. One of the most widely used models is the EPIC model (Williams, 1995). It is also available in network form (Liu *et al.*, 2007). Another method is the CROPWAT model developed by the United Nations Food and Agriculture Organization (FAO, 2010a), which is based on the method described by Allen *et al.* (1998). The AQUACROP model is used specifically to estimate crop growth and ET for water scarcity status (FAO, 2010b). An irrigation program (with the ability to determine irrigation reserve in time) is recommended to use the second method as far as possible, since it can be used for both optimal and non-optimal conditions and is more accurate (since the model is based on dynamic soil and water balance).

Conceptual guidance on the practical use of the CROPWAT program is available online (FAO, 2010a). Calculating the green, blue, and gray water footprints of a product's growth requires a great deal of data: product variables, crop maps, crop yields, soil maps, irrigation maps, application rates, fertilizers, pesticide rates, leaching-runoff section, ambient water quality standards, natural concentrations in the receiving water body, and actual concentrations of consumed water. Generally, it is preferable to extract local data on the crop's location. In some cases, it is very difficult to collect data for a specific location for the purpose of evaluation. If the evaluation can be made with an estimate of the data, it can be done with data from nearby or regional averages or from countries that may be easier to access.

9.5.5 Selection of business organizational boundaries

Trading in this context means a manufacturer of a product providing their services to other businesses or consumers; this manufacturer can be a joint stock company, or a public or non-governmental organization. The manufacturer can operate at multiple scales, for example it can operate as a special unit as a part of a company, as a complete company or a complete business segment. Also, the term 'business' can refer to a consortium or joint venture of companies or organizations aiming to provide a specific product or service. In fact, 'business' can refer to any project (such as building a piece of infrastructure) or activity (for example, organizing a major sporting event). In technical terms, a business unit is a unit of activity that converts a set of inputs into one or more outputs.

To be able to evaluate the water footprint of a business, the business must be clearly identified. It should be clear what trade boundaries should be considered. It should be possible to design intra-system trade that is clearly distinguished from its surroundings and where inputs and outputs are well known. Companies usually consist of several units. For example, a company may have different operating areas (such as factories) in different locations or different sections in one place. In terms of calculating water footprint, it is often useful to differentiate between various business units. for example, when a manufacturing company has different factories in multiple locations, it is likely that each plant operates in a different situation, and receive inputs from different locations. In such a case, it is useful to first calculate the water footprint for each business unit, then aggregate the business unit calculations to calculate the entire business.

9.5.6 Water footprint trade

This section first shows how to calculate the water footprint of a business unit, and then discusses how to calculate the water footprint of a business consisting of several business units.

The water footprint of a business unit is calculated by adding the business footprint to its supply chain water footprint:

$$WF_{bus} = WF_{bus,oper} + WF_{bus,sup} \tag{9.11}$$

where WF_{bus} is the water footprint of a business unit in volume per unit of time, $WF_{bus,oper}$ is the water footprint of business unit operations in volume per unit of time, and $WF_{bus,sup}$ is the supply chain water footprint in volume per unit of time.

These two components are the water footprint which can be directly associated with the product that is produced in the trade unit, and the water footprint of the trade unit's overheads. The term overhead is used to identify water consumption which is necessary for the continued functioning of the business; however, it does not directly relate to the production of one specific product:

$$WF_{bus,oper} = WF_{bus,oper,inputs} + WF_{bus,oper,overhead} \tag{9.12}$$

$$WF_{bus,sup} = WF_{bus,sup,inputs} + WF_{bus,sup,overhead} \tag{9.13}$$

where $WF_{bus,oper}$ is the water footprint of a business unit operation with a product produced in the business unit in terms of volume per unit of time, $WF_{bus,oper,inputs}$ is the overhead unit operations water footprint in terms of volume per unit of time, $WF_{bus,sup,inputs}$ is the chain water footprint associated directly with the product produced in the business unit in terms of volume per unit of time, and $WF_{bus,sup,overhead}$ is the overhead chain water footprint in terms of volume per unit of time.

The operational water footprint is equal to the amount of water consumed and water pollution caused by trade operations. By reviewing the guidelines provided, it is easy to observe evaporation flows from an operation, the volume of water associated with production, and the return flows to other basins compared to the water extracted. In addition, wastewater volumes and associated chemical concentrations can be considered as overhead operation water footprint. Consumption and pollution related to general water consumption activities in the business unit can be treated similarly to direct operation water footprint associated with the production process. Often, the overhead water footprint serves more than one business unit, for example the overheads for a factory with two production lines is distributed over more than two production lines. If a business is defined as referring to one of the production lines, it is necessary to calculate the share of the overhead water footprint for that production line, or multiple production lines. The water footprint of the supply chain of each business unit (volume per time) can be calculated by multiplying the different input product volumes (data available from the business itself) by the water footprints of their respective production (data obtained from each process). Suppose that different input products, originating from different sources x, are found in the supply chain footprint of a business unit as follows:

$$\mathrm{WF}_{\mathrm{bus,sup}} = \sum x \sum I\ (\mathrm{WF}_{\mathrm{prod}}[x, i] I[x, i]) \tag{9.14}$$

where $\mathrm{WF}_{\mathrm{prod}}[x, i]$ is the water footprint of production in terms of volume per unit of production of input i from source x, and $I[x, i]$ is the production volume of Input i from source x in terms of volume of production per unit of time.

The water footprint of production depends on the source of production. When production comes from the same business unit, the business accounting system knows the total water production footprint. When production originates from an out-of-business supplier, the amount of water footprint of the product is either obtained from the supplier or estimated from the supplier's specific production data.

The different production water footprints are made up of three colors (green, blue and gray) that need to be calculated separately. Therefore, the result of the water footprint of the supply chain of the business unit has three color components.

The water footprint of the product p from a business unit can be calculated as follows:

$$\mathrm{WF}_{\mathrm{prod}[p]} = \frac{E[p]}{\sum pE[p]} \times \frac{\mathrm{WF}_{\mathrm{bus}}}{P[p]} \tag{9.15}$$

where $\mathrm{WF}_{\mathrm{prod}[p]}$ is the production water footprint of input product p from a business unit in terms of volume per unit of time, $E[p]$ is the total economic value of output product p in terms of its currency per unit of time, $\sum pE[p]$ is the total economic value of all output products, $\mathrm{WF}_{\mathrm{bus}}$ is the water footprint of a business unit in volume per unit of time, and $P[p]$ is the output volume of product p of the business unit in terms of unit of production per unit of time.

It should be noted that if the trading unit delivers only one product, the equation can be shortened as follows:

$$\mathrm{WF}_{\mathrm{prod}}[p] = \frac{\mathrm{WF}_{\mathrm{bus}}}{p[p]} \tag{9.16}$$

All of the above equations are applied at the level of a business unit. Suppose a business unit consists of a number of sub-business units u, $\mathrm{WF}_{\mathrm{bus,tot}}$ is calculated by the sum of the business units' water footprints. To avoid duplicate computation, virtual water flows between different units within the business are deducted as follows:

$$\mathrm{WF}_{\mathrm{bus,\ tot}} = \sum u\ (\mathrm{WF}_{\mathrm{bus}}\ [u]) - \sum u \sum p\ (\mathrm{WF}_{\mathrm{prod}}\ [u, p] P^{*}\ [u, p]) \tag{9.17}$$

where $\mathrm{WF}_{\mathrm{bus,tot}}$ is the total trade water footprint in terms of volume per unit of time, $\mathrm{WF}_{\mathrm{bus}}\ [u]$ is the trade water footprint of unit u in volume per unit of time, $\mathrm{WF}_{\mathrm{prod}}$ [u, p] is the virtual water flows between different units of trade in terms of volume per unit of time, and P^* $[u, p]$ is the annual output volume of product p from business unit u to other business units within the same business.

9.6 CASE STUDIES

9.6.1 The water footprint of tourism in Spain

Cazcarro *et al.* (2014) evaluated virtual water trade (both blue and green components) of agricultural, industrial and tourism services in Spain. In Spain, more than 10% of Gross Domestic Product (GDP) originates from tourism services. The method used in this study was a combination of input–output analysis and process analysis. In order to evaluate agricultural products' virtual water, the authors used both water content in products and also virtual water content in primary and transformed products.

Out of the 6.9 km^3/year of Spain's water resource consumption for tourism, 3.2 km^3/year are for resident (local) tourists and 3.7 km^3/year for non-resident (foreign) tourists, thus implying Spain's position as a net virtual water importer (i.e. some of them are net virtual water imports due to foreign tourists). The numbers shown are significant, despite the fact that specific tourism activities do not have an intense impact on direct water consumption as they mainly relate to services. However, the virtual water content is significant, as Spaniards are often tourists themselves.

High and concentrated tourism may exceed levels of water assimilation, as is the case in Spain in rural tourism and in winter sports. Spain has more than 50 million tourists, leading to a high tourist water footprint. In addition, the biggest tourist destinations in Spain are sunny regions and beach areas, which are located in arid and water-stressed areas. This necessitates the formulation of appropriate policies to reduce direct and indirect water consumption.

To estimate tourism water footprint, Cazcarro *et al.* (2014) considered costs and sales relationships in the national economy. The water needed to meet demand from both non-resident and resident tourists is about 6.985 km^3. They found that, to meet the demands of foreign tourists in Spain, there is a need for about 3.737 km^3 from both blue and green water footprints, whilst Spanish water resource requirements are only about 1.570 km^3.

They suggested that the use of developed irrigation technologies in Spain is one of the solutions which could reduce the water footprint. It should be noted that price increases, economic incentives, environmental levies and altruism are the main solutions and drivers for the implementation of any type of tourism policy from a water perspective.

9.6.2 Regional footprints and interregional interactions of COD discharges in China

Wang *et al.* (2017*a*, *b*) evaluated the domestic chemical oxygen demand (COD) footprints for individual regions of China and their interactions in 2010 by using a multi-regional input–output model (MRIO). COD is one of the main indicators of pollutants in agricultural, industrial and household sector wastewaters and its reduction has become an important requirement in China. Therefore, understanding the spatial pattern and factors which influence regional COD discharge in China are vital to efforts for reducing regional discharges.

In terms of interregional interactions of COD, Tianjin, Beijing and Shanghai transferred a large amount of their COD discharge to other regions. It should be noted that interprovincial flows of virtual agricultural COD were about 68.8% of the total flows, which were transferred from eastern and western regions over long distances. COD discharged by the agriculture sector compared to freshwater resources in Henan, Hebei and Ningxia was excessive, which may lead to a more severe water scarcity in these regions. Net COD exporters bore the COD discharge caused by their own as well as the demands of other regions. Also, they found that major flows of virtual industrial COD occurred in the eastern and central regions, while the northeastern and northwestern regions exhibited smaller flows.

The authors divided COD flow patterns into two categories of agricultural and industrial discharge, through which the former flow transfers crossed long distances between the eastern and western regions of China. A comparison of water resources characteristics and their regional COD flows suggests that China's policymakers should design and implement a general national plan to reduce the country's total COD discharge based on supply and demand, especially in net exporter regions that have fresh water scarcity problems. In addition, reducing demand for contaminated products and increasing demand-side reuse and recycling is crucial.

9.7 SUMMARY

In this chapter, various water footprint studies were first reviewed. Then, common terms such as blue, green, gray and white water footprints, etc., were explained. By definition, it was found that a water footprint is a multidimensional indicator that indicates the amount of water consumed by the source as well as the amount of contaminated water by type of contamination. Water footprint divisions were then introduced, including both direct and indirect. Subsequently, the necessity of research and evaluation, and the purpose of evaluating water footprints, was stated. It was also determined that water footprint studies may have different purposes and can be used in different contexts. Each goal needs its own perspective and analysis and, in short, the purpose of assessing water footprints is to analyze how human activities relate to specific products, with water scarcity and pollution issues, and to see how activities and products can be optimized in a sustainable manner with respect to water. Next, how to calculate blue, green, gray, etc. water footprints was described. Finally, with all this in mind, some of the challenges ahead were discussed.

Looking at the concepts of virtual water and water footprints as well as their application and the problems facing water resources in the world, the main challenges and basic needs of water footprints can be identified. There are a number of practical issues that are undeniable when evaluating water footprints. Often, a key question is what to do when the data is not available? A key challenge in developing more accurate guidance is what contractual data can be used when accurate local estimates are not available. In this context, it is necessary to create a database with contractual water footprint estimates for different processes and products which are different for various production areas (such as countries). But is such a database useful for evaluating the footprint of consumers and manufacturers? Trade data is available, but often there are no details about production and supply chains. The question is, what data should be included in an analysis and what data should not be considered? The answer is that some components do not have a significant contribution to all water footprints, and their calculation is not worth the extra return compared to the supply chain. Therefore, more practical experience is needed to calculate the water footprint for different products in order to develop practical guidelines for establishing a rule to exclude some cases from water footprint analysis.

An issue that has yet to be addressed is how water footprint changes over time. For example, irrigation in agriculture depends on the pattern of precipitation in the same year. In addition, water productivity may change from year to year under the influence of other factors (including factors that are not specific to water) that leads to variability in water footprint throughout a year. So, it is obvious that changes in water footprint from one year to another cannot simply be seen as a case of improving the structure or deterioration of water consumption. Therefore, if water footprint data shows a multi-year average, a more meaningful picture can be provided.

Other questions to consider include the statistical period appropriate for analysis: should it be five years, ten years, or more? When can a trend be analyzed over time? In addition, some input data may be obtainable over a very long period (for example, for conventional weather data, it is standard to be recorded for 30 years), while other input data can be obtained annually or only for a duration of five years, on average. As a result, guidelines need to be developed.

Finally, it is worth noting that another important issue in assessing water footprints is data uncertainties, reinforcing the fact that results must be interpreted carefully. Uncertainty analysis is appropriate but time constraints often do not allow for sensitivity analysis and uncertainty analysis to be carried out. It is useful to have at least some surface indexes to show the great degree of uncertainty in different types of water footprint calculations. At present, studies of uncertainty in water footprint have been neglected.

The calculation of water footprints allows for a distinction between green, blue and gray water footprints. Gray water footprint can be divided into calculations for specific pollutants. One of the challenges in calculating a gray water footprint is the development of guidelines to define maximum permissible and natural concentrations, since such concentration standards are not available for all pollutants (in which case there should be instructions to provide usable contract values).

REFERENCES

Ababai B. and Ramezani-Etedali H. (2016). Estimation of Water Footprint Compartments in National Wheat Production. *Water and Soil*. **29**(6), 1458–1468 (In Farsi).

Allan J. A. (1993). Fortunately, there are substitutes for water otherwise our hydro-political futures would be impossible. Proceedings of the Conference on Priorities for Water Resources Allocation and Management, Southampton, July 1992. Overseas Development Administration, London, UK. pp. 13–26.

Allen R. G., Pereira L. S., Raes D. and Smith M. (1998). Crop evapotranspiration: Guidelines for computing crop water requirements. Food and Agriculture Organization (FAO) Irrigation and Drainage Paper 56. FAO, Rome.

Bazrafshan O., Ramezani-Etedali H., Gerkani-Nezhad-Moshizi Z. and Shamili M. (2019). Virtual water trade and water footprint accounting of saffron production in Iran. *Agricultural Water Management*, **213**, 368–374.

Bicknell D. (1998). SAP to fight drug firm's $500M. suit over R/3 collapse. *Computer Weekly*, 3 September.

Blas A., Garrido A. and Willaarts B. (2018). Food consumption and waste in Spanish households: water implications within and beyond national borders. *Ecological Indicators*, **89**, 290–300.

Cai B., Wang C. and Zhang B. (2017). Worse than imagined: unidentified virtual water flows in China. *Journal of Environmental Management*, **196**, 681–691.

Cazcarro I., Duarte R. and Chóliz S. (2016). Downscaling the grey water footprints of production and consumption. *Journal of Cleaner Production*, **132**, 171–183.

Cazcarro I., Hoekstra A. Y. and Chóliz J. (2014). The water footprint of tourism in Spain. *Tourism Management*, **40**, 90–101.

Chini C. M, Konar M. and Stillwell A. S. (2016). Direct and indirect urban water footprints of the United States. *Water Resources Research*, **53**(1), 316–327.

Cui X., Wu X., He X., Li Z., Shi C. and Wu F. (2018). Regional suitability of virtual water strategy: evaluating with an integrated water ecosystem-economy index. *Journal of Cleaner Production*, **199**, 659–667.

Delpasand M., Bozorg-Haddad O. and Loáiciga H. A. (2020). 'Integrated virtual water trade management with considering countries' self-sufficiency in production of strategic agricultural and industrial products.' *Science of the Total Environment*, 140797.

European Union (2000). Directive 2000/60/EC of the european parliament and of the council of 23 October 2000 establishing a framework for community action in the field of water policy. EU, Brussels. See: http://data.europa.eu/eli/dir/2000/60/2014-11-20 (accessed April 2021).

Fan J. L., Wang J. D., Zhang X., Kong L. S. and Song Q. Y. (2019). Exploring the changes and driving forces of water footprints in China from 2002 to 2012: A perspective of final demand. *Science of the Total Environment*, **650**, 1101–1111.

Feng L., Chen B., Hayat T., Asaedi A. and Ahmad B. (2015). The driving force of water footprint under the rapid urbanization process: a structural decomposition analysis for Zhangye city in China. *Journal of Cleaner Production*, **163**, 322–328.

Ferng J.-J. (2001). Using composition of land multiplier to estimate ecological footprints.

Food and Agriculture Organization (FAO) (2010a). 'CROPWAT 8.0 model' FAO, Rome.

Food and Agriculture Organization (FAO) (2010b). 'AQUACROP 3.1.' Fao, Rome. See: www.fao.org/nr/water/aquacrop.html (accessed 27 May 2021).

Hoekstra A. Y. (ed.) (2003). Virtual water trade: Proceedings of the International Expert Meeting on Virtual Water Trade, Value of Water Research Report Series No 12. UNESCO-IHE, Delft, Netherlands.

Hoekstra A. Y. (2008a). The relation between international trade and water resources management. In: Handbook on trade and the environment, K. P Gallagher (ed.), Edward Elgar Publishing, Cheltenham. pp. 116–125.

Hoekstra A. Y. (2008b). 'The water footprint of food.' In: Water for food, J Förare, (ed.), The Swedish Research Council for Environment, Agricultural Sciences and Spatial Planning, Stockholm. pp. 49–60.

Hoekstra A. Y. (2008c). Water neutral: Reducing and offsetting the impacts of water footprints. Value of Water Research Report Series No. 28. UNESCO-IHE, Delft, Netherlands.

Hoekstra A. Y. (2017). Water footprint assessment: evolvement of a new research field. *Water Resources Management*, **31**(10), 3061–3081.

Hoekstra A. Y. and Chapagain A. K. (2007a). The water footprints of Morocco and the Netherlands: global water use as a result of domestic consumption of agricultural commodities. *Ecological Economics*, **64**(1), 143–151.

Hoekstra A. Y. and Chapagain A. K. (2007b). Water footprints of nations: water use by people as a function of their consumption pattern. *Water Resources Management*, **21**(1), 35–48.

Hoekstra A. Y. and Chapagain A. K. (2008). Globalization of water: sharing the planet's freshwater resources. Blackwell publishing, *Oxford*, **109**(9), 3232–3237.

Hoekstra A. Y., Chapagain A. K., Aldaya M. M. and Mekonnen M. M. (2011). The water footprint assessment manual: setting the global standard. Earthscan, London, UK.

Lenzen M. (2001). A generalized input-output multiplier Calculus for Australia. *Economic Systems Research*, **13**(1), 65–92.

Lenzen M. and Murray S. A. (2003). The ecological footprint, issues and trends. University of Sydney Press, Sydney, Australia. p. 6.

Li H., Liu G., Yang Z. and Hao Y. (2016). Urban gray water footprint analysis based on input-output approach. *Energy Procedia*, **104**, 118–122.

Li H., Yang Z., Liu G., Casazza M. and Yin X. (2017). Analyzing virtual water pollution transfer embodied in economic activities based on gray water footprint: A case study. *Journal of Cleaner Production*, **161**, 1064–1073.

Liu J. G., Williams J. R., Zehnder A. J. B. and Yang H. (2007). GEPIC: modelling wheat yield and crop water productivity with high resolution on a global scale. *Agricultural Systems*, **94**(2), 478–493.

Luo P., Yang Y., Wang H., Gu Y., Xu J. and Li Y. (2017). Water footprint and scenario analysis in the transformation of chongming into an international eco-island. *Resources, Conservation and Recycling*, **132**, 376–385.

Madugundu R., Al-Gaadi A., Tola E., Hassaballa A. A. and Kayad A. G. (2018). Utilization of landsat-8 data for the estimation of carrot and maize crop water footprint under the arid climate of Saudi Arabia. *Plos one*, **13**(5).

Megdal S. B. (2018). Invisible water: the importance of good groundwater governance and management. *Nature Partner Journal (npj) Clean Water*, **1**, 1–5.

Ngo T. T., Le N. T., Hoang T. M. and Luong D. H. (2017). Water scarcity in Vietnam: A point of view on virtual water perspective. *Water Resources Management*, **32**(11), 3579–3593.

Salmoral G. and Yan X. (2018). Food–energy–water nexus: A life cycle analysis on virtual water and embodied energy in food consumption in the tamar catchment, UK. *Resources, Conservation & Recycling*, **133**, 320–330.

Shtull-Trauring E. and Bernstein N. (2017). Virtual water flows and water-footprint of agricultural crop production import and export: A case study for Israel. *Science of the Total Environment*, **622**, 1438–1447.

Sun S., Fang C. and Lv J. (2017). Spatial inequality of water footprint in China: A detailed decomposition of inequality from water use types and drivers. *Journal of Hydrology*, **553**, 398–407.

Tian X., Sarkis J., Genga Y., Qiana Y., Goad C., Bleischwitze R. and Xu Y. (2018). Evolution of China's water footprint and virtual water trade: A global trade assessment. *Environmental International*, **109**(9), 393–410.

Wang J., Li L., Li F., Kharrazi A. and Bai Y. (2017a). Regional footprints and interregional interactions of chemical oxygen demand discharges in China, *Resources, Conservation and Recycling*, **132**, 386–397.

Wang Y. B., Liua D., Cao X. C., Yang Z. Y., Song J. F., Chen D. Y. and Sun S. K. (2017b). Agricultural water rights trading and virtual water export compensation coupling model: A case study of an irrigation district in China. *Agricultural Water Management*, **180**, 99–106.

Williams J. R. (1995). The EPIC model. In: Computer models of watershed hydrology, V. P. Singh (ed.), Water Resources Publications, Highland Ranch, CO. pp. 909–1000.

Ye Q., Li Y., Zhuo L., Zhang W., Zhang Q., Wei Y. M., Wang J. W., Ding X. and Mid Z. (2018). Optimal allocation of physical water resources integrated with virtual water trade in water scarce regions: A case study for Beijing, China. *Water Research*, **129**, 264–276.

Chapter 10

The water, food and energy nexus

Shima Kheirinejad[1], Omid Bozorg-Haddad[2] and Veera Gnaneswar Gude[3]*

[1]*M.Sc. Student, Department of Irrigation & Reclamation Engineering, Faculty of Agricultural Engineering & Technology, College of Agriculture & Natural Resources, University of Tehran, Karaj, Tehran, Iran. E-mail: Shima.kheyri20@ut.ac.ir*
[2]*Distinguished Professor, Department of Irrigation & Reclamation Engineering, Faculty of Agricultural Engineering & Technology, College of Agriculture & Natural Resources, University of Tehran, Karaj, Tehran, 31587-77871, Iran. E-mail: OBHaddad@ut.ac.ir*
[3]*Associate Professor, Civil & Environmental Engineering Department, Mississippi State University, MS, USA. E-mail: gude@cee.msstate.edu*
**Corresponding author.*

ABSTRACT

The water, food and energy nexus is a critical component for sustainable development as global population and industrialization escalate. Agriculture is responsible for the majority of freshwater consumption worldwide, while one quarter of the world's energy is spent on food production and consumption. The connections between such vital areas necessitates a profound and integrated approach to securing the water, food and energy sectors across the world. Such an integrated approach should be based upon understanding the nexus between the three individual sectors, and on coordinating the interactions between them. As the global population is expected to reach 8 billion (8×10^9) by 2030, demands for essential services and higher living standards are becoming prevalent, and the need for conscious protection of vital resources – without which, meeting those demands and desires would be impossible – is more palpable than ever. Considering the impact of water and environmental crises on food and energy security,

Editor: Omid Bozorg-Haddad
doi: 10.2166/9781789062144_0175

the integrated management of water, food and energy with the collaboration of all stakeholders could result in a significant check on any detrimental changes. Due to the critical importance of food, water and energy security, policies should be implemented to conserve and protect these essential resources. Therefore, it is very important to understand the logic governing this issue, and, in this chapter, definitions and logical approaches that govern the concept of the water, food and energy nexus, the potential crises ahead, and the effective management solutions, tools, and methods used in this field are all discussed, using case studies and examples.

Keywords: Water, food, energy nexus.

10.1 INTRODUCTION

Achieving sustainable water, food, and energy supplies is a grand challenge. To address it, the United Nations (UN) developed its sustainable development goals (SDGs). Water, food, and energy supplies are identified as vital actors to reduce poverty by assuring adequate resources to sustain and improve livelihoods, which simultaneously leads to eco-system preservation and to using the eco-system to provide services necessary for livelihood. Given the water and environmental crises threatening the security of food and energy in some countries, it can be implied that SDGs can guide policies for sustainable development.

The interconnectedness of the three vital resources – water, food and energy –means that any decision to improve the critical condition of any of these resources will affect the other two. Since the security of all three resources is essential for human survival, care must be taken not to cause any adverse impacts in the other two sectors in order to improve the critical situation in one of them. Therefore, studies that have investigated the simultaneous interrelationships of all three resources are discussed in this chapter. These studies were made at different scales, using a variety of tools, and in some cases included other critical elements in addition to the three vital resources in the nexus.

10.1.1 Improvement of existing frameworks and tools

Since the concept of the water, food and energy nexus is examined at different scales and for different purposes, and this nexus is directly and indirectly related to other factors such as environment, climate change, land use, etc., different methods and frameworks are used for its evaluation. The main frameworks that are used in this field are examined in this section.

The Climate, Land, Energy, Water Strategies (CLEWs) model was developed by the United Nations on the eve of the Rio+20 Conference in 2012, and its performance was improved in the following years with the help of affiliate institutes. According to the International Renewable Energy Agency, CLEWs is the most complete framework for examining nexus synergies, but requires a large amount of input data which is not available, especially in developing countries, or is not easily usable in the framework

(Ferroukhi *et al.*, 2015). Another hole in this framework is the lack of consideration of economic aspects (Kaddoura & El Khatib, 2017).

CLEWs consists of three modules: energy modeling, water modeling and land modeling. In fact, no new tools were developed, but modeling tools were linked together. The tools used in the energy module are LEAP, MEAD, OSeMOSYS and MESSAGE; in the water module, the tools are WEAP and PODIUM; and in the land and environment module, the tool used is AEZ. By combining the two models LEAP (an energy analysis and planning model) and WEAP (a water evaluation and planning model) with an integrated binary combinations approach (for water and energy), a triple combination nexus was developed (Water in the West, 2013).

The first version of the Nexus Tool was developed in 2013 at Purdue University, and then its graphical interface was created and presented in a web-based form with the support of the Qatar Energy and Environment Research Institute. After making modifications, the second version was introduced as Tool 2, and Qatar was then examined as a case study using it. The strength of this tool is its accessibility and ease of presentation, as it has created conditions for modeling without access to detailed data. Unlike the CLEWs framework, it also takes into account economic factors. However, this simplification of partnerships and synergies between the sectors, which simplifies the modeling process, can also be interpreted as a weakness of this tool. For example, in the section on environmental impacts, only carbon emissions are considered, while the reduction of groundwater levels in arid regions such as Qatar (which was the target area of this method) is in fact very important from an environmental point of view.

In 2013, the LIPHE4 Institute, in collaboration with the Food and Agriculture Organization (FAO) of the UN, presented a report proposing the use of the MuSIASEM framework, which is based on fund-flow modeling, for nexus evaluation. MuSIASEM uses complexity theory techniques to analyze data at different scales. It should be noted that to achieve the minimum data required for this framework, extensive cooperation between local experts and government agencies is required.

Subsequently, in 2014, the FAO published two reports on nexus assessment. The first report describes the interrelationships and the concept of a nexus, and then introduces three areas that can be effective in studies to identify a water, food and energy nexus: (1) data collection and data mining; (2) developing a scenario; and (3) examining the reaction to possible options. In its second report, the FAO focuses more on the third of these three areas. In this FAO framework, a three-dimensional matrix of relationships between water, food and energy resources is formed to define a conceptual model of water, food and energy in each region. Next, in order to evaluate the sustainability of water, food and energy resources, the framework introduces a three-dimensional matrix of various indicators and suggested that the values of the indicators should be taken from a global database that is available to everyone. Of course, it should be noted that, due to the limited data available for calculating all indicators, this nexus rapid evaluation method is carried out with a limited number of indicators.

After formation of these main frameworks for modeling a nexus and its evaluation by various indicators, researchers looked at any existing gaps and tried to improve the performance of the modeling. For example, Smajgl *et al.* (2016), in a study of the Mekong basin, stated that most of the studies on the water, food, and energy nexus were centered on water, and this led them to consider the interrelationship between water, food and energy dynamically with equal weighting of the water, food and energy sectors, and finally compare the two methods (water-centric, and integrated conceptualisation of the nexus). Finally, they developed a balanced system and examined the advantages of multilateral studies over independent studies.

Investigating the water, food and energy nexus is an integrated approach, which, if used in policymakers' decisions, can optimize cooperation and management of this interrelationship. Of course, a framework appropriate to the purpose and area of the study should be sought. Kaddoura and El Khatib (2017) worked on the development of modeling tools. They reviewed the tools available for nexus modeling to help find the most appropriate tools that meet the modeling needs of policymakers, professionals, and organizations. Their study showed that, in order to advance the modeling of the nexus approach, more consensus needs to be developed among tool developers. Theoretically, complete modeling tools make it possible to create policies that lead to maximum synergistic productivity among nexus sectors.

Nexus simulation models are instrumental for decision makers and policymakers due to the existence of multiple stakeholders in planning for each of the resources and the pursuit of multiple goals (economic criteria, minimizing environmental damage and achieving maximum social welfare with adequate supply of water, food and energy), as well as the complex and interactive interaction between these resources. Due to the existence of different stakeholders, mathematical optimization is of paramount interest in planning to achieve optimal conditions in the three vital sectors of water, food and energy. For example, to minimize economic and environmental costs such as carbon dioxide emissions, Zhang and Vesselinov (2017) developed a single-objective linear model for the water, food, and energy nexus.

The effect of the water, food and energy nexus on cultivation pattern optimization is also an important issue, which has been studied with the aim of increasing the economic benefits of water- and energy-related products and reducing the use of water and energy sources. Egypt was used as the study area to demonstrate this method (El Gafy *et al.*, 2017). In addition, optimization of resource efficiency has been considered necessary for sustainable agriculture, and in this regard, an integrated model was suggested for the water, food and energy nexus by combining an ecological model, economic model and a regional climate model (Tian *et al.*, 2018).

Due to the presence of different stakeholders of dam reservoirs, a simulation and optimization model based on a nexus approach, with two objectives of maximizing monthly hydroelectric power generation of the dam and minimizing annual deficit irrigation of the dam reservoir downstream, was presented by Stamou and Rutschmann (2018). The purpose of this study was to illustrate the conflicts between the objectives of the three sources of water, food and hydroelectric power in reservoir management. The study area of this model was upstream of the Nile River.

10.1.2 The water, food and energy nexus at common water boundaries

As mentioned above, nexus analysis can have different uses at different scales. For example, it can be useful to examine the relationships between neighboring countries with shared water resources from a nexus perspective (food, energy or water exchanges in consideration of sustainable development), whilst also using the nexus perspective to allocate common waters to them in order to create a healthy competition to enhance resource efficiency in production from water resources, energy and agricultural products.

The water, food and energy nexus plays an important role in water policy and planning decisions, especially for basins that cross national borders. Yang *et al.* (2016) considered the Brahmaputra river basin in south Asia as a study area and presented a new hydro-economic model, with analysis of past scenarios and using the nexus concept. Similarly, Keskinen *et al.* (2016), when considering the importance of the water, food and energy nexus for human beings, stated that the concept had not yet been completely or fully tested in practice, and, therefore, they discussed the nexus on an international scale. Three regions of central Asia, south Asia and the Mekong region and their border river basins were studied, with the regions being chosen because of: (1) the diversity of views and scales; and (2) the importance of the political and governmental players. The results of this study show that due to the relationship between water allocation and the water, food and energy nexus, the nexus has the potential to complement existing approaches in the field of transboundary river basins.

It should also be noted that the Middle East requires special attention due to rapid industrialization, escalated environmental concerns, unstable political relations and cross-border conflicts. Cross-border water plays an important role in the sustainable development of Middle Eastern countries. Among them, Iran, Iraq and Turkey have important border river basins, in which any new developments can threaten these countries with many potential risks and benefits. The factors influencing the security of the water, food and energy nexus in the Middle East are water scarcity, migration, economic growth, urbanization, population growth, poverty and political stability. Therefore, it is important to examine the dynamics of nexus security-related concerns in this region. The results show that most countries in the Middle East face insecurity of global resources due to poor planning or mismanagement strategies. According to the Pardee RAND report (see Willis *et al.*, 2016), the water, food and energy security indexes for Iran, Iraq and Turkey were estimated as 0.68, 0.65 and 0.75, respectively in 2019. The results show that the population has increased significantly in the last two decades, strengthening the potential for conflict over water, especially over transboundary water resources. Therefore, with a comprehensive interpretation, water, food and energy nexus systems should be established for Iran, Iraq and Turkey, through which the performance of all stakeholders can be examined and the common challenges of these countries can be solved (Zarei, 2020).

10.1.3 The water, food and energy nexus at urban scale

The efficiency of using the nexus approach for the management of vital resources has also received much attention at urban scales. In this regard, it has been suggested that the nexus approach of water, food and energy is essential to address issues related to sustainability and the impact of human behavior on urban metabolism (Walker *et al.*, 2014). Walker *et al.*'s method is based on the utilization of Multi-sectoral Systems Analysis (MSA). Using the MSA method in a London study area, they marked several goals, including: estimating the flux of resource (nutrients, water and energy) input, output and circulating in the city's watershed system; identifying synergies and contradictions resulting from various combinations of different innovations in the water sector; and evaluation of the economic benefits associated with the implementation of technologies in terms of fertilizer production, energy generation and reduction of greenhouse gas emissions. The results indicate that the selection of the best innovation depends on the resource that is intended to be improved. Separation of urine can recover 47% of the nitrogen in food consumed in London, as well as generate 33 million dollars a year through fertilizer production. Collecting food waste in sewage, along with growing algae in wastewater treatment plants, could increase the amount of carbon emissions from renewable energy by 66%, in addition to generating 58 million dollars a year through fuel production.

In order to analyze the water, food and energy nexus (internal and cross-border interactions) and its incompatibilities from an urban perspective in Delhi, an attempt was made to define a system (Ramaswami *et al.*, 2017). The effects of a water footprint in this cycle were considered in this study and also the data show that more than 75% of water consumption is used for food production through agriculture, and more than 75% of energy is used for food production through burning fuels.

10.1.4 The water, food, energy and environment nexus

The water, food, energy, and environment nexus is the foundation of integrated and systematic solutions for the sustainable development and improvement of communities, and a sole path for each will have consequences in other sectors. Although the three systems of energy, water and food form the three sides of the concept of the nexus, the environment is affected by all three, so the environment is part of each of these individual systems. The concept of nexus, in addition to covering the interconnectedness of systems, is an integrated solution to interconnected challenges.

In this regard, the water footprint in the water, food, energy and environment nexus was studied and the relationship between different water sources (blue and green water) with the nexus was discussed. The results show that, in most water management planning, only water related to agriculture, industry and drinking water is considered, while considering other uses of blue and green water resources (such as water required for wild food production, automobiles, fossil fuels, hydropower plants, creating tourism, and forest-related uses that lead to food and energy production) is also essential.

At the University of Miami, Wong (2015) paid special attention to clean and renewable energy sources, namely: water, wind and solar power, and stressed the importance of using these three resources because of the respect that is placed on the environment in the production process. The relationship between water and energy has also been a very important issue in the past where a large volume of water is used for cooling industrial devices. The study also identified the areas most affected by the challenges in this nexus (sub-Saharan Africa, Japan, the Punjab, and the Arabian Peninsula).

The relationship between food, energy and water for long-term sustainability in Brazil, Russia, India, China and South Africa has also been studied (Ozturk, 2015). This study showed the ecological indicators related to long-term sustainability through the relationship between energy and food in these countries. The issue of sustainability with the theory of the Environmental Kuznets Curve (EKC) and biodiversity, which requires the assignment of appropriate resources to ensure food security among these countries, was studied using a point panel modeling in a Generalized Method of Moments system to obtain estimates of desired parameters. The EKC hypothesis was tested in these countries, and the results showed that despite the existence of an actual U-curve between carbon dioxide emissions and economic growth for Brazil, India and South Africa, it is still possible to develop newer integrated economic and environmental policies.

The water, food and energy nexus has also been examined from a livelihood perspective (Biggs *et al.*, 2015), given that water, food and energy security are essential elements for reducing poverty by providing adequate resources to sustain and improve livelihoods in a fair manner. According to this study, it is necessary to consider living conditions in order to achieve sustainable development. The authors also explored the concept of 'environmental security for livelihoods', which included examining the balance between the supply of natural resources and human demand to promote sustainability.

A study conducted in the UK calculated energy, water and food consumption for the years 1997–2013, using input–output analysis techniques to examine the interaction between the effects of energy, water and food at different points in the length of supply chains, from the extraction of materials and the burning of energy to the point of final consumption. The study suggests that strategies that were not intended to reduce environmental impact should not harm the social and economic well-being of Britain and its trading partners (Owen *et al.*, 2018). Also to assess water and food security with respect to ecological considerations, this time in the Gavkhuni Basin which is located in the center of Iran, a simulation model with temporal and spatial separations based on the water, food and energy nexus approach was prepared (Ravar *et al.*, 2020). The main components of the model developed in this study are population units, water, agriculture and energy. The evaluations of this study over a 10-year period indicate that the combined policies proposed by the agricultural and environmental sectors have been very effective in nexus system remodeling as well as estimating the environmental demand of the Gavkhuni wetland. Implementing a combination of changing cultivation patterns and increasing crop productivity, together with policies to control groundwater exploitation, not only increases surface water resources security (by 4%) and groundwater resources security (by 5%), but also reduces water for food production (by 18%) and energy for water (by 26%).

Water scarcity can create environmental, social and economic challenges, and the solution to this problem is to provide an integrated approach that considers the basin as an interconnected system (Bakhshianlamouki *et al.*, 2020). In their study, Bakhshianlamouki *et al.* developed a dynamic system model to simulate the water, food and energy nexus in the Urmia lake basin, which was used to evaluate the impact of proposed measures to restore the lake. In this model, in addition to considering the effects of climate change, the effect of various measures such as increasing irrigation efficiency, increasing return flow, water transfer within the basin, retirement of agricultural lands and vivification of a part of the lake on natural resources and socio-economic status of the basin were analyzed. The results of their studies show that increasing the lake level to the proposed ecological level by 2040 is possible following a comprehensive restoration approach. The results also show that the retirement of 20% of irrigated wheat lands to control water demand, if accompanied by a 20% increase in yield, will not reduce wheat production in the basin for 80% of irrigated and 50% of rainfed fields.

10.1.5 Other effective factors in the nexus

A survey system was designed based on issues related to the water, food and energy nexus. The results of the survey, which related to the Yellow river basin, Lake Winnipeg in Canada, and a number of basins in India, were presented to the relevant experts and the experts ranked the direct and indirect factors influencing the water, food and energy nexus, including: climate change, political and economic changes, etc. (Lawford *et al.*, 2013).

Other studies show that phosphorus is connected to water, food and energy security with a highly fragile link and must be managed accordingly. Phosphorus is one of the major challenges facing water, food and energy security for the growing global population. Accordingly, the distribution of rock phosphate as fertilizer to provide modern nutrition and food systems is a great social achievement in overcoming inequality. Jarvie *et al.* (2015) argue that in the United States, the success of phosphorus distribution and the reorganization of agricultural systems disrupted local phosphorus cycles and inadvertently created instability. In solving these problems, strategic phosphorus management can help to achieve sustainability, and minimize disturbances and build cooperation to improve resilience among water, food and energy security nexus components (Jarvie *et al.*, 2015).

In addition to the abovementioned factors, there is another well-established part of the water, food and energy nexus: soil. Soil is a very important issue in food security and energy supply, and has been well studied by several researchers but still its impact has not been discussed much in political decisions. It is imperative to consider erosion and soil fertility reduction because, with all the scientific advances, there is still no way to compensate for surface soil losses (Hatfield *et al.*, 2017).

10.1.6 Results extracted from review articles

Because many studies have been made of the water, food and energy nexus, studying the reviews completed in this field can provide useful information about the work already undertaken and the knowledge gaps that still exist in this broad area. The following is a summary of some of these reviews.

Al-Saidi and Elagib (2017) reviewed the water, food, and energy nexus, in which they attempted to identify a wide range of tools for internal analysis of issues related to this nexus and to create justification lines for discussing it, and they examined inconsistencies such as the failure of management strategies and the growing relevance of resources with their increasing crises and scarcity. Endo *et al.* (2017) studied Asia, Europe, Oceania, North America, South America, the Middle East, and Africa. According to this report, North America and Oceania tended to focus on one particular type of communication, namely water and energy (46%) and water and climate change (43%), while Africa focused less on water and energy (7%). Other regions were interested in balance in any kind of communication. Albrecht *et al.* (2018) reviewed 245 studies on the water, food, and energy nexus, and some of the results they mentioned include: (1) less than one fifth of the studies used both quantitative and qualitative methods; (2) there is limited use of social science methods (about one quarter); (3) three quarters of the studies examined the nexus using quantitative methods; and (4) there is less use of special and renewable methods for studying the nexus (one third of articles).

10.1.7 Existing challenges

The growing population needs a reliable supply of water, food and energy, yet communication and cooperation between these resources are simultaneously challenged. Chang *et al.* (2016) summarized the methodology of related support, challenges, and opportunities through a strong water, food and energy link (quantitatively). They also addressed simultaneously the link between each two out of the three aspects, in different scales and for a wide range of factors (energy, water, land use, environment, pollution and greenhouse gases), and they believe that improved decisions produced by studying results of these comparisons are possible if the boundaries, definitions, approaches and methodologies used are harmonized.

It should also be noted that 90% of water resources are used for energy and food production, and due to climate change and rapid population growth, water has become a limiting factor for economic progress and providing security in terms of food and energy supply in the future. Institutions tasked with allocating water resources are not connected, and a lack of planning has led to major problems (Mohtar & Daher, 2016). Mohtar and Daher (2016) showed that there are many gaps between underlying science and political decisions. These gaps can be filled by integrating scientific data and policies with learning tools that can meet political goals. The importance of maintaining the physical security (for example, against terrorist attacks and disease) of these resources has also been discussed.

The connection between water, food and energy is not a new or innovative concept. The definition of the link between water, food and energy varies, especially among authors. Therefore, the integration of research and the achievement of policy integration is not always possible and is not necessarily desirable (Wichelns, 2017).

With advances in interdisciplinary science, a new understanding of the interrelationships between water, food and energy systems has been provided, which has confronted us with crises for coordinated management to achieve economic and environmental sustainability in the water, food and energy cycle. Scanlon *et al.* (2017) presented the state of the cycle and discussed achieving management by reducing demand and increasing resources and relocations. They also provided solutions including combining human and institutional behavior in models, as well as partnerships between universities, industry, and government to develop policy-related data, and modeling on a regional rather than local (more general) scale.

Studies have been conducted on the relationship between government and decisions based on the nexus, and the results show that the gaps in this issue have not been clearly identified, and the cause of their impact and how to confront them should be studied; improving the performance of this link requires reviewing the boundaries of nexus analysis against other sectors (Weitz *et al.*, 2017). Similarly, examining the water, food and energy nexus at different scales reveals the challenges in this field. Garcia and You (2016) concluded that uncertainties appear at all scales, which should be taken into account in decisions, and finally suggested solutions to shape these challenges and improve the performance of this nexus.

The main challenges of the water, food and energy nexus in Brazil have also been identified and discussed (Mercure *et al.*, 2019). For each, data analysis from economic and biophysical modeling sources in conjunction with an overview of the legislative and policy perspectives highlights government weaknesses in growing challenges. It is noted that environmental and economic change at the global and local levels (in Brazil) have complex dependencies, and that by analyzing these dependencies, a scientific approach can be set to achieve sustainability

10.1.8 Summary

Studies show that due to the need to preserve three vital resources that are interconnected in nature, different methods have been designed and implemented for their correlational evaluation. Some of them are more detailed but require more data and, by contrast, those that are easier to access and use support less data and therefore fewer details. Of course, it should be noted that the purpose of studying the water, food and energy nexus and its evaluated scale is very effective in the breadth of the variety of data selected to form a correlation model framework. For the development of these models, the relationship of each of these three resources with the environment has been seen, because the environment is affected by all three resources. Of course, there are also challenges in evaluating the water, food and energy nexus, which are mainly related to the breadth and diversity of the data used, which can be controlled to some extent by creating links between universities, industry and governments.

10.2 DEFINITIONS

In this section, the concepts used in the text are explained.

10.2.1 Sustainable development

Sustainable development is a concept that has emerged from the negative environmental and social consequences of unilateral development approaches since the industrial revolution and the changing attitude of humanity to the concept of growth and development. This concept attempts to adopt a novel perspective towards development, avoiding past mistakes and making a wholesome development. Sustainable development is a process focused on an optimal future for human societies, in which the living conditions and resources provide the human needs without harming the integrity, harmony, and beauty of the vital systems. Sustainable development provides solutions for destructive structural, societal and economic patterns in order to prevent the annihilation of natural resources, the destruction of environmental systems, pollution, climate change, an excessive increase in population, injustice and low living conditions of current and next generations.

10.2.2 Millennium development goals (MDGs) and sustainable development goals (SDGs)

The Millennium Development Goals (MDGs) are comprised of eight mutual goals, over which a consensus was reached back in 2000 at the United Nations (UN). The goals were to be accomplished by 2015. In 2000, the Millennium Manifesto was accepted by 189 countries and signed by 147 officials. The goals were broken down into 21 sub-goals measurable with 60 indices. The goals include: (1) eliminating poverty and starvation; (2) basic education for all; (3) perpetuating gender equality and empowering women; (4) reducing child death; (5) improving mothers' health; (6) fighting AIDS, tuberculosis and other diseases; (7) ensuring environmental stability; and (8) promoting worldwide participation in development (United Nations, 2013, 2016).

With the end of the MDGs' time period, the issue of providing Sustainable Development Goals (SDGs) at the Rio+20 Summit in Rio de Janeiro, Brazil, was raised in 2012 and, as a result, these goals replaced the MDGs. The Rio+20 document was prepared in six chapters, of which Chapter 5 deals with the frameworks required for action and follow-up, including thematic and interdisciplinary issues. MDG targets became more complete and were expressed in 17 SDGs. These goals also specifically address issues related to water, food, and energy (such as Goals 2, 6 and 7).

10.2.3 Water security

In the MDGs, water security is defined as accessible safe drinking water, which is considered a human right. Furthermore, the definition is not all about security, it is also about water accessibility across all range of human consumption.

10.2.4 Food security

The Food and Agriculture Organization (FAO) defined food security as access to sufficient, safe, and nutritious food to satisfy nutrition needs, leading to a healthy life. Appropriate food is considered a human right (FAO, 1996).

10.2.5 Energy security

The UN defines energy security as access to clean, reliable and cost-efficient services in order to cook, illuminate, communicate, and produce (UN, 2010).

10.3 BASICS AND LOGIC

The interrelationships and links between the three vital parts, that is, water, food and energy, reveal the perspective of the dynamic system in studies related to this subject. Therefore, before dealing with modeling methods and the systemic perspective governing them, the details of these relationships will be explained. In the following, the different nexuses of each vital source of water, food and energy are discussed in more detail.

10.3.1 The water and energy nexus

The water and energy nexus has been very well studied in the past decade (Gude, 2015). This nexus shows how much water is used for energy transformation and production and how much energy is used for washing, storage, waste treatment, water transportation, and delivery. The types of energy include electricity, energy produced by fossil fuels such as oil and natural gas, and energy spent on extraction, treatment and on transportation, heating and cooling. All types of energy production require the use of water for the processing of raw materials used in the facility, construction and maintenance of the plant, and in order to provide cool water to thermal power plants. There are also various types of power plants which extract energy from water.

Hydroelectric power plants use water to produce electrical energy. Most hydroelectric power plants receive their required energy from the potential energy of the water behind a dam, from which the amount of energy produced depends on the volume of the water behind the dam and the height difference between the source and water outlet. In a tidal power plant, the plant receives its energy from the water height difference during the day; these power plants can act as a supporter of peak consumption. Some hydropower plants use the water's kinetic energy, with no need to build a dam, and the turbines of such power plants act as a water wheel. Such energy is considered to be a relatively novel branch of liquid movement science.

Pumped power plants are another type of hydropower plant, the duty of which is to support the electricity network during peak hours. In such a powerplant, water is displaced between two levels at different times during the day: when the demand for electricity is low, it pumps water to a higher basin, converting electrical energy to gravitational energy; during peak

Figure 10.1 Water and energy nexus.

hours, the water flows from the higher basin to a lower one, turning a turbine and producing electricity. These power plants improve load factor and reduce costs.

Burning fossil fuels such as sulfur dioxide, nitric acid, monoxide carbon and lead (in coal) to produce energy leads to a staggering amount of pollution. By contrast, using water to produce energy does not result in pollution. However, despite their clean and renewable energies, these power plants have their own limitations. The suitable places to build a hydropower plant are limited. In addition, the majority of hydropower plants are far from population centers, hence additional costs are needed to transfer the energy. Other limitations include heavy dependence on the input water and, since the amount of water in dams depends on precipitation, energy production can be affected by low precipitation. Using water to produce energy is not limited to hydroelectric energy, since it can also be used to produce electrical and liquid fuels. For instance, to produce one liter of oil, 40 liters of water are needed (Figure 10.1). Other sources of energy such as wind and geothermal energies also need a huge amount of water.

Just as energy can be used for waste treatment, it is also true that industrial and agricultural wastewater can be used to produce energy. Therefore, there is an interwoven relationship between water and energy, and gaining knowledge about it can lead to reduced environmental pollution and costs through lower resource usage.

10.3.2 The food and energy nexus

10.3.2.1 Using food waste

Annually, a considerable amount of food waste results from home, industrial and agricultural consumption and, on a daily basis, a gargantuan amount is disposed of in the environment. Considering that not all disposal methods are in accordance with environmental laws, this significant amount of waste can create devasting harm to the environment. If the waste is solid, a good portion of it will be buried and, if it is liquid, it will be poured into rivers, seas or public wastewater. There are a few ways to eliminate waste such as burning, burying and completely breaking it down. Every one of these methods, however, have their own specific limitations. In addition, the laws passed by governments to protect the environment impose their own limitations. Therefore, it is critical to produce energy from waste.

Using waste to produce clean and renewable energies such as bioethanol, biodiesel, bio-oil, biogas, syngas, vapor, and electricity, not only leads to significant reduction of environmental problems but also reduces the cost of buying energy and disposing of the waste.

10.3.2.2 *Methods of converting food waste into energy*

Various types of food waste can be converted into different forms of energy such as heat, biogas, syngas, bioethanol, biodiesel, and bio-oil. This conversion can be accomplished by biological, thermochemical and chemical processes.

Thermochemical conversion includes material breakdown through heat, converting the material to gas, burning, and liquidation by water under high pressure (Hanssen & Schakenda, 2013; Kranert *et al.*, 2012; Monier *et al.*, 2010). Through thermochemical conversion at a high temperature in the presence of oxygen or air, organic food waste breaks down into smaller liquid or gas molecules.

Anaerobic digestion and fermentation are two important biological processes for converting waste into energy. During anaerobic digestion, the micro-organisms in organic waste break down, producing biogas, which is a gas mix consisting of methane and carbon dioxide. During fermentation, micro-organisms such as yeasts ferment simple sugars, converting them into ethanol.

Gasification is a process in which waste is partially oxidized with a gasification agent to produce fuel gas, which is a mixture of gases like carbon monoxide, carbon dioxide, methane, and hydrogen. This pyrolysis is an oxygen-absent process that also produces solid coal and liquid asphalt (Serrano *et al.*, 2013).

10.3.2.3 *Biofuels*

Biofuels are produced from the residues of plants, wood waste, pulps from agricultural products, sugarcane (*Saccharum officinarum* L.), cereal, oil plants, vegetables, and animal waste. They are a good alternative to fossil fuels as their lack of sulfur makes them less pollutant. In addition to being a limited resource, fossil fuels are expensive and heavily pollutant. Biofuels are produced as solids, liquids and gases from biomass. In addition to being cheaper than fossil fuels, these fuels reduce greenhouse gas effects and carbon dioxide emissions, increasing energy security, since biofuels can be produced locally, reducing dependency on foreign countries. As countries use their own energy resources, dependency reduces and make them immune to foreign influences.

Such fuels are divided into four categories: first, second, third and fourth generation biofuels (Aro, 2016; EASAC, 2012). First generation biofuels are produced from food products like sugars, oil, animal fat and starch using current technologies. These fuels include biodiesel, bio alcohol, ethanol, and biogases. Second generation fuels are produced from agricultural waste, paper factories, wood products, forest waste, etc. The difference between these two generations lies in their cost, performance and carbon dioxide emission. Third generation biofuels are produced from algae or other biomass sources. The fourth generation of biofuels is produced from engineered plants or biomass, being capable of better energy performance and higher growth potential in non-agricultural or low-water lands. The first and second generation fuels are more prevalent, with two prominent first generation fuels being ethanol and a methyl-ester (biodiesel), which are produced from food products. Ethanol is obtained from sugarcane fermentation, whilst biodiesel is made by chemically reacting vegetables (oil, corn, soy or turnip) and alcohol. These fuels interact well with diesel, hence the name 'biodiesel'. After the development of the first generation fuels, concerns emerged about their raw materials (and possible competition with other food products) and their effect on biodiversity and land use (because of over-cultivation). Therefore, the second generation fuels started to emerge which reduce the emission and production of carbon dioxide significantly. However, second generation fuels have also faced a backlash because of their direct and indirect influence on the price of food products. Another challenge for these fuels lies in the fact that a staggering amount of water is needed to convert biomass to fuel, hence costs increase and productivity decreases – which should be overcome by technological advancements.

Based on predictions from the International Energy Agency (IEA), biofuels will make up 7% of light and heavy transportation fuels by 2030. Countries vary in terms of the state of development and use of these fuels. However, the share of biofuels in road transportation fuel consumption in the United State and the European Union is projected to reach 14% by 2040 (Cozzi & Gould, 2015). The big oil companies, who are dependent on big energy consumers, play an active role in providing this energy. The active companies in the fields of oil, agriculture, chemical materials, engineering, and financial services are tempted to enter this field as these fuels are good at providing energy and reducing greenhouse gases. The success and profitability of these companies depend on factors such as price and accessibility of raw materials, regulations and laws, and on technologies to convert raw materials into biofuel. Despite not having a significant role in providing energy, biofuels are getting more investment in the west and their share is higher than other renewable resources.

Low-scale ethanol refineries are more cost-efficient in terms of building and maintenance in comparison with oil refineries. With planning and various other new approaches, countries can detach themselves from absolute dependence on oil, reducing environmental and energy security costs. Investing technologies in the biofuel industry and commercializing it will promote sciences such as biology, biotechnology, and nanotechnologies. These sciences can make noticeable differences in biofuels with low-scale innovations, eliminating growth obstacles and accomplishing a transformation from fossil fuels to biofuels.

10.3.3 The food and water nexus

Water scarcity can lead to reduced food production, affecting food security. The latest irrigation methods can increase production and agricultural outputs in dry and semi-arid environments, offering numerous economic and societal benefits. Water is vital for agriculture as it provides the world's future food security. Today, water scarcity is a crucial concern in some parts of the world. In addition, general concerns are increasing about how the environment will be affected. New investments in irrigation infrastructure and drinking water management can decrease the effects of water scarcity and satisfy water demands for producing food products. As a result of the water crisis, the UN has concluded that the most important obstacle against food production in the upcoming decade will be water scarcity; not a lack of agricultural lands. For instance, Australia is one of the biggest food producers in the world but the latest drought has significantly reduced agricultural production. Irrigation, as the biggest water user, will be the first use that's affected by water scarcity. The increased costs of developing new water resources, land destruction in irrigation areas, decreased groundwater, water pollution, and destroyed eco-systems are just some of the challenges faced as a result of water scarcity. Under current methods of using water, the ever-increasing population and their meat-dependent diets have led to increasing limitations on water resources. For instance, according UN papers in 2013, the data on water demands are intriguing. Approximately 700 million people in 43 countries are experiencing water scarcity. By 2025, 1.8 billion people are expected to live in countries or regions with absolute water scarcity, and two-thirds of the world's population could be affected by water stress (United Nations, 2013). It is predicted that some of the most populated regions such as those around the Mediterranean Sea, and in India, China, and Pakistan will encounter intense water scarcity, and regions such as the south-west areas of the USA and Australia will be vulnerable to water scarcity. The problem of water accessibility is one of the major crises faced today, with ramifications for numerous other global issues including poverty, starvation, eco-system destruction, climate change, desertification, and even for the peace and security of the world.

10.3.3.1 World food demand and supply

The current world food production stems from 1.6 billion hectares of cultivated lands, forming 12% of the total land. Rainfed agriculture makes up 80% of the total agricultural lands in the world, producing 60% of the major food products. Irrigation-based agriculture covers only 19% of the total agricultural lands in the world, but produces 40% of agricultural products.

10.3.3.2 Water scarcity

Today, there is fierce competition over water resources in different locations, regions, and countries, so much so that, in 2013, water supply was declared one of the five major crises in the world over the following 10 years (World Economic Forum, 2013). All over the world, water scarcity is reflected in the reduced per capita water used to produce food products. The major irrigation water resources will be under heavy pressure as populated cities, states, and countries need more water from rivers and lakes. Therefore, a main concern for preserving water resources will be preserving underground water resources.

10.3.3.3 Climate change

Climate change is an important threat to food security and world peace, as it changes both the supply of and demand for water. It affects the production of crops and the supply of food materials, and leads to high costs of adapting to climate change. Currently, climate change effects on world production of food are minimal but, in terms of geography, they are unevenly distributed and much damage has already occurred in dry and humid regions of Africa and South Asia, especially in countries which already have a high rate of poverty. Some climate change analyses demonstrate that climate change will affect food scarcity, water accessibility, hunger, and starvation, and that it will widen the gap between production and consumption among developed and developing countries.

Climate change can affect rainfall, runoff, and water accessibility for irrigation in the majority of regions and countries of the world. Decreased rainfall alongside increased temperatures will result in crops needing more water as it evaporates rapidly. Therefore, water scarcity in many regions and countries is intensifying, affecting food production. Today, countries located in the south Sahara and some parts of South Asia are encountering the biggest problems in terms of water scarcity and food security, and are hence disposed to malnutrition, poverty and even starvation. Uncertainty regarding climate change and how irrigation systems should adapt to the changes lead to complicated problems in need of solutions. Identifying quick solutions to combat long-term climate change and its effect on food security is of the highest importance. There are numerous ways to overcome the challenge, including strengthening the interwoven relationship between water, food security, and energy security, researching climate change further and educating populations about new water plans and strategies, reforming managerial weak points in the food security area, encouraging vegetarianism and consuming less meat, and changing lifestyles to promote low energy consumption.

10.3.4 The water, food, and energy nexus

There are many interactions between water, food, and energy, all of which can be investigated in a myriad of contexts. This nexus and the relationships within it have already been analyzed. In this section, an overall perspective is taken to demonstrate the relationships between water, food, and energy.

As shown in Figure 10.2, water is inseparable from food and energy as it provides the means to produce biofuels, hydroelectric and thermoelectric power, and food products. Energy is used to develop and employ water resources. For instance, energy plays a crucial role in desalination, pumping underground water, in food production processes, transportation processes, and in ensuring the fertility of crops. In terms of the effects of land or food crops on water, we can refer to the effects of increased agricultural activities on runoff, deposition, water quality, intensified droughts, and higher flood likelihood. Mining activities can also reduce the fertility and quality of land. Furthermore, land is used for mining activities, biofuel production, wind energy, solar energy, and biomass – which are the effects of land and food on energy.

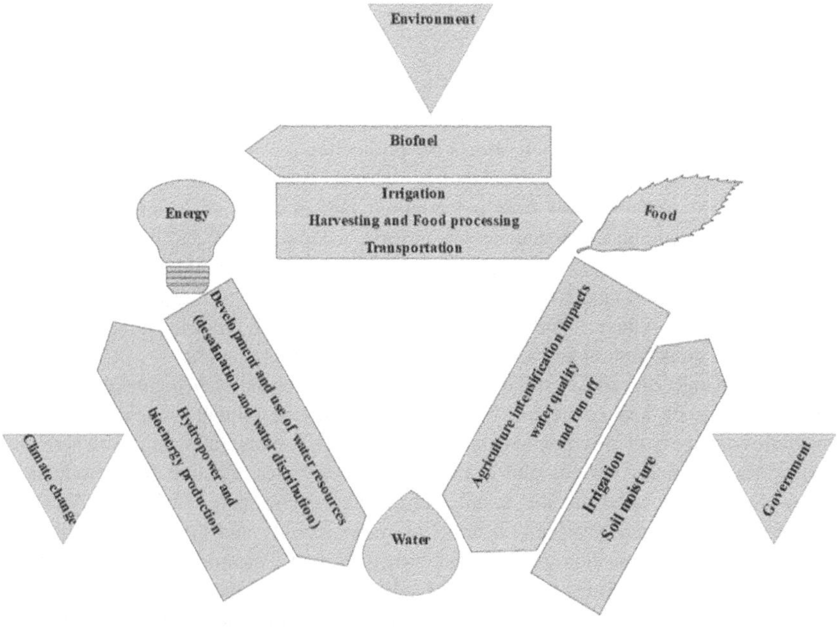

Figure 10.2 The water, food, and energy nexus.

Figure 10.2 shows the externalities affecting the water, food and energy nexus, including climate change, the environment and government. Climate change can affect weather patterns and precipitation levels, whilst government plays a role in facilitating the implementation of the needed infrastructure. Protecting the environment is one of the most important factors in this dynamic system because it is a necessary issue for humans to keep on living.

10.4 USING THE NEXUS APPROACH TO ENSURE WATER, FOOD AND ENERGY SECURITY

Using worldwide statistics, researchers have identified several factors responsible for high demand in food and nutrition, drinking water, and energy in the upcoming decades including population growth, economic development, and technological advancements. Therefore, stress is being imposed on these sectors, increasing the importance of the water, food, and energy nexus concept. Today, agriculture, using approximately 70% of total freshwater (Khokhar, 2017) is considered to be the biggest consumer of water in the world. Water is used in agriculture, food production, transportation and various forms of energy. In order to have a managerial system that leads to sustainable resources, satisfaction of vital needs, and protects the environment, it is crucial to decipher the sophisticated relationship between water, food, and energy. Among the consumers of water, agriculture is considered to be the most important food producer and energy supplier. Therefore, the line between harvesting resources and production of agriculture products needs to be straddled. To accomplish this goal, researchers have offered several special interdisciplinary frameworks to strike a balance between consumption and production – one of which is the water, food, and energy nexus. The nexus approach is an overall vision of sustainability that strives to balance the various goals, interests and needs of people and the environment by quantitative modeling of water, food and energy relations. Supplying food requires almost 30% of total world energy consumption (Scanlon *et al.*, 2017). Population growth and changing eating habits are also expected to increase food demand by 60% by 2050 (Fróna *et al.*, 2019). In terms of energy consumption, by 2050, predictions indicate an 80% increase (OECD, 2012). Energy is needed for production, transportation, and distribution of food. In addition, we need energy to extract, pump, transfer, collect, and treat water. Stemming from higher demands and lower resources of water, food, and energy all over the world, investigating the crises involved in meeting these vital needs has become the top priority for politicians. At local, regional and international scales, many attempts have been conducted to arrive at the best preventative policies, avoiding crises in water, food and energy resources, reaching a high economic efficiency and minimizing environmental harm. In recent years, the water, food and energy nexus has become known as the most essential way to better understand the intricate relations in multi-dimensional resource systems.

10.5 RELATIONSHIPS

To model this dynamic system according to the different goals and spatial scales specified at the beginning of the research, various tools and models can be used, such as WEAP, LEAP, and Vensim. In this process, the governing exchange relationships between the variables and parameters of the system must be specified, and then indicators for the decisions and management of each part of the system defined. Some of the indicators introduced by El Gafy *et al.* (2017) are presented below.

10.5.1 Water and energy consumption indices

Index 1: water consumption index ($W_{c,t}$), relating to water consumption in every hectare of product c in time t.
Index 2: energy consumption index (E_c), relating to energy consumption in every hectare of product c.

Energy consumption for producing food products can be categorized directly or indirectly. Direct energy consumption includes fuel consumption for producing electricity and driving machines and equipment. Indirect energy consumption includes energy consumption to produce fertilizers and chemical materials used in the farm. E_c is obtained from Equation (10.1):

$$E_{c,t} = \sum q_h h_{(c,t)} + q_m m_{(c,t)} + q_d d_{(c,t)} + q_p p_{(c,t)} + q_s s_{(c,t)} \tag{10.1}$$

in which q_h, q_m, q_d, q_f, q_p, q_s, and q_w are the energy equivalents of human labor (J/h), machinery (J/h), diesel oil (J/L), fertilizer (J/kg), pesticides (J/kg), seeds (J/kg), and irrigated water (J/m^3) inputs in crop c production; $h_{(c,t)}$, $m_{(m,t)}$, $d_{(d,t)}$, $f_{(c,t)}$, $p_{(c,t)}$, $s_{(c,t)}$, and $w_{(c,t)}$ are human labor (h/ha), machinery (h/ha), diesel fuel (L/ha), electricity (kWh/ha), fertilizer (kg/ha), pesticides (kg/ha), seeds (kg/ha), irrigated water (m^3/ha) inputs in crop c production at time t.

10.5.2 Productivity indices of water and energy volumes

Index 3: The productivity of water volume at time t based on (ton/m^3) is obtained from Equation (10.2):

$$W_{pro,t} = \frac{Y_{c,t}}{w_{c,t}} \tag{10.2}$$

in which $Y_{c,t}$ is the performance of product c at time t based on (ton/ha) and $w_{c,t}$ is the consumed water for every hectare of product c at time t based on (m^3/ha).

Index 4: The productivity of energy volume at time t based on (ton/J) is obtained from Equation (10.3):

$$E_{pro,t} = \frac{Y_{c,t}}{E_{c,t}} \tag{10.3}$$

in which $E_{c,t}$ is the consumed energy for every hectare of product c at time t based on (J/ha).

10.5.3 Economic productivity indices of water and energy

Index 5: Economic productivity from irrigation water at time t based on ($\$/m^3$) is obtained from Equation (10.4):

$$W_{EV,t} = N_{c,t} - \frac{C_{c,t}}{w_{c,t}} \tag{10.4}$$

in which $N_{c,t}$ is the cultivated product in every hectare of product c at time t based on ($/ha) and $C_{c,t}$ is the additional cost for every hectare of product c at time t.

Index 6: Economic productivity of energy for product c at time t based on ($/J) is obtained from Equation (10.5):

$$E_{EV,t} = N_{c,t} - \frac{C_{c,t}}{E_{c,t}} \tag{10.5}$$

10.5.4 The water, food, and energy nexus index (WFENI)

El Gafy *et al.* (2017) also introduced the water, food, and energy nexus index (WFENI). This index gives an overall picture of efficient management in the water, food and energy nexus for policymakers. WFENI is obtained from Equation (10.6):

$$WFENI_t = \sum_{i=1}^{n} w_i X_i \Big/ \sum_{i=1}^{n} w_i \tag{10.6}$$

The WFENI indicators were normalized to exclude the inflection of different dimensions, applying the Min–Max normalization technique in Equations (10.7 and 10.8). Equation (10.7) is used when the $\text{Min}(x_i)$ of the indicator is the least preferred value and $\text{Max}(x_i)$ is the most preferred value, where Equation (10.8) is used for the opposite situation. Equations (10.7 and 10.8) are as follows:

$$x_i = \frac{x_i - \text{Min}(x_i)}{\text{Max}(x_i) - \text{Min}(x_i)} \tag{10.7}$$

$$x_i = \frac{\text{Max}(x_i) - x_i}{\text{Max}(x_i) - \text{Min}(x_i)} \tag{10.8}$$

in which x_i is the normalized index and x_i is the value index. $\text{Min}(x_i)$ and $\text{Max}(x_i)$ show the lowest and highest index values, respectively; w_i is the weight used for each index and n is the number of WFENI indices. The highest value in the best case scenario is 1 and the lowest value in the worst case scenario is 0.

10.5.5 The total consumption of water and energy

The total consumed water per product in one hectare is obtained from Equation (10.9):

$$W_t = \sum_{c=1}^{v} A_{c,t} \times w_{c,t} \tag{10.9}$$

in which $A_{c,t}$ is the cultivated area of product c at time t.

The total consumed energy per product in one hectare is obtained from Equation (10.10), in which v is the number of the studied product:

$$E_t = \sum_{c=1}^{v} A_{c,t} \times E_{c,t} \quad (10.10)$$

10.6 NEXUS CASE STUDIES

10.6.1 Lake Winnipeg in Canada, Yellow River in China, and Indian basins

Lawford *et al.* (2013) discussed the major factors influencing the water, food and energy nexus, and how they are understood in various basins. Their findings stem from a poll conducted by Global Catchment Initiative in alignment with the Global Water System Project in basins in the Yellow River in China, Lake Winnipeg in Canada, and several basins in India. This study demonstrates how the population of a river can reap the benefits of the water, food and energy nexus more efficiently through improved management of water resources. According to the analysis of this study, the water, food and energy nexus can prove a useful tool in moving the goals of sustainability forward. The analysis includes a ranking of the issues related to a river in the context of the water, food and energy nexus, approved by experts. Summaries of the status of the three vital resources – water, food, and energy– in the basins discussed by Lawford *et al.* (2013) are presented below.

10.6.1.1 Lake Winnipeg in Canada

Lake Winnipeg basin, which is a basin with a high land to surface water area ratio (1/40), covers an area of 9,500,000 km^2. A runoff with an area of 245,000 km^2 finds its way into the lake. With the exception of Southern Alberta (in the Saskatchewan River basin), irrigation is not used broadly, as the majority of agricultural lands are rainfed. Agriculture in the basin has provided the means for exporting products. In Manitoba alone, 7.7 billion dollars of agricultural products were exported to 178 countries in 2009. In producing these products, a significant amount of fertilizer is used, leading to residues like nitrogen and phosphorus which, along with other waste, find their way into Lake Winnipeg. The production of hydroelectricity occurs in Manitoba. The electricity capacity in rivers leading to Lake Winnipeg is 1268 MW, while at the outlet of Lake Winnipeg and the Hudson Bay there is an additional 3494 MW capacity for producing hydroelectricity. Hydroelectricity is usually delivered to consumers in Manitoba, followed by allocation to other basins. The rest of the energy needs in the basin are provided by fossil fuels such as coal. The US, and to a lesser degree Canada, promote biofuel as a part of their energy policies. These policies use some parts of agriculture to produce ethanol, resulting in higher food prices.

Given the nutrients spreading into the lake, the propagation of algae is the biggest environmental challenge faced in summer. Satellite data is used to monitor the algae's propagation and the amount of sediment in the lake. Algae shoots have been known to spread across the lake, covering an area of 15,000 km^2 (60% of the lake) in summer, clogging the beach. These algae have negatively impacted the entertainment and tourism aspects of the lake, lowering its value. A bio-economy approach backed by improved water management should be adopted to solve this issue.

10.6.1.2 Yellow River basin in China

Like other Asian basins, the Yellow River basin is under severe pressure to provide food for its ever-increasing population. The area of the Yellow River basin is 443,752 km^2 and its average runoff is approximately 47.38 billion m^3. Lake Winnipeg and the Yellow River experience a similar range of precipitation, whilst water demand is higher in the latter. In this region, water consumption increased from 10 billion m^3 in 1949 to 37.5 billion m^3 in 2006. The Yellow River basin makes up 17% of the agricultural land in China, producing over 75 million ton of agriculture products, constituting 16% of the total national production. The agricultural areas equipped with an irrigation system have increased tenfold, reaching over 7.5 million hectares in 2000. A significant portion of electricity in the basin is provided by coal power plants; however, there are important hydroelectricity power plants in the middle of the river capable of a 33.4 MW capacity. Despite having wetlands in the middle and lower portion of the river which are protected by the environment, the river lacks sufficient water to satisfy the environmental needs of the wetlands. Wetlands are varied and essential habitats, covering 70% of the Yellow River delta. Approximately 1 billion ton of sediment reaches the river. While a large part of this is moved to the ocean, the remaining 400 million ton requires a current of 14 billion m^3 to drain it (over 25% of the current). The accumulated sediment leads to a 5–10 cm increase in the river bed, hence leaving it vulnerable to flood.

10.6.1.3 River basins in India

The existence of river basins in India provide the basis for the integration of water, food and energy policies. Irrigation, a key issue in the water, food and energy nexus, plays a crucial role in agriculture policies. Groundwaters are the main source of irrigation in some states, requiring significant energy to be pumped to the surface. In the Indus basin, the use of underground waters is higher than the supply, lowering the water table level and threatening the sustainability of groundwaters and the quality of agricultural lands.

In Punjab, Earth Observations help with monitoring groundwater and lowering of the water table level (Rodell, 2005). These studies and observations should be accompanied by regional studies in order to evaluate the causes of the problem, offering possible solutions.

Increasing food and energy prices have led to an increase in financial aid in the majority of the northern region of India. This policy helps agriculture against the negative consequences of water consumption. To reduce the drainage of groundwaters, urgent implementation of political actions is needed, such as monitoring groundwater usage and giving financial aid. Apparently, in India a lack of collaboration in water, food and energy policies has prohibited usage of all three. Lawford *et al.* (2013) suggested a list of solutions for improving collaboration on nexus policies, including:

(1) improving irrigation productivity;
(2) developing hydroelectric power resources;
(3) increasing water price to reduce losses and to increase productivity;
(4) encouraging the government to pass laws and establish organizations for improving water management;
(5) creating an energy price structure for groundwater pumping to promote conservation;

(6) establishing multi-purpose projects to produce energy, regulate the irrigation laws and control flood;
(7) allocating sufficient water to preserve the ecosystem.

Investigation of the food, water, and energy nexus included ranking of factors by experts based on their influencing power, as follows:

(1) climate change: apparently, changes in temperature, precipitation, runoff and their relationship with carbon dioxide influence the majority of basins.
(2) political and economic changes: changes in water accessibility and food and energy production stemming from the world economic conditions and government policies.
(3) economic and regional development: changes in water accessibility and food and energy production stemming from regional economic and investment changes.
(4) demography: population changes lead to changes in water, food and energy utilization, impacting the balance between demand and resources.
(5) urbanization: the impact of growing cities on water, food and energy needs resulting from an increase in water infrastructure, healthcare services, and land cover changes.
(6) changes in land use: the changes stemming from transforming forests into agricultural lands and agricultural lands into industrial complexes.
(7) basin infrastructure: changes in water pathways and storage in order to satisfy the current needs of the basin.

10.6.2 Greater London

In this example, the environment – which is one of the factors heavily involved in the water, food and energy nexus and is crucial in reaching sustainable development – is discussed (Walker *et al.*, 2014).

Greater London encompasses 32 boroughs (plus the City of London), with an area of 1572 km^2. In 2009, its population was about 8.7 million and it is expected that it will reach approximately 9 million by 2030. There has not been a significant change in land use in the last decades. In 2010, land cover was dominated by the city (63%) and the rest belonged to landscape (24%), forests (6%), agricultural products (5.5%) and surface water (1.5%). It should be noted that this study investigated London changes in 2010.

London produces 1400 MW electricity, distributing 400 MW in Enfield and 1000 MW in Barking and Dagenham. This can be turned into a constant production capacity of 12,200 GW annually, covering more than 3 million consumers (based on average family consumption in 2010). This is, however, lower than 30% of the total needed electricity for industrial, business, and local consumption; therefore, the rest has to be imported. Another source of electricity is burning waste, which occurs in three centers, in Beckton (North East), Crossness (South) and Mogden (South West); the total waste produced in the aforementioned centers is 48%, 28%, and 25%, respectively; Beckton has a capacity of producing 8 MW electricity based on sewage sludge. (Following European laws on wastewater disposal demanding a decrease in disposing of wastewater in the sea, these three main wastewater treatment stations in London performed their waste disposal with energy recycling. Therefore, Walker *et al.* (2014) assumed that all the waste produced in London is burned.

England has adopted laws for decreasing carbon emissions. The purpose of decreasing carbon emissions is to improve energy productivity and decrease emissions from big public and private organizations, especially water and electricity. London has committed to decrease its carbon emissions by 60% by 2025, by comparison to 1990. In addition, besides the obvious environmental, societal and economic policies, London has adopted policies in line with climate change. One of the policies about reducing carbon emission requires at least 20% of the city's energy to be produced using renewable resources.

Based on population predictions made by the National Statistics Office in 2008, it is expected that London's population will reach 9 million by 2030, indicating a 15% increase in comparison to 2010. This has a direct impact on water and food consumption, waste production and energy demands. Noticeable attempts have been made to reduce local demand for water by increasing the efficiency of system equipment and minimizing leakage. Higher water demands and 23% leakage lead to higher pressure on water distribution networks. The water authorities believe that by 2030, daily water consumption per capita will reduce to 120 liters, and by 2050 to 110 liters.

In this example, urban mechanisms and the interactions between five energy resources (water, energy, carbon (C), nitrogen (N) and phosphorus (P)) circulating around five economic segments (water, energy, waste, paper, and forest) are discussed using Multi-sectoral Systems Analysis (MSA). For a city as big as London, a higher share for these mechanisms is expected; however, analyses demonstrate that a significant amount of N resources in urban mechanisms belong to natural gas imports for purposes other than producing electricity. In addition, the maximum amount of P entering the city is segregated at the output sludges, entering wastewater treatment stations.

When performing MSA, the benefits and conflicts among various technologies are of high importance. For London and many cities with a similar waste system, there is a myriad of equipment and methods for waste treatment (like the recovery of nutrients P and N in the upstream system, especially through urine separation toilets (UST) and downstream algae wastewater (AWW) treatment plant systems for biofuel production). Analyses show that the UST method has reduced water and energy consumption and recovered a staggering amount of nutritious content, especially N. Given the financial benefits (first and foremost, the saved cost because of reduced water consumption for flushing), UST will be a crowning achievement.

A method called 'pyrolysis of separated sewage sludge' (PSS) helps with P recovery and produces renewable fuel through the breakdown of sewage sludge. Another method known as 'consolidation and co-treatment of household organic waste' (COW) is used to recover C through recycling food waste at waste treatment stations.

MSA results show that the best use of combined technologies is to increase C concentrations through COW, and then recycle carbon-based fuels using the AWW and PSS methods, marking a fruitful collaboration between the three methods.

10.6.3 Climate change pressures in the Middle East and north Africa

In addition to an ever-increasing population, climate change is one of the most significant factors threatening the security of water, food, and energy, and this section investigates climate change effects on all factors in the nexus.

The Middle East and North Africa are among the hottest places on earth, showing high vulnerability to climate change. In an exclusive report from the Intergovernmental Panel on Climate Change (IPCC) on regional climate change, models predicted that the temperature will rise about 1 to 2°C from 2030 to 2050. Authorities and citizens in all countries have realized that, to adapt themselves to climate change, they should take the urgency of changing the current policies on energy, agriculture, water, and environment seriously. Climate Engine satellite data taken from 1979 until now show that the average maximum temperature is around 25°C in winter and 35°C in summer, with a 0.7°C increase from 1900 to 1996. The Middle East and North Africa region is noticeably becoming hotter, with the IPCC issuing warnings that the rise in temperature might go further, making some areas uninhabitable and reducing agricultural production. Cities can already sense the increase in temperature and the majority of big cities in the Middle East are facing the hottest days. The temperature rise puts a staggering amount of pressure on crop production, reducing environmental resources and increasing migration, conflicts and inadaptability. Following temperature rise, intense and irrevocable climate incidents might occur. For instance, Cyclone Phet, the most intense cyclone occurring in Oman in 2010, led to significant casualties of 44 dead and 700 million dollars of damage (Scherr, 2018). Below, we look at the immediate effects of climate change on the environment, energy, agriculture, water and well-being, reported by the World Bank in this region (World Bank, 2017).

10.6.3.1 Climate change effects on the environment

Foliage patterns indicate that climate change effects in semi-arid regions are higher than in dry ones. Managerial options such as improved management of raw materials and the integration of agricultural mechanisms can ameliorate land conditions, eliminating the stresses resulting from climate change.

10.6.3.2 Climate change effects on energy

The Middle East and North Africa (MENA) countries have the highest capacity in terms of wind and solar energies. Taking advantage of such renewable energies can work wonders for Arabian countries to reduce losses related to current energy systems. Using wind and solar energies also helps with electricity production. In addition, it is expected that in the upcoming decades the demand for such energies will increase as a result of population growth and economic development and also, as a result of the need to cool the environment.

10.6.3.3 Climate change effects on agriculture

Agricultural productivity is impeded by limiting factors such as water scarcity and environmental problems, threatening the food security of a few countries. The factors responsible for the 2°C increase in temperature are becoming more intense, resulting in a revised predicted increase of 4°C. Abnormal heat can affect local food production, worsening the livelihood of villagers, hence encouraging their migration to the cities. Granted, there are not enough estimations of the effects of climate change on fiber and food production, but studies predict an increase in some food products. Implementing managerial policies during crucial times and using multi-segmented production systems will lead to higher food security in the region.

10.6.3.4 Climate change effects on water resources

The Middle East and North Africa face the highest water scarcity in the world. Water scarcity has already been identified as a crisis in these dry areas and it is possible that the situation will get worse as climate changes progress. Considering the two-fold population growth predicted by 2070, the region will face a staggering demand for food and water, resulting in constant pressure on resources and increased societal unrest in political arenas. Fast population growth and urbanization lead to more intense pressure on natural resources. It is predicted that as precipitation reduces by 20 to 40%, the region will become 2°C hotter, becoming 4°C hotter with 60% reduction, hence the region's capability to handle the water needed and improve livelihoods will be tested rigorously. In some countries, attempts have been made to preserve or recover water, and such attempts might overcome the deficiencies, especially if they are implemented all over the region. Changes in cultivation methods and improving irrigation can lead to a tremendous decrease in water consumption.

10.6.3.5 Climate change effects on health and well-being

The concept of 'health' varies from region to region, reflecting the local economy. In countries with a higher rate of poverty, infant death is high and life expectancy is low. Climate change harms people through hyperthermia and the diseases resulting from a lack of water in the body. Water not being accessible and problems with food production (especially, if the water scarcity affects irrigation) lead to an indirect effect on health. In some areas, there are vast expanses of dry and desolate lands that absorb rainfall. Flash floods can block roads, disrupting commuting and travel. The floods of June 2009 in Yemen led to disruption of electricity, which can halt the activities of schools while, in worse situations, climate can also wreak havoc with infrastructure. As well as the well-known effects of these disruptions on the educational performance of students, breaks in sessions can also undoubtedly have ramifications for their learning efficiency. Alongside factors such as the high poverty rate in the region, it is expected that climate change will lead to environmental degradation, increased migration, and destroyed livelihoods. Studies show that all of these problems affect the well-being of children and their share of a good education.

10.6.3.6 Recommendations to alleviate climate change pressures in the MENA region

Climate change is a serious threat and one that should not be glossed over by world leaders. The policies adopted by MENA countries might provide noticeable opportunities to preserve resources, making up for climate change effects. In public sectors, having an efficient framework based on national, regional and international collaborations, together with environment-friendly ideas will be necessary to synchronize climate actions. Active participation of the private and public sectors, of human rights organization and universities in lowering the climate change effects is important. To garner a deeper understanding of the intertwined nexus between water, food, and energy, each segment should first be analyzed in its interactions with the other two, and then the relationship between all three should be investigated.

10.7 SUMMARY

This chapter has considered the complexities of the water, food and energy nexus. Every component of the nexus was explained and its importance has been emphasized. The results of research from all over the world in recent years has been discussed. Given the fact that all the three components are separately connected with each other directly and indirectly, complete understanding of the continuous relationships between the components is required to achieve sustainability. In terms of presenting efficient solutions to preserve this nexus, researchers have suggested a myriad of methods based on various indices such as climate effect, geographical situation, government perspective, the environment, etc., which, in some areas, have been quite successful. In offering solutions, it should be noted that all of them must be based on preserving the eco-system, and on protecting the earth against a multitude of environmental and societal dangers. Political mechanisms should be developed in order to implement a connective approach on various levels. Doing so leads to decreased negative political impacts, resource savings, reduced conflicts, and increased collaborations. Policies should be in line with each other, protecting the eco-system and providing safe and accessible natural resources, especially for poor people. Governments, the UN, and other international organizations should go for improved management and adaptability of the water, food and energy areas. Their attempts should be made by considering the interactions between natural resources, societal and environmental implications, and development in various contexts. In the water, food and energy nexus, economic motivations play a vital role. For instance, reforming prices to better reflect the shortage of resources, improving economic productivity and offering targeted subsidies without any negative consequences for various sectors are all fruitful economic responses. Establishing an interactive approach to issues related to the water, food and energy nexus is one of the top priorities in sustainable development. It should be emphasized that, for any society, long-term stability stems from providing clean water, soil productivity, and energy systems, which are all significantly correlated with economic prosperity.

International organizations, the private and public sectors, and other investment sectors should make sure that their research is based on demand, and invest more in studies about the water, food, and energy nexus. Subjects for these studies should include agricultural crops and products, energy production, water provision and health, and improving water quality and the eco-system. These studies should offer ways to facilitate the use of technologies and provide integrated approaches to the better use of land, water, and energy resources. Research teams should adopt a specific methodology when investigating the water, food and energy nexus, employing different methods to assess and present their findings, including risk assessments, threat indices, and an assessment of environmental and societal indices. Research teams should constantly analyze the nexus and the relationships between its three components when estimating costs and benefits. Such analyses can lead to better and more informed decisions that will support business models, as well as proving useful in helping to accomplish sustainability in the water, food and energy nexus.

REFERENCES

Albrecht T. R., Crootof A. and Scott C. A. (2018). The water–energy–food nexus: a systematic review of methods for nexus assessment. *Environmental Research Letters*, **13**(4), 043002.

Al-Saidi M. and Elagib N. A. (2017). Towards understanding the integrative approach of the water, energy and food nexus. *Science of the Total Environment*, **574**, 1131–1139.

Aro E. M. (2016). From first generation biofuels to advanced solar biofuels. *Ambio*, **45**(1), 24–31.

Bakhshianlamouki E., Masia S., Karimi P., van der Zaag P. and Sušnik J. (2020). A system dynamics model to quantify the impacts of restoration measures on the water–energy–food nexus in the Urmia lake Basin, Iran. *Science of the Total Environment*, **708**, 134874.

Biggs E. M., Bruce E., Boruff B., Duncan J. M., Horsley J., Pauli N., McNeill K., Neef A., Van Ogtrop F., Curnow J. and Haworth B. (2015). Sustainable development and the water–energy–food nexus: a perspective on livelihoods. *Environmental Science & Policy*, **54**, 389–397.

Chang Y., Li G., Yao Y., Zhang L. and Yu C. (2016). Quantifying the water–energy–food nexus: current status and trends. *Energies*, **9**(2), 65.

Cozzi L. and Gould T. (2015). World Energy Outlook 2015. International Energy Agency, Paris.

El Gafy I., Grigg N. and Reagan W. (2017). Water–food–energy nexus index to maximize the economic water and energy productivity in an optimal cropping pattern. *Water International*, **42**(4), 495–503.

Endo A., Tsurita I., Burnett K. and Orencio P. M. (2017). A review of the current state of research on the water, energy, and food nexus. *Journal of Hydrology: Regional Studies*, **11**, 20–30.

European Academies Science Advisory Council (EASAC). (2012). The current status of biofuels in the European Union, their environmental impacts and future prospects. EASAC Policy Report 19. EASAC, Brussels, Belgium.

Ferroukhi R., Nagpal D., Lopez-Peña A., Hodges T., Mohtar R. H., Daher B., Mohtar S. and Keulertz M. (2015). Renewable Energy in the Water, Energy & Food Nexus. International Renewable Agency (IRENA), Abu Dhabi.

Food & Agriculture Organization (FAO). (1996). Rome Declaration on World Food Security and World Food Summit Plan of Action: World Food Summit 13–17 November 1996, Rome, Italy. Food and Agriculture Organization of the United Nations, Rome.

Fróna D., Szenderák J. and Harangi-Rákos M. (2019). The challenge of feeding the world. *Sustainability*, **11**(20), 5816.

Garcia D. J. and You F. (2016). The water–energy–food nexus and process systems engineering: a new focus. *Computers & Chemical Engineering*, **91**, 49–67.

Gude V. G. (2015). Energy and water autarky of wastewater treatment and power generation systems. *Renewable and Sustainable Energy Reviews*, **45**, 52–68.

Hanssen O. J. and Schakenda V. (2013). Nyttbart matsvinn i Norge. Analyser av status og utvikling i matsvinn i Norge 2010–11–Rapport fra ForMat-prosjektet (Efficient uses for food waste in Norway. Analysis of status and developments in food wastage in Norway. Report from ForMat project). Østlandsforskning, Krøkerøy, Norway. (In Norwegian).

Hatfield J. L., Sauer T. J. and Cruse R. M. (2017). Soil: the forgotten piece of the water, food, energy nexus. *Advances in Agronomy*, **143**, 1–46.

Jarvie H. P., Sharpley A. N., Flaten D., Kleinman P. J., Jenkins A. and Simmons T. (2015). The pivotal role of phosphorus in a resilient water–energy–food security nexus. *Journal of Environmental Quality*, **44**(4), 1049–1062.

Kaddoura S. and El Khatib S. (2017). Review of water–energy–food Nexus tools to improve the Nexus modelling approach for integrated policy making. *Environmental Science & Policy*, **77**, 114–121.

Keskinen M., Guillaume J. H., Kattelus M., Porkka M., Räsänen T. A. and Varis O. (2016). The water–energy–food nexus and the transboundary context: insights from large Asian rivers. *Water*, **8**(5), 193.

Khokhar T. (2017). Chart: Globally, 70% of freshwater is used for agriculture. World Bank Data Blog. See: https://blogs.worldbank.org/opendata/chart-globally-70-freshwater-used-agriculture (accessed 26 March 2021).

Kranert M., Hafner G., Barabosz J., Schuller H., Leverenz D., Kölbig A., Schneider F., Lebersorger S. and Scherhaufer S. (2012). Ermittlung der weggeworfenen Lebensmittelmengen und Vorschläge zur Verminderung der Wegwerfrate bei Lebensmitteln in Deutschland. (Determination of the amount of food thrown away and suggestions for reducing the throw-away rate for food in Germany.). University of Stuttgart, Stuttgart. See: http://www.bmelv.de/SharedDocs/Downloads/Ernaehrung/WvL/Studie_Lebensmittelabfaelle_Langfassung.pdf (accessed 27 May 2021). (In German).

Lawford R., Bogardi J., Marx S., Jain S., Wostl C. P., Knüppe K., Ringler C., Lansigan F. and Meza F. (2013). Basin perspectives on the water–energy–food security nexus. *Current Opinion in Environmental Sustainability*, **5**(6), 607–616.

Mercure J. F., Paim M. A., Bocquillon P., Lindner S., Salas P., Martinelli P., Berchin I. I., de Andrade Guerra J. B. S. O., Derani C., de Albuquerque Junior C. L. and Ribeiro J. M. P. (2019). System complexity and policy integration challenges: The Brazilian energy–water–food nexus. *Renewable and Sustainable Energy Reviews*, **105**, 230–243.

Mohtar R. H. and Daher B. (2016). Water–energy–food nexus framework for facilitating multi-stakeholder dialogue. *Water International*, **41**(5), 655–661.

Monier V., Mudgal S., Escalon V., O'Connor C., Gibon T., Anderson G., Montoux H., Reisinger H., Dolley P., Ogilvie S. and Morton G. (2010). Final report–Preparatory study on food waste across EU 27; European Commission (DG ENV–Directorate C). BIO Intelligence Service, Paris.

OECD. (2012). Energy and Climate Policies and Innovation. OECD Publishing, Organisation for Economic Co-operation and Development, Paris, France.

Owen A., Scott K. and Barrett J. (2018). Identifying critical supply chains and final products: an input–output approach to exploring the energy–water–food nexus. *Applied Energy*, **210**, 632–642.

Ozturk I. (2015). Sustainability in the food–energy–water nexus: Evidence from BRICS (Brazil, the Russian Federation, India, China, and South Africa) countries. *Energy*, **93**, 999–1010.

Ramaswami A., Boyer D., Nagpure A. S., Fang A., Bogra S., Bakshi B., Cohen E. and Rao-Ghorpade A. (2017). An urban systems framework to assess the trans-boundary food–energy–water nexus: implementation in Delhi, India. *Environmental Research Letters*, **12**(2), 025008.

Ravar Z., Zahraie B., Sharifinejad A., Gozini H. and Jafari S. (2020). System dynamics modeling for assessment of water–food–energy resources security and nexus in Gavkhuni basin in Iran. *Ecological Indicators*, **108**, 105682.

Rodell M. (2005). India's water economy: bracing for a turbulent future. Report No. 34750-IN. World Bank, Washington, DC.

Scanlon B. R., Ruddell B. L., Reed P. M., Hook R. I., Zheng C., Tidwell V. C. and Siebert S. (2017). The food-energy-water nexus: Transforming science for society. *Water Resources Research*, **53**(5), 3550–3556.

Scherr L. A. (2018). Water–Energy–Food Nexus/Climate Change Impacts in the MENA Region. See: https://www.water-energy-food.org/news/water-energy-food-nexus-climate-change-impacts-in-the-mena-region (accessed 16 March 21).

Serrano A., Siles J. A., Chica A. F. and Martín M. Á. (2013). Agri-food waste valorization through anaerobic co-digestion: fish and strawberry residues. *Journal of Cleaner Production*, **54**, 125–132.

Smajgl A., Ward J. and Pluschke L. (2016). The water–food–energy Nexus: realising a new paradigm. *Journal of Hydrology*, **533**, 533–540.

Stamou A. T. and Rutschmann P. (2018). Pareto optimization of water resources using the nexus approach. *Water Resources Management*, **32**(15), 5053–5065.

Tian H., Lu C., Pan S., Yang J., Miao R., Ren W., Yu Q., Fu B., Jin F. F., Lu Y. and Melillo J. (2018). Optimizing resource use efficiencies in the food–energy–water nexus for sustainable agriculture: From conceptual model to decision support system. *Current Opinion in Environmental Sustainability*, **33**, 104–113.

United Nations. (2013). The Millennium Development Goals Report 2012. United Nations, New York.

United Nations. (2016). Sustainable Development Goals Report 2016. United Nations, New York.

United Nations Development Programme. (2010). Energy for a Sustainable Future: The Secretary-General's Advisory Group on Energy and Climate Change, Summary Report and Recommendations. United Nations, New York.

Walker R. V., Beck M. B., Hall J. W., Dawson R. J. and Heidrich O. (2014). The energy–water–food nexus: Strategic analysis of technologies for transforming the urban metabolism. *Journal of Environmental Management*, **141**, 104–115.

Water in the West. (2013). Water and Energy Nexus: A Literature Review. Water in the West, Stanford University, CA.

Weitz N., Strambo C., Kemp-Benedict E. and Nilsson M. (2017). Closing the governance gaps in the water–energy–food nexus: insights from integrative governance. *Global Environmental Change*, **45**, 165–173.

Wichelns D. (2017). The water–energy–food nexus: Is the increasing attention warranted, from either a research or policy perspective? *Environmental Science & Policy*, **69**, 113–123.

Willis H. H., Groves D. G., Ringel J. S., Mao Z., Efron S. and Abbott M. (2016). Developing the Pardee RAND Food-Energy-Water Security Index: Toward a Global Standardized, Quantitative, and Transparent Resource Assessment. RAND Corporation, Santa Monica, CA. See: https://www.rand.org/pubs/tools/TL165.html (accessed April 2021).

Wong K. V. (2015). Energy–Water–Food Nexus and Recommendations for Security. *Journal of Energy Resources Technology*, **137**(3), 034701.

World Bank. (2017). *Climate Change in the Middle East & North Africa*. See: https://www.worldbank.org/en/programs/mena-climate-change# (accessed 21 May 2020).

World Economic Forum. (2013). Global Risks 2013. 8th edn. World Economic Forum, Geneva, Switzerland.

Yang Y. E., Wi S., Ray P. A., Brown C. M. and Khalil A. F. (2016). The future nexus of the Brahmaputra River Basin: climate, water, energy and food trajectories. *Global Environmental Change*, **37**, 16–30.

Zarei M. (2020). The water–energy–food nexus: a holistic approach for resource security in Iran, Iraq, and Turkey. *Water–Energy Nexus*, **3**, 81–94.

Zhang X. and Vesselinov V. V. (2017). Integrated modeling approach for optimal management of the water, energy and food security nexus. *Advances in Water Resources*, **101**, 1–10.

Chapter 11

Water security

Sahar Baghban[1], Omid Bozorg-Haddad[2] and Ronny Berndtsson[3]*

[1]Water Engineering M.Sc. Student, Department of Irrigation & Reclamation Engineering, Faculty of Agriculture Engineering & Technology, College of Agriculture & Natural Resources, University of Tehran, 3158777871 Karaj, Iran

[2]Distinguished Professor, Department of Irrigation & Reclamation Engineering, Faculty of Agricultural Engineering & Technology, College of Agriculture & Natural Resources, University of Tehran, Karaj, Tehran, 31587-77871, Iran. E-mail: OBHaddad@ut.ac.ir

[3]Professor, Department of Water Resources Engineering and Center for Middle Eastern Studies, Lund University, Lund, Sweden

**Corresponding author.*

ABSTRACT

Today, the concept of security manifests itself in a number of forms in most scientific, social, environmental, and political issues. One of the most essential and prominent types of security issue faced today is water security, which has introduced a complex and new concept to the literature of today's world. Not only has the current water crisis in most parts of the world increased the risk of military conflicts between countries, but the value of safe water will become even more apparent in the coming decades, which is why water security will have a special importance in the modern world. Water security is a concept that has attracted particular attention in recent decades from various dimensions. Estimates show that by 2100, the planet's population will increase from the current 7.7 billion (7.7×10^9) to 11.2 billion, and that providing adequate safe water for such a population is one of the significant challenges facing humans – because the world's freshwater resources are not just for individual consumption and for the environment, but

Editor: Omid Bozorg-Haddad
doi: 10.2166/9781789062144_0205

they are also vital for the agricultural, energy, industrial, and transportation sectors. Therefore, it should be noted that the existing world water resources, both in terms of quantity and quality, are at a critical stage. However, the severity of this problem varies from region to region, with regions such as the Middle East and South Africa at greater risk of water scarcity. Water security is therefore a vital issue for all of us. This chapter first outlines the concept of water security from different perspectives, emphasizing that water security is a multifaceted issue, before finally examining solutions and policies to address to the challenges.

Keywords: Water resources, Water security, Population growth, Climate change, Terrorism.

11.1 INTRODUCTION

Water is an integral part of every ecosystem and the primary material of life used for agriculture, industry, drinking, and the environment. Today, water scarcity and pollution have seriously endangered the lives of millions of people living on the Earth, especially in developing countries, and no country can maintain its economic, political, and social stability without regard to water and related issues. Historically, many ancient civilizations have been formed along the great fertile valleys such as the Nile, Sand, Yellow River, Tigris and Euphrates. This enforces the message that we are dependent on water because, not only is water alone essential for human survival, but human food supply is also strongly related to water. For this reason, the agricultural sector accounts for an average of 70% of the freshwater resources available on Earth (Khokhar, 2017). In recent decades, population growth has posed a severe threat to the human world, with respect to water resources. Today, 80% of the world's population have access to only 20% of safe water resources, and if we take into account factors such as climate change and the rapid growth of industry, we find that available and safe water will significantly decrease in the near future. Hence, the pressure on freshwater resources will continue to increase. In general terms, there are two main approaches to global water security: a developmental approach and a risk-based approach. The developmental approach seeks to increase water security over time, as a result of a combination of policies and projects. The risk-based approach seeks to increase water security through risk management and increase the system's reversibility in view of climate change and water disasters (van Beek & Arriens, 2014). Remarkably, water risks account for almost 90% of natural hazards worldwide (Nazemi & Madani, 2018).

11.1.1 State of the art in water security studies

In 2009, the World Economic Forum (WEF) prioritized water security as a global risk, stating that 'water security is the gossamer that links together the web of food, energy, climate, economic growth, and human security challenges that the world economy faces over the next decades' (Catley-Carlson, 2009). The concept of water security has different meanings from different perspectives, but in all the definitions it is proper water management that leads to its security, expressed as an improvement in quality of life for everyone. It is a concept that brings together politicians, business leaders, water experts, and many different disciplines and interest groups with a common language. Many types of research have been undertaken in different case studies on the water security concept, in each of which different perspectives on water security have been considered. In general, the studies conducted have included urban water security, issues related to the quantity and quality of surface and groundwater resources, water security with an emphasis on the supply management of drinking, agriculture and environmental requirements whilst also addressing political and socio-economic issues, as well as the impact of climate change on water security. In this section, this research into water security and related issues will be discussed.

Vörösmarty *et al.* (2010) studied global threats to human water security and biodiversity in rivers, examining 47 major rivers in the world with the highest freshwater supply. The authors concluded that to protect all freshwater resources globally, identifying and finding threats on a wide range of scales (global to local) are necessary. The study used a combination of commonalities in human perspectives and biodiversity on water security, using a theoretical framework that examines multiple stressors and the extent of downstream impacts. The study found that nearly 80% of the world's population are at high risk for water security. Wealthy countries with considerable investments in water technology can solve many stressors, but countries with less wealth remain vulnerable. The study organized stresses as spatial stimuli under four themes: catchment disturbance, pollution, water resources development, and biotic factors. Therefore, to ensure global water security, management policies should be presented to eliminate the signs of threats to humans and biodiversity.

Gerlak and Mukhtarov (2015) investigated Integrated Water Resources Management (IWRM) and water security as complementary processes. The study reveals how IWRM can achieve water security. In this regard, the authors confirm that IWRM is a tool for planning and managing water based on technical–scientific knowledge, but water security is an argumentative method for recognizing water according to human values, ethics, and power. Therefore, these two methods are complementary.

Wegerich *et al.* (2015) investigated water security in the Syr Darya Basin – the Garlic Basin in Central Asia. This case study addresses the vital role of water supply in achieving sustainable water security and water supply management for irrigated agriculture, and considered two parts: the first on securing water supply at sea level, and challenges related to the past and current irrigation water management strategies for the irrigation area; and the second on water security approaches in Ferghana province, where water needs are changing. Security approaches highlight water, and there is a strong emphasis on the irrigation sector, which has not achieved water security since Uzbekistan became independent. The authors conclude that the existing rules of water resources management should be amended to effectively address the critical challenges of water security.

Jia *et al.* (2015) investigated improved water security in an integrated way. Their case study was conducted in the Yellow River Basin in China. Since water is a major limiting factor for human survival and socio-economic development, water security is crucial for the ecological system. An index for the water security of the Yellow River basin was developed based on three subsystems: water resources, environment, and socio-economics. An integrated water security assessment was performed based on a combination of different processes in four stages. Finally, an integrated water security assessment model was selected based on the evaluated models' rating values and weight coefficients. Based on the results of the developed integrated assessment model, they concluded that: Sichuan province can be regarded as secure; Henan, Shandong, Qinghai, and Shaanxi provinces have reached the security threshold; Shanxi, Gansu, and Inner Mongolia are relatively insecure; and Ningxia province is without water security.

Jiang (2015) studied the current situation, emerging challenges, and prospects for water security. This case study was conducted in China, where five key aspects of water sustainability were used to examine the current state of risk-based water security, including water availability, water use patterns and pollution control, water institutions and management, aquatic system health, and social vulnerability. The study analyzed future challenges to water security, including climate change, population growth, rapid urbanization, and the water–food–energy nexus. The author concluded that China has the potential to achieve long-term water security, which can be achieved through government control and a firm policy, massive investment, and ongoing water management reforms.

Wheater and Gober (2015) studied water security and the science agenda. The case study was conducted in the Saskatchewan River Basin in Canada. The freshwater environment is facing unprecedented global pressure. Excessive use of surface water and groundwater is widespread. In this study, the water science community's role in addressing water security challenges was examined, requiring a deep understanding of aquatic and terrestrial environments and their interactions with the climate system and the analysis of vulnerabilities to environmental and social change. Scientific products must emerge from a two-way exchange, with bilateral cooperation with management and political communities for solving real-world issues. The authors concluded that, as tensions in water resource systems increase, science needs to understand how we improve the system and that water science should help humans better understand all the dimensions of water systems.

Varady *et al.* (2016) investigated the conditions and evolution of adaptive water management and water security. The most appropriate and tangible water management is based on sustainability and integrated management of water resources. They cited water security in the absence of a clear water management goal and concluded that both concepts need to be measured and monitored to determine progress to assist in decision-making. They concluded that water security as a framework provides comprehensive goals and measures progress in achieving these goals concerning the relationship between systems and sectors.

Romero-Lankao and Gnatz (2016) studied the conceptualization of urban water security. They discuss existing frameworks, criteria, and aggregation methods based on national-level models and criteria. This approach focuses on the gaps between research, and considers urban dynamics with regional, national, and global perspectives while ignoring local metropolitan areas. Such approaches can help move urban-regional water security goals toward greater coherence and appropriate policy and intervention. The hybrid approach to aggregation indicators and methods highlights the multidimensional and interconnected nature of urban water security features, and shows the different levels of impact in the five dimensions of SETEG (Social, Economic, Technological, Ecological, and Governance) development. They conclude that, to create water security and sustainability, a bridge should be built between people working on urban dynamics and water security.

Kirchhoff *et al.* (2016) investigated water security and adaptive capacity. The study was conducted in arid regions of the Americas (the USA, Mexico, and Argentina). Freshwater resources are under tremendous pressure to meet the needs of humans and ecosystems in the face of climate change. This pressure has raised concerns about water resources and put improving water security on the global agenda. The study adds new dimensions to water security by proposing a theoretical model that considers the commonalities and interrelationships between water security, adaptive capacity, and adaptive water management. The results show that water security indicators should consider adaptive capacities and flexible water management structures to achieve comprehensive water security.

Gain *et al.* (2016) investigated global water security for sustainable development. Water plays a critical role in creating equitable and sustainable societies and ecosystems, and thus the United Nations recognizes water security as one of the 17 goals for achieving sustainable development. In this research, a multi-criteria analytical framework is presented to provide a global assessment of water security. Their assessment shows that countries in Africa, South Asia, and the Middle East experience shallow water security, whilst in other areas with severe water scarcity (including parts of the United States, Australia, and southern Europe), due to good performance in management, safety, quality, accessibility are better indicators of global water security. They concluded that the Global Water Security Index should be used to progress towards sustainable development.

Zeitoun *et al.* (2016) examined the complex challenges of a water security policy with mitigation and integration approaches. They studied two approaches that water security researchers use to understand the complexity of water-community policy challenges. The reductionist approach seeks to show uncertainty by calculating and linking Gross Domestic Product (GDP) to hydro-meteorological factors, biodiversity, and politics in society. The second approach is more integrated, to address a wide range of uncertainties, explicitly recognizing diversity in society and the environment. They concluded that water security should remain a term and remind us that there is no single, decisive, and unified response to the challenges of water, society, and climate.

Elgert *et al.* (2016) investigated water security improvement through a collection network in Guatemala. Urban infrastructure is often not available to supply water to communities living in and around rural areas. Given that rainwater harvesting can improve the three dimensions of water security (quantity, quality, and accessibility); they concluded that the system had improved water security; however, concerns remain about the project's potential for expansion, including system cost, water quality, and individual responsibility for water security.

Lu *et al.* (2016) investigated urban water security assessment in Chengdu, China. The Analytic Hierarchy Process (AHP) and Delphi methods were used to weigh an index to assess urban water security levels. Among nine indicators for evaluating urban water security, the values of environmental and socio-economic indicators showed that there is relatively more significant pressure on the city's social, economic, and environmental sectors. On the other hand, the Water Security Assurance Index (WSAI)' high values in water resources and environmental water pressure indices showed the model could be used to assess the extent of urban water security support quantitatively.

Veettil and Mishra (2016) studied water security assessment using blue and green water footprints in the Savannah River Basin in the United States. They applied the Soil and Water Assessment Tool (SWAT) to quantify freshwater availability (blue water and green water) in this basin. Quantitative assessment of water security using the concept of blue and green water footprints can improve water resources management at the regional scale.

Human factors (e.g., water demand) and environmental flow needs were combined to quantify water security in terms of scarcity and vulnerability indicators. The authors concluded that climatic factors control spatio-temporal distributions of blue water and, on the other hand, green water flow is influenced by intense agriculture and water bodies (e. g., reservoir); and also the amount of blue water shortage is high in the southern regions.

Cook (2016) implemented water security with limited protection resources in Ontario, Canada. Since the long-term growth and development plans for land use for location-based planning, rules, and policies make water resource management more effective, this study aims to convince people to protect water resources in land use and economic development priorities for resource extraction. This research shows that drinking water security through law and resource protection policy is limited to social water relations. Implementing drinking water security will require a fundamental overhaul of the existing trade balance relationship, and one way to examine water security policy and power is to find social water relations related to water governance.

Ludwig and Roson (2016) investigated the relationship between climate change and water security, conducting their case study in the Middle East. Climate change is affecting the Mediterranean region. Existing trends and future forecasts indicate extreme sensitivity to changes in hydrological regimes, scarcity of water resources and, consequently, threatening the amount of water available and water management. They concluded that current uncertainties in climate forecasts would lead to an incomplete understanding of the impact of climate change signals on economic mechanisms, integrated security concepts, and severe restrictions on water-related climate change decisions due to a lack of public information. The regionality and inconsistency of temporal and spatial scales of operations come from different scientific perspectives.

Maganda (2016) researched water security issues. This case study was conducted in the European Union, with specific focus on the Grand Duchy of Luxembourg. This study examined the availability of and access to water in weak and prosperous countries. It suggests that if norms such as water security or the human right to global water and health are accepted, they should be adopted by wealthy, powerful countries and regions rich in water, and their domestic and foreign policies should be guided by them. If such governments do not support these norms, implementing them will become a severe challenge, and water security debates will also reflect the imbalance of power in world affairs.

Nel *et al.* (2017) investigated strategic water resource areas for urban water security. This case study was conducted in South Africa. Strategic water resource basins are areas with relatively high natural runoff, which support the region's population or economy. The study identified 22 strategic water resources areas in South Africa that were associated with important urban centers. The authors concluded that strategic water resources make up only 8% of the land area but provided 50% of runoff. When connected to urban downtowns, they support at least 51% of South Africa's population and 64% of its economy. However, only 13% of the total land area is protected. They recommended that multiple strategies be used to protect these basins legally. Identifying strategic areas of water resources and their relationship with downstream users provides an opportunity to achieve collaborative planning in different policy areas that enable new cooperation patterns between government, business, and civil society.

Zeitoun *et al.* (2017) studied a war city's environmental status from a water supply perspective. This case study was conducted in the city of Basrah in Iraq, and evaluated the impact of the armed conflict on Basra's drinking water services from 1978 to 2013 through an 'urban warfare ecology' to delineate the implications of planning and linking to urban studies. This analysis indicates a one-step decline in service quality due to a lack of water treatment chemicals, a lack of spare parts, and, in the first place, a 'brain drain' of qualified water service personnel. As a result, professionals refuse to develop Relief-Rehabilitation-Development (RRD) ideas to understand the specific needs of a complex urban warfare space, which affects all aspects of service delivery through biological and social processes.

Powell *et al.* (2017) studied water security in times of climate change. This study is about achieving fair governance of water and its effects in support of sustainable development, based on case studies of international climate change plans and government projects. This paper is based on how situations characterized by water contention (exacerbated by the uncertainties posed by climate change) can be adapted, and the aim was to support a transformation towards both systemic and adaptive water governance. The authors conclude that the adoption of fair laws could effectively manage water resources, which would change the one-dimensionality of international or national negotiations.

Housh and Ohar (2018) investigated models for detecting cyber attacks on water distribution systems. This research presents a model-based numerical method that uses a hydraulic water simulation model to detect cyber-attacks in water distribution systems. The method uses a three-step approach in which: (1) demand is estimated based on a reading of part of the management system and data collection; (2) a hydraulic model is used to review the hydraulic data of the management system and data collection; and (3) a multi-level classification method is applied to classify the errors obtained in typical and external errors. The results showed that the proposed method increased the level of timely warning for all simulated cyber-attack events. The results obtained in this hypothetical, fully calibrated hydraulic model had just one source of uncertainty (network demand data), which was not considered in the calculations of algorithms.

Alvarez *et al.* (2018) researched nanotechnologies to increase water security. Affordable water supply for increased comfort and convenience is a big challenge because the demand for water consumption increases with increases in population and, to prevent it, reusing water is the best answer. Several nanomaterial engineering features can be designed to improve various water treatment processes. Contaminants such as salts, trace chemicals, toxins, pesticides, etc. can be removed with nanotechnologies. If nanotechnology is used in this way, it will be possible for us to use unconventional water resources, which will increase water security.

As a conclusion, much research has been done within the framework of water security, with different aims at a regional, local, and global scale. By studying this research, we can reach a common language to achieve and maintain water security. This chapter concludes with a review of water security, examining it from various perspectives, and evaluates the most critical issues that play a role in achieving it. But before that, we examine how water security can be addressed in the face of political and socio-economic challenges, population growth, and the impact of climate change. The chapter also supplies a vision of how we can better protect the quantity and quality of water resources as efficiently as possible. It also draws out conclusions for a broader consideration of water security and points out that environmental, economic and community security can all be achieved by targeting water security.

11.2 DEFINITIONS AND TERMS

11.2.1 Security

In this context, security is the set of methods, activities, and actions that ensure that an accident either never happens or reduce the likelihood of it happening and, in the event of a dangerous accident, return the situation to normal as soon as possible and at the lowest possible cost.

The definition of security varies depending on what 'security' refers to. In this regard, there are two different views on the 'security' authority, one of which considers government as the security authority and states that 'security is static in nature and oversees the government.' Another concept considers the individual as a source of security and states that 'it is only the foundation of the individual (and not the government) that can be a perfectly suitable context for security studies' (Buzan, 2008).

11.2.2 Water security

In general terms, water security is the reliable access to a reliable volume and quality of water at the right time, on a large scale, to meet the needs of water-dependent systems to the extent that it provides health, assets, and production for living things. Such water security is a fundamental precondition for human security, food, the environment, as well as for economic growth, and sustainable development is not possible without water security. Different definitions from the perspective of different experts are presented below.

According to the World Water Partnership, security at every level – from the family to the global dimension – means that everyone has enough money to access adequate drinking water for a safe, healthy, and efficient life while maintaining the environment (Whiteford & Witter, 1999). In this comprehensive view, the main goal is to meet basic human needs and provide food with sufficient attention to the environment and ecosystem, and water security without sustainable development is doomed to fail and future generations will face serious risk. However, in addition to the scientific dimension of risk, water security studies also approach another critical dimension that can lead to internal or international conflicts: the political dimension.

Many organizations have focused on water security in the context of human needs for water, stating that security is a situation which requires only a sufficient amount of quality water necessary to be provided to people at affordable price to meet the short- and long-term needs of nations. Water security is when a person has access to sufficient and affordable water to meet their needs, including drinking, washing, and living (Rijsberman, 2006).

Of course, this approach in many cases can be dangerous because the importance of the ecosystem and its integral role for both water and humans can be neglected. This is especially the case when the focus is put on agriculture. It is essential to provide adequate food for humans and this is the primary purpose of water security (Falkenmark *et al.*, 2007).

In its definition of water security, the World Bank provides a quantitative standard: the annual supply of 1 m^3 per person per day of sanitary water to meet sanitary needs and 1000 m^3 of water for productive activities in agriculture, industry, and the environment (Imbulana *et al.*, 2002).

Water security from the perspective of the World Water Assessment Program identifies four leading indicators as components of water security: water supply for basic human needs, water supply for civilized security, water supply for environmental protection, and water supply for industry and energy.

The United States Environmental Protection Agency defines water security as a tool to prevent and protect against pollution and terrorism, and US lawmakers, therefore, pay special attention to drinking water infrastructure and related security (Minamyer, 2008).

Water scarcity, water stress, water risk, and water crisis are some of the terms related to water security; but each has a different meaning (see below). In general, concepts related to water security focus on the amount and availability of water. Thus, the most prominent tools for assessing water security are the two concepts of water stress and water scarcity (Falkenmark *et al.*, 2007).

11.2.3 Water scarcity

Water scarcity refers to a situation where water demand exceeds the amount of water available. Water scarcity is related to human demand for water; accordingly, an uninhabited arid area with no human need for water is not referred to as being water-scarce. Water scarcity indicates the physical abundance of freshwater relative to the consumer population, regardless of water resource quality. Consequently, an area may have abundant water resources but the amount of water pollution is such that it is impossible to use that water, even though the area does not suffer from water shortage (Schulte, 2017).

11.2.4 Water stress

Water stress refers to the ability or inability to access clean water to meet human and environmental needs. As a result, this is a broader and more comprehensive concept than water scarcity and includes various physical aspects of water, including the quantity and quality of water and access to it. According to the Falkenmark Indicator, areas with annual water intake of between 1000 and 1700 m^3 per person experience water stress. Water stress evaluates the ratio of water use to the extent of its availability from sources such as rivers, aquifers, etc. (Falkenmark & Molden, 2008).

11.2.5 Water risk

Water risk refers to the possibility of a situation in which an institution such as a government faces water-related challenges such as water scarcity, water stress, floods, droughts, infrastructure collapse, and the like. Water risk is experienced and defined differently for each section of society and the organizations in which they are based. For example, after assessing risk issues in the Caucasus region, a wide range of risks was identified, some of which included the following (Jones *et al.*, 2009):

- the risk of persistent water shortages for the current population and for economic growth;
- increased risk of floods and droughts and their damage;
- risk of contamination of water resources both when there is too much water and when water is scarce;
- resources and options for dealing with natural disasters and technology;
- risks related to the anticipated effects of climate change, especially for regional environmental characteristics and ongoing trends of economic globalization.

11.2.6 Water crisis

A phenomenon is a crisis when it has unpredictable, dynamic, temporary, and uncertain characteristics. For water, a critical situation is one that is out of balance due to some threat to production, distribution, or consumption of water. These threats can also be of natural origin, such as floods, droughts, and earthquakes. Such a threat can also occur due to human factors such as mismanagement, water supply network wear or insufficiency of equipment, and improper consumption. Simply put, a water crisis is a situation in which access to renewable water per person per year is less than the global standard. According to the Falkenmark Indicator, countries with an annual water availability of less than 500 m^3 per capita suffer from absolute water scarcity or a severe water crisis (Liu *et al.*, 2017).

11.2.7 Terrorism

According to the United Nations (UN), immediate human victims of violence are usually randomly targeted (opportunistic targets) or selectively targeted (representative or symbolic targets) from a population, to act as message generators. Terrorism is defined by the United States Department of Defense as: 'Illegal use or threat of violence, often motivated by religious, political, or other provocative ideology, to intimidate or force governments or communities to pursue a specific goal that is primarily politically motivated' (Office of the Chairman of the Joint Chiefs of Staff, 2019). The definition of terrorism by the Federal Emergency Management Agency is 'the use of force or violence against individuals or property in violation of U.S. criminal law to intimidate, coerce or compensate.'

11.2.8 Bioterrorism

Bioterrorism is the deliberate release of highly pathogenic microorganisms or toxins into aquatic or non-aquatic sources that cause death or disease on a large scale in human, plant, and animal populations irreparable damage. Therefore, bioterrorism is a severe threat, because it simultaneously endangers both national security and public policy. Throughout history, humanity has used its advanced technology to kill and destroy innocent human beings. Recent developments have led to microbial agents and biological toxins becoming deadly weapons and threatening human life and health (Tavakoli *et al.*, 2005). There is ample evidence that biological warfare devices generally have more significant potential and more uncontrollable damage capabilities than chemical warfare devices and are not comparable in size and scope to any weapon. The development of technology and the advancement of biochemistry and biotechnology have made these weapons easier to access, and genetic engineering has played a most crucial role (Zare Bidaki & Balali-Mood, 2015).

11.3 BASICS AND LOGIC

It is predicted that by 2030 the world will have a 40% shortage of water supply and demand (Kojm, 2015). This shortage is serious, and the complex nature of water security and its relationship to climate change, population growth, and increasing energy needs will add to the difficulty of meeting this challenge. This means having little or no water for drinking, agriculture, transportation, production, health, habitats, and healthy ecosystems for humans, plants, and animals. The collaboration between businesses, governments, and civil society can make a positive difference in the global water challenge.

11.3.1 The effect of climate change on water security

Many parts of the world are currently facing severe water problems, both quantitatively and qualitatively. Emissions from human activities have significantly increased the concentration of atmospheric greenhouse gases. This increase has intensified the greenhouse effect and, as a result, has increased the average temperature of the Earth. With global warming instability, the complexities of the global water security situation are intensifying, and the risks posed by it are multiplying. According to the latest studies, 2°C of global warming will affect approximately 15% of the world's population with a drastic reduction in water resources, and consequently increase the number of people living under absolute water scarcity to 40% (Schewe *et al.*, 2014). Since surface water resources are spatially non-uniform, water resources already often need to be transferred over long distances, before being stored and released. Global warming will intensify regional disparities as well as the negative effects of dry and wet seasons. It may disrupt supply and demand of water (Biswas & Tortajada, 2016). Forecasts show that massive rainfall events in most mid-latitude areas and in humid tropics are likely to intensify and become more frequent by the end of this century as global average temperatures rise (Stocker *et al.*, 2013).

11.3.2 Effects of climate change on freshwater resources

Rising temperatures alone can directly affect access to water resources through changes in flow regime in snow-capped or glacial river basins (Biswas & Tortajada, 2016). It is noteworthy that in many areas of these rivers, the total water resulting from glaciers melting will increase in the coming decades before then decreasing (Jiménez Cisneros *et al.*, 2014). Rising sea levels, mainly due to rising temperatures through ocean warming and melting glacial masses, exacerbate onshore influx, and increase surface waters and reduce access to freshwater (Biswas & Tortajada, 2016). Higher air temperatures increase surface water temperatures, which harms freshwater ecosystems, reducing the quality of pure water, and imposing risks on water quality (Jiménez Cisneros *et al.*, 2014).

11.3.3 Adaptation to climate change in the water sector

The basic concept of adaptation means reducing the vulnerability of natural and human-made environments (exposed ecosystems and human habitats) to the risks associated with climate change, and minimizing any risks (Biswas & Tortajada, 2016). One of the first steps to adapt to future climate change is to reduce vulnerability and exposure to current climate change (Field, 2014). For example, in the agricultural sector, improving irrigation efficiency, reducing demand for irrigation water to irrigate crops are among the most effective adaptations (Biswas & Tortajada, 2016). In the industrial sector, the revision of the location of industries and agricultural products that need more water and their placement in water-rich areas, and the introduction of industrial water efficiency certificates are amongst the adaptations that can be made in the face of climate change (Biswas & Tortajada, 2016).

11.3.4 Water–food–energy nexus challenge

One of the significant challenges for Asian cities is the interconnectedness of water–food–energy network security issues, exacerbated by climate change (Biswas & Tortajada, 2016). Moreover, water security is a crucial element of this network (Beck & Walker, 2013). Water, food, and energy are the resources that are essential for directing all human activities and, in addition, these factors have very close and interconnected internal relationships with each other. Water cannot be delivered to the consumer without consuming a significant amount of energy (for spatial transfer, etc.), and the production and productivity of food and energy will not be possible without consuming a large amount of water. Safe use of wastewater for irrigation and waste-to-energy conversion approaches are among the complementary solutions needed to overcome these challenges in rural and urban areas (Biswas & Tortajada, 2016).

11.3.5 Water for food security

Water and food security are both highly vulnerable to ongoing climate change. The global average temperature is predicted to rise by 1.4–4.8°C, resulting in a significant reduction in freshwater resources and agricultural production by the end of the 21st century (Misra, 2014). The world's population is projected to grow by 30% from 2015 to 2050, with global demand for food production increasing by 70%, and in developing countries by 100% (Chang *et al.*, 2013). Meeting this demand depends on the use of innovations such as the more efficient management of water resources. South Asian agriculture is highly vulnerable to climate change, and high temperatures and changing rainfall patterns pose a severe threat to the region's food security (Ahmed & Suphachalasai, 2014). Demand management and increasing water efficiency could be two essential changes needed to overcome water scarcity. The most critical water use efficiency measures are participatory irrigation management, fully efficient irrigation systems, crop stress management, and application of improved crops (Biswas & Tortajada, 2016).

11.3.6 Water for energy security

The relationship between water and energy is entirely reciprocal: water plays a role in producing almost all energy, and energy is an essential factor in providing water and wastewater services. For example, water is used to cool power plants and various industrial units, to generate electricity, for transport, and for irrigating biofuel feedstock crops, while energy is used for pumping, collecting, cleaning, storing, disposing of, and distributing water (DoE, 2006). Today, about 70% of electricity is generated by thermal power plants, highly dependent on water resources. Therefore, for example, new technologies should be used for cooling, using air, or a combination of methods, instead of using water or to reduce its use (USNIC, 2012). The amount of water used to produce fossil fuels for crops such as sugarcane, corn, and soy is significantly high. Generally, the amount of water consumed varies depending on the type of crop and the conditions in which it is located. Therefore, by choosing the right crops – which are low water consumption products – and preparing the conditions, it is possible to produce bioenergy correctly and with minor water casualties (Chang *et al.*, 2016). More crop production, more efficient use of energy in food production and the adoption of more efficient water use operations and technologies (such as solar-powered irrigation pumps) are valuable opportunities to improve water use efficiency and help ensure the sustainability of Asian agricultural systems (Biswas & Tortajada, 2016).

11.3.7 Demand management for water security

With increasing population growth, quantitative and qualitative growth of the agricultural sector, and the expansion of urbanization, groundwater resources in most areas have exceeded allowable limits. The final costs of supplying the extra water required from surface water sources and water source pollution are increasingly accelerating. Governments and different government agencies, society, and, consequently, macroeconomic policies are all undergoing significant changes. Therefore, trying to balance water supply and demands and the evolving needs of different regions in different climatic and geographical conditions by merely relying on the dimension of supply management – relying on the construction of new facilities and hardware aspects of technology – cannot alleviate the problems faced.

It is estimated that with the total stable annual freshwater remaining constant at the current amount of 4200 billion m^3, the annual water shortage in 2030 will be about 2765 billion m^3, which is equivalent to 40% of current demand (Addams *et al.*, 2009). Reducing water demand is one of the most critical factors that will increase security and reduce water resources risk. The main goal of communities is to allocate water according to the principles of sustainable development goals. To do this, we must be able to predict the process of accessing and supplying water over the next 30 years. The impact of industrialization, urbanization, and new water pricing systems must be considered.

However, it can be predicted that if the trend of water consumption due to population growth continues, global freshwater consumption will continue to exceed the tolerance threshold of the planet. Thus, it seems that an accurate assessment of water production and

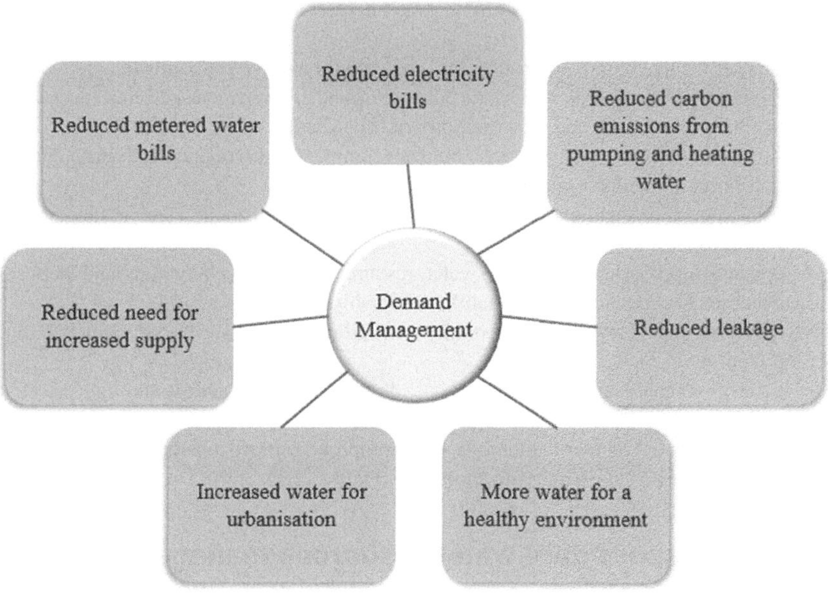

Figure 11.1 Benefits of demand management (Brears, 2016).

consumption is necessary. An assessment of water production and consumption should therefore be specified. But where, by who and and how? Water should be utilized more efficiently for greater profitability to prevent water wars and conflicts that may occur due to lack or absence of water, and this shows the importance of paying attention to water security in the world today.

In general, demand management includes a set of policies that assess existing urban water resources before planning for their supply and distribution. In addition to environmental benefits and ecosystem conservation, demand management is more cost-effective and efficient than consumption management because more efficient allocation of financial resources is not often needed (Brears, 2016). Finally, demand management leads to the fair use of water and the sustainability of water resources. The benefits of demand management are shown in Figure 11.1.

11.3.8 Urban water security

Cities play the most crucial role in sustainable development and poverty reduction., because most economical, social, national, government services, trade, and transportation activities are concentrated there and provide essential links with rural areas (Brears, 2016). Achieving water security in urban environments is of particular importance because, today, cities contain more than half of the world's population. New insights, tools, and methods are needed to understand the complex relationships and interactions in human and natural systems in urban areas at appropriate time and place scales (Nazemi & Madani, 2018).

The most important key elements in achieving urban water security are as follows (Brears, 2016):

- access to safe and affordable drinking water to meet basic needs such as health, safety, and well-being;
- support for livelihoods, human rights, and cultural and recreational values;
- protection of ecosystems in water allocation and management systems in order to maintain their health and the continuity of ecosystems;
- sufficient water resources for economic development and economic activities (energy, transportation, industry, tourism, etc.);
- collection and treatment of water used to protect human life and nature against pollution;
- common approaches to border water resources management within and between countries to promote the sustainability of freshwater resources;
- ability to cope with water uncertainty and hazards, including floods, droughts, and pollution;
- good governance and consideration of all stakeholders' interests through effective policies;
- establishment of partner institutions and appropriate planned, operated and maintained infrastructures, and capacity development (Figure 11.2).

11.3.9 Transboundary water resources management

Surface water or groundwater that flows through two or more countries is called a common water source. Rivers, seas, and lakes are examples of shared waters that may lead to

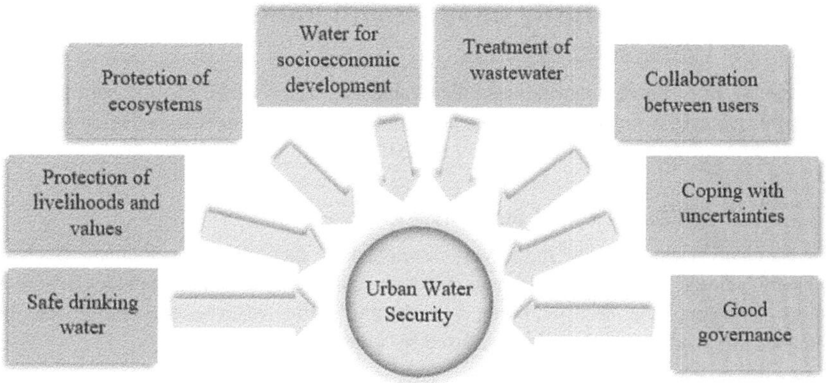

Figure 11.2 Elements of urban water security (Brears, 2016).

cooperation and the conclusion of treaties between countries. Indeed, one important issue surrounding dependence on water resources is shared water resources with neighbors. Groundwater contains the bulk of available freshwater resources and, according to statistics, 50% of the world population depend on these sources for drinking. Furthermore, a significant amount of water from these sources is used for agricultural purposes. Thus, groundwater resources are among the most critical of shared resources (Shahbazbegian *et al.*, 2016). The competition between states for the acquisition and exploitation of shared waters has led to numerous disputes, threatening national peace and security when unique legal and political mechanisms are not found to resolve them.

Water has always been a source of conflict, especially cross-border water resources, where different countries may clash over a common water source. Lack of proper management of border water resources and lack of cooperation between different communities have had adverse effects on social, economic, and environmental issues (Wheeler, 2014).

In general, many experts and actors in the Middle East, South Africa, and East Asia warn of water scarcity as a significant and growing problem in their region, leading to severe conflicts, hence the importance of water security studies. Often, national political, social, and economic development plans do not consider neighbors' rights to shared water resources.

Differentiation of shared water resources becomes a problem when there is a conflict between water resource exploitation for different purposes. For example, using a river's water for agriculture in one country may conflict with generating hydropower from the same river in another. Alternatively, one country may discharge waste into a lake, while another country uses the lake's water as a source of drinking water.

The amount of freshwater on the planet is small and relatively constant. Factors such as population growth, development of industrial activities, agriculture, and urbanization have led to an increase in the exploitation of these waters and in the conflicts of interest of governments in using these resources – a factor which can disrupt achievement of the goal of water security and consequently also disrupt peace and security. Therefore, quantitative management of water resources, which guarantees sufficient water resources to meet human needs, is one of the critical dimensions of security of access to water.

11.3.10 Water security assessment indicators

11.3.10.1 Water security index

A Water Security Index is an indicator that calculates water security to measure physical dimensions, using a model focused on household water consumption behavior in urban areas in developing countries (Shrestha *et al.*, 2018).

11.3.10.2 National water security index

The National Water Security Index was developed by the Asian Development Bank to compare national security performance in Asia and the Pacific and includes five key factors: economic water security, home water security, urban water security, environmental water security, and disaster resistance caused by water (Siddiqi, 2016).

11.3.10.3 Water insecurity index

The Water Insecurity Index is used on a regional scale in India and includes six key factors: water resource security, water access security, water use security, water capacity security, environmental water security, and water security in times of stress and climate change (Kher *et al.*, 2015).

11.3.10.4 Water poverty index

The Water Poverty Index consists of five components related to water: resources, access, capacity, consumption, and environment. The Water Poverty Index provides a better understanding of the complexities involved in the optimal management of water resources. Therefore, it can be used to assess the status of water resources in different regions and rank them in terms of physical, social, economic, and environmental factors related to water scarcity (Lawrence *et al.*, 2002).

11.4 IMPORTANCE AND NECESSITY OF WATER SECURITY

The issue of water security covers a wide range, from quantitative and qualitative water issues to cross-border waters and diplomatic relations between countries. Water security has a special place in many of the world's environmental summits; during the Second World Water Forum in The Hague in March 2000, for example, the Ministers' Final Statement emphasized the importance of water security and prioritized work on a global scale (Soussan & Harrison, 2000):

- ensuring the protection of related ecosystems and coasts, improving freshwater resources, promoting sustainable development and political stability;
- ensuring everyone has access to safe and sufficient water at a reasonable cost to enjoy a healthy and dynamic life; and
- supporting vulnerable people from water hazards.

The severe increase in business risk over the past decades is one of the critical issues that was not addressed in the above statement. A survey of business leaders at the World Economic Forum showed that water security is recognized as one of the most critical business risks

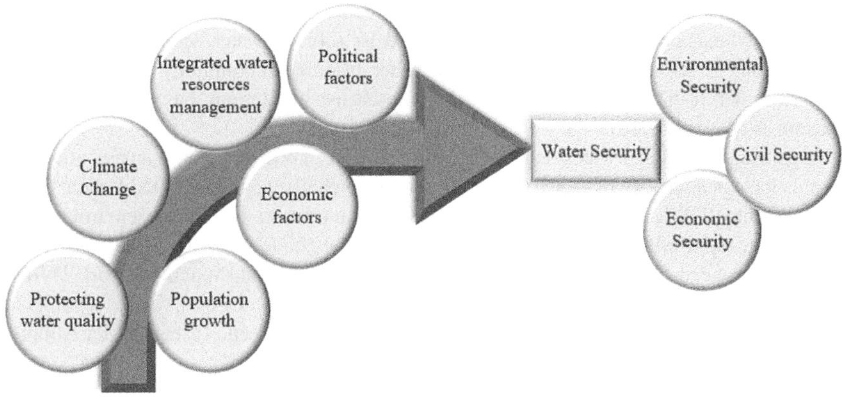

Figure 11.3 Drivers of water security.

in terms of impact intensity (Catley-Carlson, 2009). If human casualties occur in a water-related terrorist incident, the psychological effects on society will be devastating. Such psychological effects can spread quickly and last for a long time because, in such attacks, a society's sense of security is damaged, people's trust in water industry officials is weakened, and it takes a very long time to compensate.

Lack of adequate water quality and reasonable access due to insufficient water governance in previous decades has made water security a significant risk in all kinds of business. That is why, in addition to scientific and research aspects, legal, social, and political aspects should all be considered in water security studies. In general, in hostilities, some of the methods that an enemy might choose to accelerate conflict and defeat a target include 'terrorist' acts such as contaminating drinking water, destroying infrastructure and water facilities intended to cut off water supply services, or causing security crises to endanger water resources. In recent decades, water security has always been the main topic of many terrorist, military, and political attacks (Figure 11.3).

11.4.1 Integrated water resources management

In today's world, water as a socio-economic commodity is of particular importance, given the growing need for water in all aspects of human life. With global warming and the undeniable impact of climate change, especially in arid and semi-arid regions with more limited renewable water resources and where water can sometimes be a rare commodity in some communities, achieving water security is becoming essential and necessitates water resources management with a comprehensive and accurate approach.

Integrated water resources management is a process that promotes the coordinated development and management of water, land, and resources so that economic and social well-being can continue to operate legally without compromising the sustainability of vital ecosystems and the environment. With the increase of issues associated with the threat of water resources, various developments have arisen in the scientific and political fields that could be tools to achieving acceptable quantity and quality of water for the health of individuals and the environment (Grey & Sadoff, 2007). Since the mid-1990s, these issues have been considered in the context of 'water security'. Since then, Integrated Water Resources Management (IWRM) has been used to achieve water security with proper management, governance, and policies. The water resources community has agreed that complex water resources problems require comprehensive solutions based on new ideas (Nazemi & Madani, 2018).

Regarding the issue of water security and the concept of IWRM, which is an essential step towards the sustainability of water resources, one main challenge is Integrated Urban Water Management (IUWM). The concept of IUWM is meant to counteract environmental degradation, rapid urban population growth, and climate change (Brears, 2016). The main difference between IWRM and IUWM is in their spatial scale and functional part; IWRM is at the river basin level (which can include urban areas), while IUWM can be considered as a subset of IWRM and related to urban water management (urban water supply, sewerage, and storms) (Brears, 2016).

11.4.2 Achieving water security to maintain sustainable development

In sustainable development, goals should be developed in three main directions:

(1) to provide social growth in which the needs of all people are met;
(2) to provide adequate protection for the environment and to ensure careful consumption of natural resources;
(3) to ensure development leads to sustainable economic growth and prosperity.

Sustainable development and sustainable water management are related to the need for water for development. Sustainable water management involves allocating water between different purposes and different uses. Achieving water security by reducing its destructive potentials and increasing its productive potentials has always been the goal of human society and is a significant challenge for many countries worldwide. The development and management of water resources are essential for wealth creation, risk reduction, and poverty reduction, and in this regard, many developing countries need large investments and water management and infrastructure at all levels (Grey & Sadoff, 2007).

Sustainable development, on the other hand, is balancing development with the environment. Of course, sustainable development focuses not only on environmental aspect but also on social and economic aspects. Sustainable development is the intersection of society, economy, and environment. Today, many natural ecosystems have been destroyed, and many other ecosystems have become fragile and vulnerable due to the improper exploitation of natural resources, especially water resources. Proper planning and management of water resources is one of the most important goals of sustainable development and ensures the survival of human communities.

In general, we must keep in mind that sustainable development is a comprehensive approach, and addressing one sector without considering other sectors will not solve problems; indeed, new problems will be raised by addressing only one sector and overlooking its adverse effects on other sectors. Therefore, by any definition and evaluation, development and growth cannot be achieved without water security, and ensuring the sustainable development of the world depends on access to a sufficient amount of quality water.

11.5 METHODOLOGY

With the industrial revolution, human societies changed the global water system significantly. These changes have taken place when not enough knowledge was at hand to comprehensively analyze this complex system and the results of human intervention in it. Analysis of water security in the late 20th and 21st centuries requires attention to management and policy aspect, as well as engineering and technology knowledge.

The concept of water security first appeared in 1990 and has changed significantly since then. In the 1990s, the term water security was explicitly related to military security and rarely referred to environmental security. However, in 2000, the Second Assembly of the Global Water Partnership provided a comprehensive definition of water security in which easy access to water, cost, human needs, and environmental health were also considered. Many scholars and politicians have considered this term and attributed different meanings and definitions to it. Some have specifically linked water security to the field of water and the environment, while others have considered an interdisciplinary approach to water security (Cook & Bakker, 2012).

Nevertheless, the essential foundations for forming the concept of water security can be found in the state of the world's water resources in the 20th and 21st centuries. Advances in industry, agriculture, technology, and urbanization have enabled modern man to achieve maximum efficiency in producing agricultural and industrial products by using all available environmental capacities – one of the most important of which is available water resources. For this reason, water, like oil and gold, has gradually become one of the main components of power and progress in a country. On the other hand, with huge consumption of the world's water resources and global warming in the last two centuries, this resource is now in danger of shortage and crisis, so that water security has found its place in global security. Lack of attention to water security is the basis of many conflicts and disputes between countries.

11.5.1 Water security from different points of view

One of the most critical issues in recent years, which has attracted many researchers, environmental activists, and politicians worldwide, is water security and related issues. Today's limited quantities of water resources are affected by a series of complex relationships of geophysical, geopolitical, social, political, and cultural conditions. It is not possible to study each part without considering other parts and factors (Falkenmark *et al.*, 2007).

There are numerous examples of water disputes and security conflicts, ranging from the restriction of transboundary water resources (e.g., Turkey, Iraq, and Syria) to the pollution of standard surface and groundwater (e.g., the Danube in Europe, which flows through ten countries and so is considered a shared water resource). This is why some researchers believe that in the third millennium AD, water will play a key role in forming political, regional, and global relations (Shahbazbegian *et al.*, 2016). Therefore, a means of ensuring water security on a political and international scale and of preventing conflicts between countries through bilateral or multilateral cooperation between countries, and the use of active diplomacy to resolve disputes over common water resources, is required. Undoubtedly, the political dimension of water security studies is crucial.

Threats to drinking water resources is one of the most critical infrastructure problems posed by military attacks. The policy of military strikes, especially in vulnerable areas, is usually to control water resources. Examining examples and historical records proves that the pollution of international water supply sources is not a new issue and has been seen throughout history. Water supply systems have always been a preferred target of regional terrorism. Terrorists have used chemical, biological, and radiological agents to attack water systems as a target. However, dam destruction or pollution in peacetime is most often due to natural disasters (Gleick, 2006).

11.5.2 Water security and dams

One of the most critical issues in the discussion of water security in military conflicts relates to attacks on dams, and a number of examples are described below:

- New Hampshire Dam; locals attacked the dam in New Hampshire in 1850 because they were dissatisfied with its impact on the river water level, as it supplied water to downstream factories (Gleick, 2006);
- Cahora Bassa Dam; this dam was targeted by the Mozambican National Resistance in 1980 during conflicts over the independence of Mozambique, Rhodesia, South Africa, and Zimbabwe, leading to its destruction (Gleick, 2006);
- Inga Dam; in 1998, Attacks on Inga Dam during efforts to overthrow President Kabila, disrupted electricity supplies from Inga Dam and water supplies to Kinshasa.

A number of different natural disasters have affected dams, including:

- Sefidrood Dam; this is one of four large concrete dams in Iran, and was destroyed by a 7.4 magnitude earthquake. Iran is located on the Alpine-Himalayan earthquake-prone bel and defense–security issues and earthquake engineering knowledge should clearly be seriously considered in the design and construction of dams in the country, especially given that most of the dams in Iran have been built or are under construction in earthquake-prone areas. Dam safety and stability against earthquakes is an essential aspects of dam design.
- Aswan Dam; located on the Nile River in Egypt, in 1998–1997 the Aswan Dam witnessed the spread of the RVF virus amongst livestock. This led to the further spread of the disease amongst humans and livestock, killing people and thousands of livestock living on the dam lake shores. The disease had never occurred in this part of Egypt before (Pepin *et al.*, 2010).
- Konya Dam; located on the Quina River in India, in 1967 a magnitude 6.5 earthquake caused severe damage to the Konya Dam and its associated facilities.
- Zipingpu Dam; located in China, in 2008 a 7.9 magnitude earthquake broke the Zipingpu Dam and about 80,000 civilians were killed (Kerr & Stone, 2009).
- New Orleans Levee; in 2005, due to Hurricane Katrina, the New Orleans Levee and floodwalls in New Orleans, USA, were broken, and 80% of the land was submerged. The destruction caused by this failure is estimated to be greater than the damage caused by the storm (Congleton, 2006).
- Brumadinho Dam; the Brumadinho Dam is located on the Ru river in southern China. In 1976, the dam broke due to a severe storm, killing 145 000 people and leaving about half a million homeless (Penuel & Statler, 2010).

- Mosul Dam; located in Mosul city in Iraq, the Mosul Dam was built in 1984 on the Tigris and Euphrates rivers. According to a US military engineering warning to the Iraqi government, the dam is being threatened by karst disappearing from its foundations. If breached, it could create a 20 meter wave of water and destroy entire downstream areas, including Mosul city, with a population of about 2 million people. Results of an investigation by the General Inspectorate for Reconstruction of Iraq show that if the dam is destroyed, the city of Baghdad may also be at risk of flooding (Filkins, 2016).

11.5.3 Water security in water supply networks

Water supply networks, pumping stations, valves, and measuring devices may provide targets for terrorist attacks, given their wide availability, insufficient protection, ease, and the high casualties and financial losses that could be incurred. Physical, chemical, biological, and cyber-attacks are among the potential methods that can be used by terrorists. Water in a network could be polluted and disrupted, potentially affecting thousands of lives. Various scenarios can be envisaged:

- Physical destruction of networks; this action may lead to the interruption of water supply due to the breakdown of facilities and equipment such as pumps, computers, buildings. However, no such case has occurred so far.
- Intentional chemical contamination; in 1972, New York City's drinking water supply was reported to contain nerve gas (Gleick, 2006). In 1977, a reservoir of water in North Carolina became contaminated with bacterial agents when the safety caps on a tank were removed, and the water was contaminated and made unusable. After this attack, water had to be supplied by trucks (Gleick, 2006). In 1984, members of the Raineeshee religious group targeted a city water supply tank with Salmonella in The Dalles, Oregon, resulting in more than 750 cases of illness (Gleick, 2006).
- Cyber attacks; in 2000, in Queensland, police arrested a man trying to drain the Maroochy Shire wastewater system using computers and radio transmitters and diverting it to parks and rivers (Gleick, 2006). In 2002, US intelligence agencies obtained evidence of an Al-Qaeda planned attack on urban water resources (Gellman, 2002).
- Bioterrorist attacks; in 1973, a German biologist threatened to pollute Germany's water supply with biological substances unless the government paid him $8.5 million. In 1972, two members of the neo-Nazi group known as the 'Order of the Rising Sun' were arrested in Chicago for plotting to carry out a plan to pollute water in several cities, including Chicago and St. Louis. They had several biological agents, including botulism, meningitis, anthrax, and 30–40 kg of typhoid (Gleick, 2006). In 1991, the town of Kelowna in British Columbia, Canada, was threatened with contamination by biological agents through an anonymous letter, but no group or individual was identified as the culprit.

Currently, factors such as limited freshwater resources, lack of proper access to water, and reduced quality have led to a crisis of safe and reliable water access in aquatic and terrestrial ecosystems. Undoubtedly, water scarcity (drought) or its excess (flood), in addition to being affected by desertification and land degradation, also present the potential for human resource risk.

Many human conflicts are somehow rooted in access to water. Water is subject to reduced quality due to human activities and exploitation, even in areas with abundance. This reduction in quality can significantly affect the conservation of aquatic biodiversity. Increased organic matter and nutrients, industrial wastes, and persistent chemical pollutants in water resources have caused significant problems such as increasing nitrate levels in groundwater, heavy metals in river and lake sediments, and depletion of oxygen in water sources (Bogardi *et al.*, 2012). In the study of water security, many mechanisms and processes should be considered for each region (see Figure 11.4).

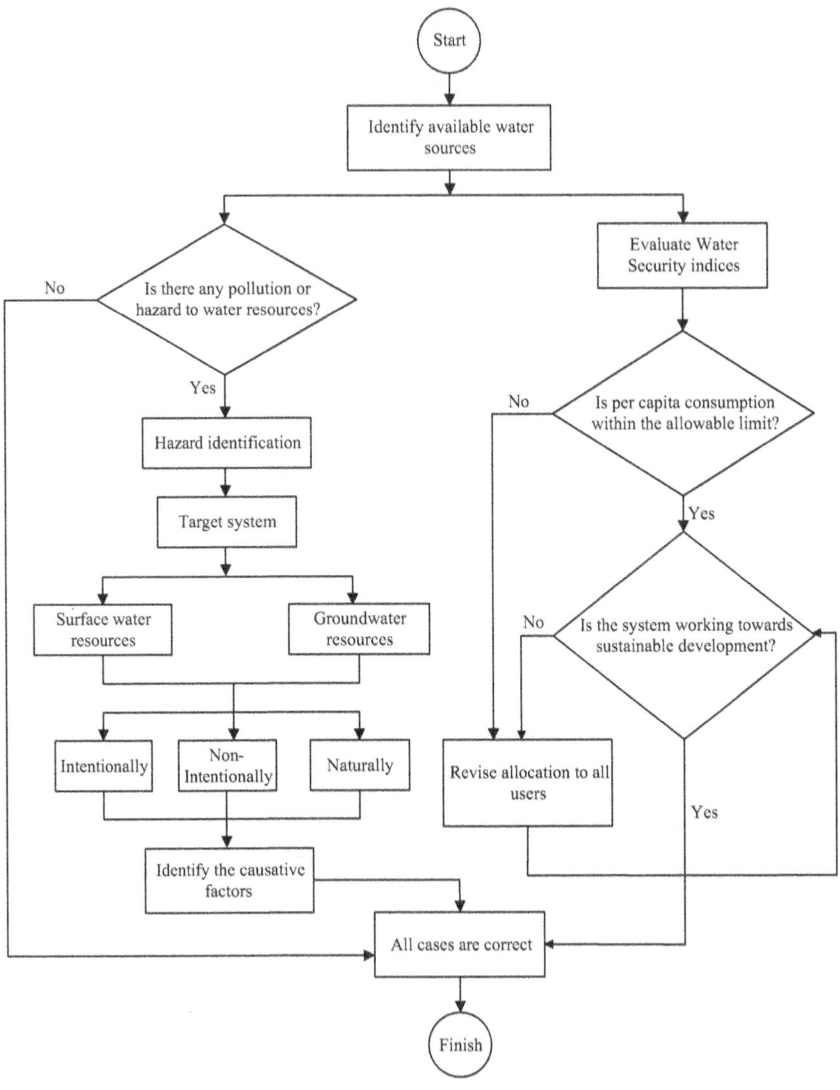

Figure 11.4 Overview of the methodology for developing water security.

11.6 EXAMPLES

11.6.1 Water security in Central Asia and the Amu Darya

Central Asia is a vast landlocked continental region in which water resources are inadequately distributed. Its densely populated urban settlements, such as Bukhara, Samarkand, and Merv, are among the world's most critical areas in terms of water issues. The drying up of the Aral Sea in the region has made it an international issue. The water needs of five countries – Tajikistan, Kyrgyzstan, Uzbekistan, Turkmenistan, and Kazakhstan – are met by two major rivers, the Amu Darya and the Syr Darya. Tajikistan and Kyrgyzstan are on the upper reaches of these rivers, whilst the republics of Uzbekistan, Turkmenistan, and Kazakhstan are on the lower reaches.

The conversion of desert lands into vast cotton fields during the Soviet era and the diversion of the Amu Darya and Syr Darya to irrigate these arid lands led to the establishment of new population centers. The Amu Darya and Syr Darya originate in Tajikistan and Kyrgyzstan, respectively, and meet the water needs of the downstream countries Turkmenistan, Uzbekistan, and Kazakhstan.

Before the collapse of the Soviet Union, Turkmenistan, Uzbekistan, Kyrgyzstan, and Tajikistan were in one economic zone, whilst Kazakhstan was in another independent zone, interacting in coordination with the central government (Moscow). Tajikistan and Kyrgyzstan did not have energy resources such as oil and gas; thus, they took supplies of their required hydrocarbon energy from Uzbekistan, Turkmenistan, and Kazakhstan and, in return, they supplied water for agriculture, industry, and urban consumption to the downstream countries (Amir Ahmadian & Naseri, 2013). After the collapse of the Soviet Union, the regional, political, and economic relations between these countries wholly changed, so that these countries have repeatedly considered the issue of sharing and joint use of resources. Tajikistan and Kyrgyzstan, which were among the Soviet Union's most impoverished countries during the Soviet era, became weaker after independence due to insufficient income sources. However, due to the existence of oil and gas resources, the downstream countries canceled the exchanges in the energy basin with the upstream countries after independence and also changed the exchange prices of their energy resources to the level of free-market exchanges (Amir Ahmadian & Naseri, 2013).

Due to the lack of sufficient revenue sources to purchase the energy needed, the upstream countries implemented a plan to sell water to the downstream countries, and due to the lack of sufficient reservoir dams to implement this plan, they sought to build new dams. In all existing countries except Kazakhstan, where the industry and services sectors are strong, the agricultural sector has a large share in the population's employment. Due to the existence of vast agricultural lands and the lack or absence of water for irrigation of agricultural lands, daily livelihood, water security, food security, and job security are putting millions of low-income people in crisis in the region, which has created tensions between the downstream and upstream countries.

China's demand for both small and large lakes in the region, including Lake Balkhash in Kazakhstan (downstream), and German support for projects such as the reconstruction of small hydropower plants in Kyrgyzstan (upstream) have exacerbated the water crisis in these countries. The downstream countries whose agricultural production is limited to cotton, wheat, and rice are opposed to the current policies and plans in the upstream countries due to their depleted water resources and the threat to their water and food security and economic stability.

Various attempts have been made to reach a compromise between the upstream and downstream countries, but all of them have failed. Due to the strong military power of the lower countries, military maneuvers have always been one of the main tools of pressure on the upstream countries, and because of this, the possibility of conflict and regional threats increases. For this reason, the water security situation in this region is very critical and severe, and if special measures are not taken to solve this problem, this issue could lead to a more significant crisis than the current situation (Rittger *et al.*, 2017).

11.6.2 The effects of climate change on water conservation and security in Uganda

For many developing countries, achieving the goal of safe, affordable and accessible water for all is a challenge, especially as populations grow and climate change is predicted. Therefore, due to the inadequacy of public water supply systems to reach remote rural areas, water supply systems are increasingly becoming decentralized, especially in rural areas. As a result, for most developing countries, rainwater harvesting is gradually becoming a significant water source to improve domestic water security. In this study, water conservation and water security are compared between observed scenarios based on historical data and six different climate models (BCC-CSM1.1, ACCESS1.0, CNRM-CM5, HadGEM2-CC, and HadGEM2-ES) for the periods 2025–2055 and 2060–2090.

This case study was conducted in the Kabarole area of western Uganda because it has the highest rainfall in the country and, as a result, has a high ability to harvest rain. The analysis is based on two dry seasons – December, January, February (DJF) and June, July, August (JJA) – and two rainy seasons – September, October, November (SON), and March, April, May (MAM).

In this study, average daily water savings is determined as the sum of water harvested during the simulation period divided by the total number of days in the same period. Water security is determined as the number of days a tank of a particular volume is empty in a simulation period divided by the number of days in the same period.

The average per capita household water consumption in this area is about 16 liters, which according to WHO recommendations should be at least 20 liters; in this study, per capita, household water consumption per person per day was considered to be 50 liters. The average number of people in each family was five.

Although there are many variations in the models' predictions, this study showed that due to forthcoming climate change, daily water saving and water safety increase in DJF and MAM and decrease in JJA and SON. MAM will have the most significant impact on forecasting water savings daily to 50% by 2070. The models generally concluded that households with a roof of more than 50 m^2 and a reservoir of 5 m^3 would be less affected by climate change if their water treatment was effective and their water security was better (Kisakye & Van der Bruggen, 2018).

11.6.3 Achieving water security in Egypt through desalination

Freshwater is an essential source of life. Its availability increases the quality of life and the economy of the community. Seawater (or saline) desalination has been performed for over 50 years and is a well-established means of water supply in many countries. The gradual

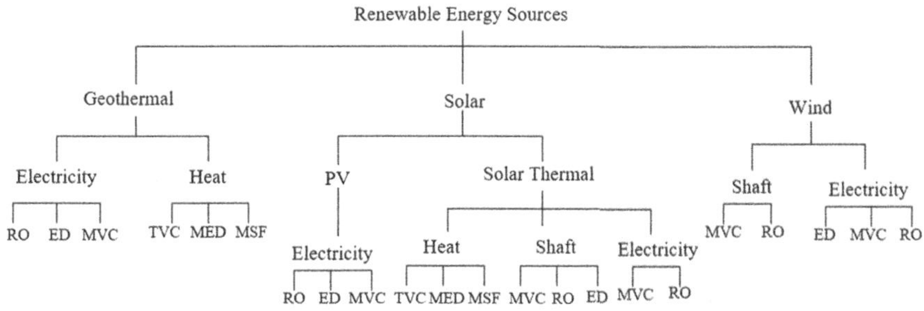

Figure 11.5 Renewable energy sources available for desalination technologies. PV = photovoltaic solar; RO = reverse osmosis; ED = electrodialysis; MVC = mechanical vapor compression; TVC = thermal vapor compression; MED = multi-effect distillation, MSF = multi-stage flash distillation (El-Sadek, 2010).

warming of the planet exacerbates countries with arid climates with difficulty meeting their water needs, so the desalination process and the reuse of water for urban water supply can significantly contribute to those countries' water resources. It is now technically and economically possible to produce large quantities of high-quality water from desalination processes.

The National Water Policy in Egypt has three objectives: (1) to promote the protection of water quality; (2) to make optimal and efficient use of available resources; and (3) to strengthen international cooperation for the protection and increase of water supply in the Nile Basin (El-Sadek, 2010).

The diversity of available water resources is vital for achieving water security. Water resources include surface water, groundwater, return water, and direct runoff from rainfall. Desalinated water can be added to these groups (Gerlak *et al.*, 2018). Desalination is a process that has recently received a great deal of attention from scientists, water resource planners, policymakers, and other stakeholders. The main impetus for this attention given to desalination is the dramatic technological advancements being made in desalination processes, resulting in a much lower cost than before. Due to the over-exploitation of available water resources, the desalination process, because of its lower cost, is becoming a viable option for drinking water supply (El-Sadek, 2010) (Figure 11.5).

Because the cost of using fossil fuels to supply the energy required for the desalination process is very high, renewable energy sources offer a cost-effective option and do not lead to air pollution. In addition to maintaining water security, the security of energy resources is not threatened (El-Sadek, 2010).

11.7 SUMMARY

Water is an important human necessity. Demand for scarce resources, such as for water, has always been the source of national and international tensions. However, exploiting a scarce resource such as water has profound economic, security, and geopolitical implications that governments cannot ignore. Today, water resources worldwide are in a critical situation, yet this deterioration varies from region to region. Areas such as the Middle East, sub-Saharan Africa, and South Africa are most at risk of prolonged droughts and water shortages. Rapid population growth, climate change, and water mismanagement, especially in developing countries, are all factors that threaten water security. The nature of water security implies access to sufficient good quality water for personal and domestic use, for energy production, industry, transportation, land ecosystem sustainability and biodiversity, as well as to ensure food security. Indeed, today, the relationship between water security and food security is increasingly used in development literature, and the consumption of more than 70% of water resources for food production is an actual example of this close relationship. Therefore, it can be concluded that in recent years, the concept of water security has gone beyond military and war concerns, and now has multiple meanings. Given the global importance and the interconnective role of water, water management should be at the top of political and social programs and priorities, and undoubtedly access to the ideal state of water security requires the protection and development of a knowledge of risk awareness and preparedness to face any risks. Changing lifestyles as a result of engineering concerns, policy implementation and purposeful management, within a coordinated and effective legal framework, is essential for successful water resources management.

REFERENCES

Addams L., Boccaletti G., Kerlin M. and Stuchtey M. (2009). Charting our Water Future: Economic Frameworks to Inform Decision-Making. McKinsey & Company, New York.

Ahmed M. and Suphachalasai S. (2014). Assessing the Costs of Climate Change and Adaptation in South Asia. Asian Development Bank, Mandaluyong, Philippines.

Alvarez P. J., Chan C. K., Elimelech M., Halas N. J. and Villagrán D. (2018). Emerging opportunities for nanotechnology to enhance water security. *Nature Nanotechnology*, **13**(8), 634.

Amir Ahmadian B. and Naseri M. (2013). Water crisis in Central Asia with emphasis on regional conflict over water issue. *Journal of Central Eurasia Studies*, **6**(12), 1–20.

Beck M. B. and Walker R. V. (2013). On water security, sustainability, and the water–food–energy–climate nexus. *Frontiers of Environmental Science & Engineering*, **7**(5), 626–639.

Biswas A. K. and Tortajada C. (eds). (2016). Water Security, Climate Change and Sustainable Development. Springer, Singapore.

Bogardi J. J., Dudgeon D., Lawford R., Flinkerbusch E., Meyn A., Pahl-Wostl C., Vielhauer K. and Vörösmarty C. (2012). Water security for a planet under pressure: interconnected challenges of a changing world call for sustainable solutions. *Current Opinion in Environmental Sustainability*, **4**(1), 35–43.

Brears R. C. (2016). Urban Water Security. Wiley, United Kingdom.

Buzan B. (2008). People, States & Fear: an Agenda for International Security Studies in the Post-cold War Era. Lynne Rienner Publishers, No. 1, Boulder, Colorado.

Catley-Carlson M. (2009). The bubble is close to bursting: a forecast of the main economic and geopolitical water issues likely to arise in the world during the next two decades. Draft for Discussion at the World Economic Forum Annual Meeting, Geneva, Switzerland.

Chang Y., Ries R. J. and Wang Y. (2013). Life-cycle energy of residential buildings in China. *Energy Policy*, **62**, 656–664.

Chang Y., Li G., Yao Y., Zhang L. and Yu C. (2016). Quantifying the water–energy–food nexus: current status and trends. *Energies*, **9**(2), 65.

Congleton R. D. (2006). The story of Katrina: New Orleans and the political economy of catastrophe. *Public Choice*, **127**(1–2), 5–30.

Cook C. (2016). Implementing drinking water security: the limits of source protection. *Wiley Interdisciplinary Reviews: Water*, **3**(1), 5–12.

Cook C. and Bakker K. (2012). Water security: debating an emerging paradigm. *Global Environmental Change*, **22**(1), 94–102.

DoE U. S. (2006). Energy Demands on Water Resources: Report to Congress on the Interdependency of Energy and Water. US Department of Energy, Washington DC.

Elgert L., Austin P. and Picchione K. (2016). Improving water security through rainwater harvesting: a case from Guatemala and the potential for expanding coverage. *International Journal of Water Resources Development*, **32**(5), 765–780.

El-Sadek A. (2010). Water desalination: an imperative measure for water security in Egypt. *Desalination*, **250**(3), 876–884.

Falkenmark M. and Molden D. (2008). Wake up to realities of river basin closure. *International Journal of Water Resources Development*, **24**(2), 201–215.

Falkenmark M., Berntell A., Jägerskog A., Lundqvist J., Matz M. and Tropp H. (2007). On the Verge of a New Water Scarcity: a Call for Good Governance and Human Ingenuity. Stockholm International Water Institute (SIWI), Sweden.

Field C. B. (ed.). (2014). Climate Change 2014–Impacts, Adaptation and Vulnerability: Regional Aspects. Cambridge University Press, Cambridge.

Filkins D. (2016). A Bigger Problem than ISIS? www.newyorker.com (accessed 27 May 2021).

Gain A., Wada Y. and Giupponi C. (2016). Measuring global water security towards sustainable development goals. *Environmental Research Letters*, **11**(12), 2–13.

Gellman B. (2002). Cyber-attacks by Al Qaeda feared. Washington Post. Jun 27 2002: A01.

Gerlak A. K. and Mukhtarov F. (2015). Ways of knowing water: integrated water resources management and water security as complementary discourses. *International Environmental Agreements: Politics, Law and Economics*, **15**(3), 257–272.

Gerlak A. K., House-Peters L., Varady R. G., Albrecht T., Zúñiga-Terán A., de Grenade R. R., Cook C. and Scott C. A. (2018). Water security: a review of place-based research. *Environmental Science & Policy*, **82**, 79–89.

Gleick P. H. (2006). Water and terrorism. *Water Policy*, **8**(6), 481–503.

Grey D. and Sadoff C. W. (2007). Sink or swim? *Water security for growth and development. Water Policy*, **9**(6), 545–571.

Housh M. and Ohar Z. (2018). Model-based approach for cyber-physical attack detection in water distribution systems. *Water Research*, **139**, 132–143.

Imbulana K. A. U. S., Droogers P. and Makin I. W. (eds). (2002). World Water Assessment Program case study, Ruhuna Basins, Sri Lanka. International Water Management Institute, Colombo, Sri Lanka.

Jia X., Li C., Cai Y., Wang X. and Sun L. (2015). An improved method for integrated water security assessment in the Yellow River basin, China. *Stochastic Environmental Research and Risk Assessment*, **29**(8), 2213–2227.

Jiang Y. (2015). China's water security: current status, emerging challenges and future prospects. *Environmental Science & Policy*, **54**, 106–125.

Jiménez Cisneros B. E., Oki T., Arnell N. W., Benito G., Cogley J. G., Döll P., Jiang T., Mwakalila S. S. (2014). Freshwater resources. In: Climate Change 2014 – Impacts, Adaptation and Vulnerability: Part A: Global and Sectoral Aspects: Working Group II Contribution to the IPCC Fifth Assessment Report, Intergovernmental Panel on Climate Change (eds), Cambridge University Press, Cambridge, doi: 10. 1017/CBO9781107415379

Jones J. A., Vardanian T. and Hakopian C. (eds). (2009). Threats to Global Water Security. Springer Science & Business Media, New York.

Kerr R. A. and Stone R. (2009). A human trigger for the great quake of Sichuan? *Science*, **323** (5912), 322–322.

Kher J., Aggarwal S. and Punhani G. (2015). Vulnerability of poor urban women to climate-linked water insecurities at the household level: a case study of slums in Delhi. *Indian Journal of Gender Studies*, **22**(1), 15–40.

Khokhar T. (2017). Chart: globally, 70% of freshwater is used for agriculture. *World Bank Data Blog*. See: https://blogs.worldbank.org/opendata/chart-globally-70-freshwater-used-agriculture (accessed April 2021).

Kirchhoff C. J., Lara-Valencia F., Brugger J., Mussetta P. and Pineda-Pablos N. (2016). Towards joint consideration of adaptive capacity and water security: lessons from the arid Americas. *Current Opinion in Environmental Sustainability*, **21**, 22–28.

Kisakye V. and Van der Bruggen B. (2018). Effects of climate change on water savings and water security from rainwater harvesting systems. *Resources, Conservation and Recycling*, **138**, 49–63.

Kojm C. A. (2015). Change and continuity: the national intelligence council, 2009–2014. *Studies in Intelligence*, **59**(2), 1–12.

Lawrence P. R., Meigh J. and Sullivan C. (2002). The Water Poverty Index: An International Comparison. Department of Economics, Keele University, Staffordshire, UK.

Liu J., Yang H., Gosling S. N., Kummu M., Flörke M., Pfister S., Hanasaki N., Wada Y., Zhang X., Zheng C. and Alcamo J. (2017). Water scarcity assessments in the past, present, and future. *Earth's Future*, **5**(6), 545–59.

Lu S., Bao H. and Pan H. (2016). Urban water security evaluation based on similarity measure model of Vague sets. *International Journal of Hydrogen Energy*, **41**(35), 15944–15950.

Ludwig R. and Roson R. (2016). Climate change, water and security in the Mediterranean: Introduction to the special issue. *The Science of the Total Environment*, **543**(Pt B), 847.

Maganda C. (2016). Water security debates in 'safe' water security frameworks: Moving beyond the limits of scarcity. *Globalizations*, **13**(6), 683–701.

Minamyer S. (2008). Effective crisis communication during water security emergencies. *Journal American Water Works Association*, **100**(9), 180–184.

Misra A. K. (2014). Climate change and challenges of water and food security. *International Journal of Sustainable Built Environment*, **3**(1), 153–165.

Nazemi A. and Madani K. (2018). Urban water security: emerging discussion and remaining challenges. *Sustainable Cities and Society*, **41**, 925–928.

Nel J. L., Le Maitre D. C., Roux D. J., Colvin C., Smith J. S., Smith-Adao L. B., Maherry A. and Sitas N. (2017). Strategic water source areas for urban water security: making the connection between protecting ecosystems and benefiting from their services. *Ecosystem services*, **28**, 251–259.

Office of the Chairman of the Joint Chiefs of Staff. (2019). DOD dictionary of Military and Associated Terms. The Joint Staff, Washington DC.

Penuel K. B. and Statler M. (2010). Encyclopedia of Disaster Relief. SAGE, California, USA.

Pepin M., Bouloy M., Bird B. H., Kemp A. and Paweska J. (2010). Rift Valley fever virus (Bunyaviridae: Phlebovirus): an update on pathogenesis, molecular epidemiology, vectors, diagnostics and prevention. *Veterinary Research*, **41**(6), 61.

Powell N., Kløcker Larsen R., de Bruin A., Powell S. and Elrick-Barr C. (2017). Water security in times of climate change and intractability: reconciling conflict by transforming security concerns into equity concerns. *Water*, **9**(12), 934.

Rijsberman F. R. (2006). Water scarcity: fact or fiction? *Agricultural Water Management*, **80** (1–3), 5–22.

Rittger K., Bair E., Racoviteanu A., Brodzik M. J., Armstrong R. L., Jodha Khalsa S. and Dozier J. (2017). Quantifying the individual contributions of melt from snow and glaciers in High Mountain Asia river basins: Syr Darya, Amu Darya, Indus, Ganges, and Brahmaputra. EGU General Assembly Conference Abstracts, Vienna, Austria, p. 11559.

Romero-Lankao P. and Gnatz D. M. (2016). Conceptualizing urban water security in an urbanizing world. *Current Opinion in Environmental Sustainability*, **21**, 45–51.

Schewe J., Heinke J., Gerten D., Haddeland I., Arnell N. W., Clark D. B., Dankers R., Eisner S., Fekete B. M., Colón-González F. J. and Gosling S. N. (2014). Multimodel assessment of water scarcity under climate change. *Proceedings of the National Academy of Sciences*, **111**(9), 3245–3250.

Schulte P. (2017). What do 'Water Scarcity', 'Water Stress', and 'Water Risk' actually mean? The CEO Water Mandate. See: https://ceowatermandate.org/posts/water-scarcity-water-stress-water-risk-actually-mean/ (accessed 15 March 21).

Shahbazbegian M. R., Turton A. and Mousavi Shafaee S. M. (2016). Hydropolitical Self-Organization theory; system dynamics to analyse hydropolitics of Helmand transboundary river. *Water Policy*, **18**(5), 1088–1119.

Shrestha S., Aihara Y., Bhattarai A. P., Bista N., Kondo N., Futaba K., Nishida K. and Shindo J. (2018). Development of an objective water security index and assessment of its association with quality of life in urban areas of developing countries. *SSM – Population Health*, **6**, 276–285.

Siddiqi Y. (2016). Asian Water Development Outlook 2016: Strengthening Water Security in Asia and the Pacific. Asian Development Bank, Report No. 9292575449. Asian Development Bank, Mandaluyong, Philippines.

Soussan J. and Harrison R. (2000). Commitments on Water Security in the 21st Century. Analysis of Pledges and Statements Made at the Ministerial Conference and World Water Forum, March 2000, The Hague. University of Leeds, Leeds.

Stocker T. F., Qin D., Plattner G. K., Tignor M., Allen S. K., Boschung J. and Midgley P. M. (2013). Climate change 2013: the physical science basis. Contribution of Working Group I to the Fifth Assessment Report of the Intergovernmental Panel on Climate Change. Cambridge University Press, New York.

Tavakoli H. R., Sarafpour R. and Samadi M. (2005). Water, food and bioterrorism. *Journal of Military Medicine*, **7**(1), 75–82.

United States National Intelligence Council (USNIC). (2012). Global Trends 2030: Alternative Worlds. USNIC, Washington DC, USA.

van Beek E. and Arriens W. L. (2014). Water Security: Putting the Concept into Practice. Global Water Partnership, Stockholm.

Varady R. G., Zuniga-Teran A. A., Garfin G. M., Martín F. and Vicuña S. (2016). Adaptive management and water security in a global context: definitions, concepts, and examples. *Current Opinion in Environmental Sustainability*, **21**, 70–77.

Veettil A. V. and Mishra A. K. (2016). Water security assessment using blue and green water footprint concepts. *Journal of Hydrology*, **542**, 589–602.

Vörösmarty C. J., McIntyre P. B., Gessner M. O., Dudgeon D., Prusevich A., Green P., Glidden S., Bunn S. E., Sullivan C. A. and Liermann C. R. (2010). Global threats to human water security and river biodiversity. *Nature*, **467**(7315), 555.

Wegerich K., Van Rooijen D., Soliev I. and Mukhamedova N. (2015). Water security in the Syr Darya basin. *Water*, **7**(9), 4657–4684.

Wheater H. S. and Gober P. (2015). Water security and the science agenda. *Water Resources Research*, **51**(7), 5406–5424.

Wheeler S. A. (2014). Insights, lessons and benefits from improved regional water security and integration in Australia. *Water Resources and Economics*, **8**, 57–78.

Whiteford S. and Witter S. G. (1999). Water security: the issues and policy challenges. *International Review of Comparative Public Policy*, **11**, 1–25.

Zare Bidaki M. and Balali-Mood M. (2015). Bioterrorism and biological warfare, from past to the present: a classic review. *Journal of Birjand University of Medical Sciences*, **22**(3), 182–198.

Zeitoun M., Lankford B., Krueger T., Forsyth T., Carter R., Hoekstra A. Y., Taylor R., Varis O., Cleaver F. and Boelens R. (2016). Reductionist and integrative research approaches to complex water security policy challenges. *Global Environmental Change*, **39**, 143–154.

Zeitoun M., Elaydi H., Dross J. P., Talhami M., de Pinho-Oliveira E. and Cordoba J. (2017). Urban warfare ecology: a study of water supply in Basrah. *International Journal of Urban and Regional Research*, **41**(6), 904–925.

Chapter 12

Food security from a water perspective

Shima Kheirinejad[1], Omid Bozorg-Haddad[2*] and Vijay P. Singh[3]

[1]M.Sc. Student, Department of Irrigation & Reclamation Engineering, Faculty of Agricultural Engineering & Technology, College of Agriculture & Natural Resources, University of Tehran, Karaj, Tehran, Iran. E-mail: Shima.kheyri20@ut.ac.ir

[2]Distinguished Professor, Department of Irrigation & Reclamation Engineering, Faculty of Agricultural Engineering & Technology, College of Agriculture & Natural Resources, University of Tehran, Karaj, Tehran, 31587-77871, Iran. E-mail: OBHaddad@ut.ac.ir

[3]Distinguished Professor, Regents Professor, Caroline and William N. Lehrer Distinguished Chair in Water Engineering Department of Biological and Agricultural Engineering & Zachry Department of Civil & Environmental Engineering Texas A and M University 321 Scoates Hall, 2117 TAMU College Station, Texas 77843-2117, USA

*Corresponding author.

ABSTRACT

Access to enough food to eliminate hunger is a fundamental right of society. Lack of food is an obstacle to social, political, economic, and cultural development of society. Investment in agriculture, support of education, and health development in the community can lead to food security. Water is fundamental to agriculture and hence to food and nutritional security. Water is also vital for plants and livestock. Agriculture has the largest share of water consumption, accounting for about 70% of all freshwater earmarked for human use, and good quality water is needed for production of a wide variety of non-food products, such as cotton, rubber, and industrial oils. In 1948, the Universal Declaration of Human Rights affirmed everyone's right to adequate food.

Editor: Omid Bozorg-Haddad
doi: 10.2166/9781789062144_0235

However, accessing adequate food in rural areas in many developing countries depends on access to natural resources, including water. On 28 July 2010, the UN General Assembly declared access to clean drinking water and sanitation as a human right. Significant changes in policy and management across the entire agricultural production chain are necessary to ensure the best use of available water resources to meet the growing need for food and other agricultural products. The Food and Agriculture Organization (FAO) works with countries around the world to improve food security. It has been assisting member states in cooperation with public and private financial institutions since 1964, and has implemented numerous programs to invest in agriculture and rural development. In recent years, emergency aid to meet the urgent needs of people in Economic Cooperation Organization (ECO) member states has been provided through these programs in times of crises, such as earthquakes, floods, droughts, and avian influenza.

Keywords: Food security, Hunger, Water security, Virtual water, Irrigated agriculture.

12.1 INTRODUCTION

One of the most important aspects of human life and health is nutrition. The availability and supply of healthy and sufficient food, ease of access, family income and purchasing power are important factors for ensuring food security. Another important factor needed to ensure food security is nutritional security, which depends on knowledge in families of how to allocate funds to prepare the best (nutritionally rich) available food and how to share it.

In order to ensure food security within a country and across the social system, organizations and institutions must collaborate and coordinate with a food safety organization in the production or import of food, imparting education, and raising public awareness. The local food safety authority must be aware of the type, quantity, and price of food. It must ensure that all people have access and purchasing power, and it should sound an alarm in the event of a crisis. Food security has been a major challenge in agriculture since ancient times, and in developing countries it is a function of two factors: population growth and food production.

Another challenging issue for food security for the growing population today is water security, an emerging concept and a challenge. Water security means 'reliable access to acceptable quantities and quality of water for health, livelihood and production' and is one of the key factors of sustainable development. Inefficient utilization of water resources drains water reserves, undermines economic development, and eventually becomes a serious threat to long-term food security. Water plays a key role in food security through its multiple effects on health, nutrition, energy generation, pollution abatement, agricultural production, and food processing.

With population growth, climate change, competition over water resources, and environmental quality concerns, water supply is a major issue in food production. A review of water resources shows that water security around the world is at risk. It is expected that the world's population will grow to 9–10 billion ($9–10 \times 10^9$) by 2050, and more than 50% of the world's population will live in areas that suffer from water shortages for at least one month a year (United Nations, 2018).

Therefore, water scarcity and water quality are major challenges for future food security. While agriculture is expected to remain the largest consumer of fresh water resources worldwide, industrial and domestic demands are also increasing rapidly. Further, the reliability of agricultural water supply will be reduced without significant improvements in water resources management and investment policies. Strong competition in different sectors due to water scarcity, coupled with a decline in agricultural water supply reliability, puts food supply under pressure and raises concerns for food security. The agricultural sector faces a complex set of challenges, such as the need to produce larger amounts of food with greater quality while using less water per unit of production, providing healthy living for rural people with available resources and opportunities, using clean technologies to ensure environmental sustainability, and achieving an efficient way to improve local and national economies. It is worth mentioning that farmers are at the core of any changing process and hence must be encouraged and guided through appropriate incentives and governmental measures introduced to preserve natural ecosystems and to minimize possible negative effects. In addition, there is a need for targeted investments in infrastructure renovation, organizational restructuring, and technical capacity building for farmers and water managers.

12.2 WATER SECURITY OR FOOD SECURITY

Since water is important for both food security and public health, its deficiency can lead to social unrest and conflict. Food security is a vital requirement for sustainable development, but it cannot be achieved without water security. As water security cannot be ignored for food security, problems caused by rising water stress cannot be ignored. Hence, policymakers need to plan appropriately in order to avoid problems that arise from water crisis and its adverse effects on food security.

In the future, water shortages will not only be primarily due to simple water scarcity but will also be due to unplanned use of water. Accordingly, water governance has now become a key concern in rising water scarcity, in local and transboundary water conflicts, and with regard to global climate change.

12.2.1 State of food security in the world

In 2015, after a relatively long period of effort to reduce hunger and malnutrition, 784 million people were affected by this crisis around the world. In 2017, with an increase in hunger and malnutrition, the figure had risen to 821 million. These statistics prove the practicality and necessity of continuing the study of this issue (Moraru *et al.*, 2020). This decline is more pronounced in developing regions, despite significant population growth. In recent years, slower and less pervasive economic growth, as well as political instability in some developing regions such as Central Africa and West Asia, have hampered the development of food security programs (McGuire, 2015).

2015 was the end of the Millennium Development Goals (MDG) Monitoring Period. In developing regions, the overall proportion of malnutrition in the population declined from 23.3% in 1990 to 12.9% in 1992. Some areas, such as Latin America, Eastern, and Southeast Asia, the Caucasus and Central Asia, and North and West Africa, have made rapid progress. In addition, improvements have been made in South Asia, Oceania, the

Caribbean and South, and East Africa, but have been slower in achieving the MDG goal (1C) of halving the proportion of people with chronic malnutrition.

In developing regions, two indices of the MDGs – the prevalence of undernourishment (PoU) and the proportion of underweight children under five years of age (CU5) –both declined from 1990 to 2017. In some areas, including East Africa, Southeast Asia, and South America, malnutrition rates declined at a faster rate than underweight children, indicating that there is still a need to improve diet quality, hygiene, and access to clean water, especially for populations. Economic growth is a key factor in reducing malnutrition, but it must be inclusive and create opportunities to improve the livelihood of the poor. The development key is to increase the productivity and income of small farm-holder families.

Social protection systems have played a significant role in boosting progress towards the goal of eliminating hunger and poverty in a number of developing countries. Social protection is achieved by promoting economic security and access to adequate nutrition, health care, and better education, which directly contribute to the reduction of poverty, hunger, and malnutrition. By improving human capacities and reducing the impact of economic damage, social protection enhances the ability of the poor to access employment.

In a report in 2009, the FAO said that to feed 9 billion people in the world by 2050, production has to increase by 70%, requiring the removal of current obstacles such as agricultural land constraints, water scarcity, high energy prices, investment reduction in agricultural research, and increasing food waste (FAO, 2009).

From the 21st century food security perspective, food production is facing problems. Competition over agricultural land and water resources, high energy prices, and climate change suggest that with fewer resources, more food needs to be produced. Sustainable growth in agriculture is a vital factor in feeding the world for decades to come and, as we know, agricultural output has a direct relationship with water resources.

After taking stock of the food security situation globally, we now consider various experts who have been working on water-based food security issues in recent years and give a snapshot of their work.

Rockström (2001) pointed to one of the biggest food security challenges associated with green water security. He believed that the greatest challenge to biophysical water is whether freshwater is sufficient to sustain global food production and natural ecosystem services or not. He also focused on small-scale drought and agricultural disasters.

Rosegrant and Cline (2003) examined the challenges and policies related to world food security, which will be a global concern for the next 50 years and beyond. They reported that crop yields had recently declined in many areas due to reduced investment in research and infrastructure and increased water scarcity. Climate change is a major contributor to providing food security in many areas and, they pointed out, with increased investment and political reforms, food security in developing countries will improve significantly.

Yang *et al.* (2003) pointed out water resources thresholds and their effects on food security. They calculated per capita renewable water resources by 2030 for Asian and African countries (assuming full availability of water) using population forecasting figures and related uncertainties. They also stated that a numerical account of the relationship between water scarcity and grain imports was important in forecasting future food demands and formulating relevant national and international policies. Population growth as one of the influential factors of rising water demand and decline of the per capita availability of water has led more people to experience water crisis. Thus, it is logical to expand the demand for food imports to compensate for water shortage. Studies indicate that a threshold decline has occurred. Most of the countries with lower thresholds were rich in oil, thus were able to pay for grain. However, in the next 30 years, many poor and populated countries may be on the verge of rapid population growth and groundwater depletion. Food insecurity and starvation caused by water scarcity can be exacerbated because the grain imported by these countries may not be profitable.

White *et al.* (2007) discussed the impact of climate change on food and water security in the Arctic, pointing out that lakes, rivers, and lagoons in the Arctic landscape are not typically connected to groundwater, hence only lakes deeper than two meters and rivers with significant flow retain liquid water. They also refer to the biological conditions and suggest that humans in the Arctic rely on surface water, so the important issue is understanding the impacts of climate change on food and water security in the Arctic, although predicting water changes and future climate will cause significant uncertainty. This uncertainty can be used along with impact analysis to help the community with planning and preparedness through a risk assessment of climate change.

Godfray *et al.* (2010) addressed the issue of food security and the challenge of nutrition. They argued that continued population growth and consumption meant that global food demand would continue to rise for at least another 40 years, and there was no easy solution for the sustainable nutrition of 9 billion people. A wide range of options, including alternatives covered by Godfray *et al.* (2010), should be followed simultaneously. For example, agricultural land and water used for aquaculture and fisheries could be managed in ways specifically designed to reduce the negative impacts on biodiversity. However, it must be kept in mind that enormous challenges exist in achieving food security, such as sustainable food production while controlling greenhouse gas emissions, reduction in water consumption, and meeting the MDGs. Climate change is another threat, but the world can produce more food if it is managed more effectively.

Hanjra and Qureshi (2010) examined the overall role of several emerging forces in redefining global food security. These forces include climate change, water scarcity, energy crisis, population growth, and credit crisis. Studies have shown that feeding the population in the year 2050 would require about 12,400 km^3 of water, of which only 6800 km^3 of water is used today. However, about 3300 km^3 of water scarcity remains, even after improving agricultural irrigation efficiency, water management, and updating dryland agriculture. This will lead to a food gap unless today's coordinated actions, such as confronting climate change, land and water conservation, reduction in food system energy efficiency, modernizing irrigation infrastructure, supplying domestic foods, reforming international food trade, and other global challenges, are addressed. They also pointed out that there were restrictions on conveyance capacity in some regions. In addition, increase in food production should be fulfilled in regions that are now being cultivated under improved water and energy efficiency. In order to eliminate institutional, sovereign, and financial constraints needed for future food security of people by 2050, an unprecedented global co-operation would be needed.

Turral *et al.* (2011) summarized the anticipated impacts of climate change on water availability for agriculture and examined the implications of local and national food security. They analyzed the impacts of climate change on a set of major endangered agricultural systems and made it faster to implement 'certain' strategies that would result in positive development to make agricultural systems more flexible. It is hoped that policymakers and planners can use this report to tailor their adaptation responses when considering both water in agriculture and competing demands from other users.

Bakker (2012) examined water security and its related challenges and opportunities and stated that water security was recognized as an acceptable level of water-related hazard to humans and ecosystems. Therefore, water security is attracting increasing attention. On the other hand, he stated that due to the challenges and research opportunities, it was estimated that 80% of the world population was at a high risk of environment-related diseases. However, to date, little progress has been made in integrating academic research on water security with the needs of policymakers and experts. Hence, if researchers could play an important role in resolving the global water crisis, significant changes in budget, education, research, and scientific motivational structures would be needed.

Cook and Bakker (2012) provided a comprehensive review of water security from both scientific and policy perspectives. The use of the term water security has increased significantly over the last decade across different disciplines. They compared definitions and analytical approaches to water security in the natural and social sciences showing that the methods and scales thereof are distinct and sometimes inconsistent. As a result, an integrated approach to water security raises issues of governing management, thus promising a new approach to water management.

Brauman *et al.* (2013) discussed food security, its definition, and how it is measured. The term 'food security' has been used to define different things and this study discussed various meanings associated with this concept and suggested this to be a useful measure of household and individual well-being. If food security is the target, an estimate of food access should be combined with an estimate of proper access to clean water. Furthermore, the paper points out that anthropometric measures are probably more appropriate than food safety considerations targeting child nutrition policies and programs.

In examining the relationship between water and food, virtual water is effective to reduce water scarcity, because it creates a link between water, food, and trade. Zhang *et al.* (2018) considered agricultural trade from 2001 to 2015 with the aim of understanding the Belt and Road Initiative from the perspective of agricultural virtual water trade. In addition, more than half of the countries' virtual water trade with China was in line with a virtual water strategy that helped alleviate water crisis. Trade in agricultural products with China is definitely in the interest of both countries in terms of virtual water. These findings can also provide resources for optimizing business structures.

Kang *et al.* (2009) provided a comprehensive survey of the impacts of climate change on productivity using climate, water, and agricultural models. Existing studies show that climate change models with the higher spatial resolution are able to predict the future of climate. It is anticipated that water availability will increase in some parts of the world, which has its own impact on water use efficiency and water allocation. If the level of irrigation expands or if irrigation intensifies, crop production can increase, but this can also increase the amount of environmental damage. As climate change affects the soil-water balance, soil evaporation and heat transfer changes occur, therefore future crop growth could vary. It is estimated that crops are affected by climate change in different regions; in some areas, crop production will increase, and in others, it will decrease, depending on the area and irrigation program. Existing modeling results indicate that increasing rainfall results in increased crop yield, and maximum crop yields vary with temperature and temperature sensitivity. If the available water declines in the future, soils with high water retention capacity will have a better performance to reduce the effects of potential drought and maintain crop yields. With rising temperatures and rainfall fluctuations, access to water and crop production will likely decline in the future.

Despite the challenges in establishing food security, efficient solutions must be sought. Another way to ensure food security and produce more food is to use fertilizers. However, it should be noted that the use of artificial fertilizers is worrisome, because although they contribute to increased yield, they lead to a series of environmental problems, especially in areas facing water shortages. For example, the North China Plain (NCP), one of China's main food production areas, has helped increase food production by over-exploiting groundwater for irrigation as well as over-using nitrogen (N) fertilizers. It has also led to water shortages and groundwater pollution. Therefore, potential conflicts between strategies that ensure water security, food security, and water pollution reduction in the NCP are being explored in this area. Wang (2019) state that the transition from food-oriented strategies to sustainable resources and environment development and management is essential.

Food insecurity is doubly important because it affects children around the world, and malnutrition makes them more susceptible to various diseases in many developing countries. Various countries, such as South Sudan, Yemen, Syria, Kenya, Uganda, and Libya, have been hit by a severe food insecurity crisis resulting in a humanitarian emergency. However, several EU countries and policymakers have been involved in addressing the food crisis (by supporting agriculture, starting resource recycling, developing water–energy–food initiatives, and introducing water sustainability and food security strategies). Another way out of the food security crisis is genetically modified crops, which can help solve this problem by 25–40%. To prevent disruption of crops and incorporate new dynamic technologies to ensure global food security, it is necessary to minimize pathogenic infections through horizontal gene transfer (Pachapur *et al.*, 2020).

It should be noted, however, that in addition to the challenges discussed above, the increasing prevalence of co-morbidity between humans and livestock may also be a threat to both resources (water and food) due to such factors as deforestation and climate change, and increase the potential for future epidemics. Immediate provision of water and sanitation against epidemics, such as Covid-19, can save lives, but in the medium term, steps need to be taken to strengthen water security. Medium-term measures, in addition to ensuring sustainable access to water with acceptable quality, can help countries avoid some of the economic and human costs that result from epidemics (Cooper, 2020).

It should also be noted that the latest data show that the food security of 135 million people is classified as a crisis level or worse. That number could almost double before the end of 2020 due to the effects of COVID-19. The epidemic could push some 49 million people into extreme poverty by 2020. Each 1% reduction in global gross domestic product (GDP) is expected to result in 0.7 million underdeveloped children. The COVID-19 crisis threatens the food and nutrition security of millions of people, many of whom have already suffered. In the long run, we face disruptions in the functioning of food systems, which have serious consequences for health and nutrition. By taking concerted action, we can not only prevent some of the worst effects, but also do so in a way that shifts to more sustainable food systems that are better balanced with nature and support healthy diets (United Nations, 2020).

12.3 DEFINITIONS AND IDIOMS

12.3.1 What is food security?

According to the United Nations (UN), food security is access (physically, socially, and economically) to adequate, safe, and nutritious food for all people at all times, which of course corresponds to food preferences and nutritional needs for an active and healthy life (FAO, 1996). This will be achieved only if the family's food basket is properly prepared. In other words, it should be sufficient for the family members and cooked properly so that nutrients reach the cells and organs of the body.

12.3.2 Key elements of food security

There are three elements to food security: availability of food, access to food, and stability of food intake (FAO, 1996):

(1) food availability: includes food supply existing within national borders, through domestic production and food imports;
(2) food access: physical and economic access to resources for providing nutritious food items is needed;
(3) stability of food intake: stability and consistency is required in the nutritional value needed to sustain the community.

12.3.3 Factors affecting the food security of a community

Various factors can affect the food security of a community, including family income and families' tastes and nutritional knowledge of how to budget to supply the most valuable variety of food and how to share that diet among family members.

The collaboration of organizations and institutions to provide this important principle in all countries is also important. The principle responsible organization should always be aware of the diet being offered, including its quantity and price, and should make sure that everyone has physical access to it, ensuring that it is affordable given each individual's income. In case of crisis, this organization is responsible for raising the alarm and undertaking security measures.

12.3.4 Food security threats

Food security and nutritional threats can occur for a number of reasons, including if food is overpriced, or if there are problems of supply as a result of defective food distribution systems. A country or region's food security can also be threatened by pollution or water crisis.

12.4 BASICS AND LOGIC

12.4.1 Investigation of agricultural water resources

The water cycle on Earth has a relatively stable volume. It is estimated that the amount of water on Earth is about 1.39 billion km^3 and agriculture consumes around 7130 km^3 (Graham *et al.*, 2010; IWMI, 2007). Food production and distribution are dependent on the existence of water. Agriculture without the use of soil has been recently introduced; however, agricultural production without water is beyond expectation. Food supply relies on the internal resources of each country, necessarily depending on the amount of its available water as direct trade in water is not a common practice for agricultural production.

In addition to its vital role in sustaining the livelihood of individuals, water has a widespread role in economic, social, and cultural functions of societies. Water footprints can also be found in the production of every single human commodity, as water is needed to produce a variety of industrial and service products.

12.4.2 The role of water in food security

The food security of a country depends on its water resources. Since the relationship between food security and water resources in each country and region is complex, the ratio of harvested water from renewable resources to water from other sources indicates the degree of food security and highlights the risk of food supply for that country or region. Since the water cycle in the world is a closed cycle and has a certain amount of water, the imbalance between water supply and food demand affects the supply of food around the globe, and sets the competitive stage for food supply for most countries in the world.

Higher food demands from countries that import nutritious ingredients change prices in the world market and affect both the importing and exporting countries' economies. Increasing demand, followed by increased production and exports, puts pressure on the water resources of the countries that export food, and causes environmental problems and decline in the quantity and quality of water suitable for food production. Therefore, all countries in the world are tied up together in water issues regarding food security. All regions and countries are more and less interconnected in a chain and cannot exclude themselves from this interaction.

12.4.3 Harvesting water resources for food security

Harvesting water for food production is done as a matter of course, and it is rational to consume water for food production along with drinking needs over other uses. Establishing sustainable food security in each country requires that water harvesting for food production should be prioritized over other sectors' consumption. At the same time, the efficiency of water usage in food production should be maximized.

Different climatic conditions in different parts of the globe and water availability are the most important factors affecting water harvesting and consumption. Other factors, such as technology level, cropping pattern, population density, economic issues, and socio-cultural characteristics also affect the quantity and quality of water harvesting.

In order to compare efficiencies of water systems of different countries, comprehensive information regarding climate, population, consumption patterns, and production strategies in relation to food, agriculture, environment, foreign trade, and economy is needed. Based on its source, water harvesting can be undertaken from groundwater sources, rivers, lakes, ponds, or directly from precipitation. Groundwater and onshore water resources are referred to as blue water, and rainwater (directly) is called green water in the water resources literature.

12.5 THE NEED FOR FOOD SECURITY

Food security depends on sustainable agricultural growth and development, and will not be achieved unless there is an emphasis on preserving, restoring, and balancing the basic resources of water, soil, vegetation, environment, and the air.

Although food security is considered to be a multifaceted element which is influenced by several factors, the most important and influential of these is perhaps production; in other words, producing adequate food, at the right time, is the most important factor to ensure food security. This imposes great pressure on agriculture and on the livelihoods of individuals, especially as climate change has made rainfall much more volatile than previously experienced.

12.6 MAJOR CHALLENGES IN GLOBAL FOOD SECURITY

In its 2011 report, the U.S. International Food Policy Research Institute outlined the challenges of food security (Kuyvenhoven, 2012):

(1) severe climatic shocks, especially human-made: Climate change is causing an increase in the number of people with malnutrition, from 40 to 170 million people worldwide. While higher temperatures in higher latitude regions can be associated with higher crop yields, in lower regions (where most developing countries are located), negative consequences are expected;
(2) food price instability: From the price of fertilizer and energy consumption on the farm, to transportation costs and the use of agricultural products for fuel production (which may be the most important cause of rising food prices in recent years), a wide range of different factors can contribute to food price instability;
(3) water scarcity: Half a billion people live in countries with severe water scarcity, and by 2050 this number will reach 4 billion. This is not only due to climate change but also due to the excessive use of water from rivers, lakes, and underground resources;
(4) starvation due to war and unrest: About one and a half billion people, and about 42% of the world's poor, live in regions facing wars.
(5) biofuels: The use of biofuels was initially considered as a way to reduce dependence on fossil fuels and prevent the emission of carbon produced by them, but a strong link with agricultural markets and even land-use patterns has also emerged as a significant issue. First-generation biofuels currently in commercial use are most strongly associated with agricultural markets and land use, although there are important by-products that can be used as feed and therefore offset the effects. Given the areas of scientific uncertainty that still remain about the impact of biofuels on food security and the environment, the decision to use them is complex and is in some cases controversial.
(6) the presence of new actors (private sector, humanitarian organizations, green organizations, and developing countries such as India, China, Brazil, etc.) : In recent years, for example, emerging economies such as Brazil, India and China have had unprecedented influence on the global food system, and the same is true of some charities as well as private sector initiatives.
(7) over-exploitation and erosion of land: About 24% of the world's agricultural land is affected by erosion, and 1% of the total land used under cultivation has been lost, which amounts to 20 million tons of cereal production annually.
(8) political developments: Political developments in the Middle East region due to the Islamic awakening (Arab Spring) make countries in this region food insecure.

To resolve these challenges, a number of solutions can be proposed. To change dominant dietary patterns towards environmental sustainability and to preserve the environment, we need to look at food in a different way and move beyond just food waste reduction and lead dominant nutritional patterns towards environmental sustainability, whilst also promoting public health.

Increasing per capita intake of vegetables and legumes is a nutritionally rich and healthy option. Approximately just one-tenth of water consumed for red meat production is used to produce legumes, while vegetable protein intake prevents the spreading of many common chronic diseases. Low consumption of vegetables and legumes is the main nutritional factor for mortality.

12.7 METHODS

12.7.1 Water resources and consumption adaptation indicator

In the 21st century, especially in developing countries, the demand for water to meet the needs of the growing population has increased substantially. This is especially important, of course, in areas of the world where there are water shortages. As mentioned above, the spatial distribution of water in the world is very uneven; therefore, there are significant differences in access to water and water consumption per capita in various regions of the world.

Important indicators regarding the status of water resources are:

(1) the ratio of water harvested to the total water available; and
(2) the water productivity index.

One of the predominant indicators of water resources status in each country is the ratio of extraction to available water (water potential index). The larger the index, the greater the pressure on the country's water resources, and the greater the water stress in a country. If the index is in the range of 1–10, there is no water stress, if the index is in the range of 10–20, there is low water stress, if the index is in the range of 20–40, there is moderate stress, and if the index is over 40, there will be severe water stress.

Another indicator of water resources consumption is the water productivity index, which can be measured, for example, as the ratio of GDP to the total annual freshwater harvest (Vladimirova *et al.*, 2018). The value of this indicator varies according to the conditions and economic structure of different countries. The higher value of the indicator in developed countries is due to the higher prevalence of technologies and higher efficiency and productivity, so that they produce greater value per unit of water than do developing countries. However, most developed countries have fewer water constraints, being located at higher latitudes, and hence face fewer problems. Therefore, promoting water productivity and efficiency in countries with less water is important, and it is essential that appropriate policies and measures are taken.

12.7.2 Relations and calculations

Wheat is a staple food worldwide, produced in both rainfed and irrigated farm lands. The amount of virtual water required for both irrigated and rainfed wheat is addressed in this section.

The amount of virtual water used is calculated using gross requirement, precipitation, and crop, according to water efficiency. By definition, virtual water is the amount of water used to produce one kilogram of agricultural material (Allan, 2003). Since the water harvested for agriculture is considerably high, recent efforts to be self-sufficient for the production of strategic crops have put even more pressure on water resources. The virtual water source can be green water (part of the evapotranspiration that comes directly from rainfall during the growing season) or blue water (part of the evapotranspiration that comes from surface or underground water sources). If rainfall occurs at the desired location during the growing season, effective precipitation is stored directly in the root zone and is used by the plant; this is defined as green water. Although the source of blue water supply is also precipitation, it occurs during the non-growing season. Therefore, the water must be harvested from the surface or underground sources during the growing season to meet the plant's water needs. For this purpose, this type of water source is called blue water.

A spatial analysis of virtual water content is needed for different crops with the aim of identifying the best cultivation pattern in a region with respect to virtual water; not only does this provide an optimal use of available resources and capacities, but it also reduces water consumption and increases agricultural production.

The amount of water required for irrigated wheat can be calculated from Equation (12.1), and the virtual water needed for irrigated wheat and rain-fed irrigation are calculated from Equations (12.2 and 12.3), respectively:

$$\mathrm{CWRi} = \frac{\mathrm{ET}}{\mathrm{Ea}} \tag{12.1}$$

$$\mathrm{VWCi} = \frac{\mathrm{CWRi}}{\mathrm{Yi}} \tag{12.2}$$

$$\mathrm{VWCr} = \frac{\mathrm{Re}}{\mathrm{Yr}} \tag{12.3}$$

in which: ET is the net irrigation water required (m^3/ha); Ea is the efficiency of water used in Irrigation; CWRi is the water requirement for irrigated wheat (m^3/ha); Yi is the average yield of irrigated wheat (kg/ha); VWCi is the amount of virtual water consumed by irrigated wheat (m^3/kg of product); VWCr is the amount of virtual water consumed by rainfed wheat (m^3/kg product); Yr is the average rainfed wheat yield (kg/ha); and Re is the effective rainfall (m^3/ha). Figure 12.1 shows a flowchart which can be used for calculating the net irrigation water required and the amount of virtual water consumed by desired plants. Then, the amount of virtual water consumed by agricultural production is calculated.

The amount of virtual water is calculated based on the amount of precipitation from meteorological data and also on the amount of precipitation reported in NETWAT software. However, there is an inconsistency in the values produced by NETWAT Software in comparison to meteorological data for the same years.

It should be noted that agriculture may change water resources in other ways instead of just satisfying crop water needs. For years, declining water quality (through pollution produced by the agricultural sector) has negatively affected public health and ecosystems. It is important to consider water quality in water resources management. Accordingly, systematic monitoring of water quality, pollution control, and wastewater treatment are considered to be necessary and urgent programs in this field.

Gray water footprint is an indicator of freshwater pollution and is defined as the volume of freshwater that is required to absorb the load of pollutants based on water quality standards in the environment. The calculation of this index is important, to take into account environmental considerations. Gray water footprint is calculated by dividing the pollutant load (in mass/time) by the difference between the ambient water quality standard for that

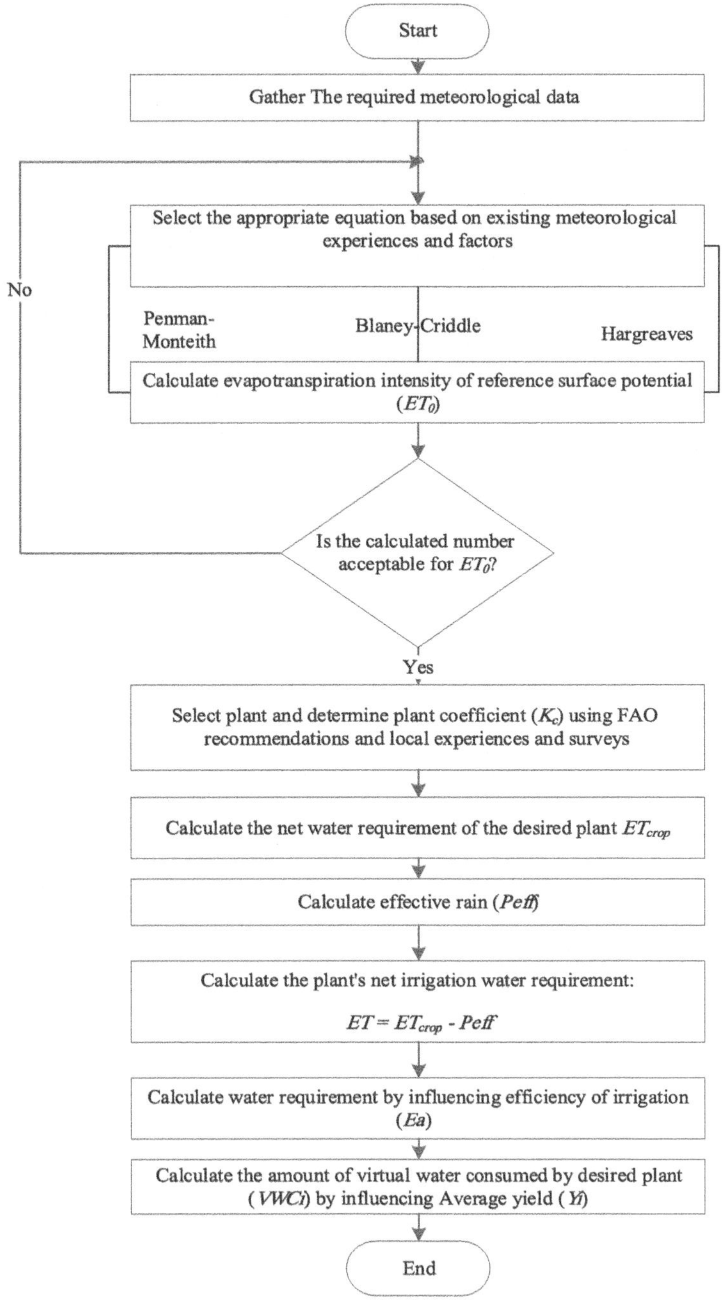

Figure 12.1 General scheme of various steps for calculating the amount of virtual water consumed by desired plants.

pollutant (maximum acceptable concentration, in mass/volume) and its normal concentration in the receiving water body (in mass/volume) (Hoekstra *et al.*, 2009).

12.8 EXAMPLES

Since China is one of the most populous countries worldwide, food security is consequently a greater risk, so it provides a good view of the state of food security, and a solution to face a more populous future with more water and food demand. Khan *et al.* (2009) reviewed the situation of water and food resources and the factors affecting them in China, which is summarized below. Agriculture in Latin America and the Caribbean (LAC) has also been considered as a further case, given that LAC agriculture is deemed a strategic area for rural development and poverty relief policies, as well as having a major role in subduing local and global food insecurities (Willaarts *et al.*, 2014).

After examining the two study areas' tangible food and water securities and their relevance, the situation of water and food security in the world is briefly discussed.

12.8.1 The state of water and food resources in China

Food security is one of the top issues on the Chinese political agenda. China's food security is affected by several factors, including political factors, population growth, urbanization, industrialization, land-use changes, water scarcity, income growth, transmission of nutrition, and disruption in global energy and food markets. Sustainable growth in agricultural productivity and sustainable relationships with global food suppliers are twin anchors of food security. Shortages in China's domestic food production could also affect international food markets. The turmoil in global energy markets can affect food prices and global supply costs, which in fact can affect food security and poverty. Policies to safeguard food supply are essential. China needs to improve its policies by investing in infrastructures, such as irrigation, drainage, storage, agricultural research, and institutional reforms exemplified by linkages between water and other development-related sectors, such as population, energy, food, and environment, as these factors together determine the future of food security and poverty reduction in China. Climate change creates a new level of uncertainty in water management and requires rapid research to develop measures to ameliorate water-related stresses.

Crude petroleum prices hit $100 a barrel in early 2008. Increasing petroleum prices and government subsidies in some countries increased food consumption for biofuels, and increased fertilizer costs and irrigation. In 2007, the prices of corn and vegetable oils rose by an unprecedented 33% and 50%, respectively, and these price shocks shifted to other food grains, with wheat prices rising by about 30%. Structural changes in the global grain markets also increased the rice price by almost 40% from December 2005 to 2007, thereby jeopardizing food security.

The turmoil in global energy and food markets could affect food security and poverty reduction in China. The sudden rise in food prices can greatly affect urban and non-agricultural homes, as well as farmers in rural areas, and also affect rural families in different ways. Price changes can have a variety of impacts on a variety of households and regions due to the heterogeneity in consumer behavior and sources of income, as well as their possible consequences for compensation policies. The cost of protecting the livelihoods and food security of these vulnerable households could be significant.

Population growth, urbanization, industrialization, income growth as well as increased consumption and lifestyle changes by global forces and market integration pose challenges to maintaining food security in China. Investing in new infrastructure, policies, and institutions in China are essential to achieving national food security goals and maintaining poverty reduction. In this regard, water security must be ensured, and efforts should be made to prevent the loss of fertile land for urbanization and industrialization. Given these challenges, the role of land and water resources, policies, and institutions in promoting food security and poverty reduction in China is discussed further below.

12.8.1.1 Economic reforms and food security

China launched its economic reform program more than 40 years ago when the government introduced the family responsibility system in agriculture. Prior to the reforms, much of China was a peasant farming community, and rapid economic growth was possible, as large portions of agricultural labor were surplus. Economic reform often creates major benefits and disadvantages, and the distribution of impacts varies across social groups, with implications for social justice and reform. The Chinese peasants were a very important group from the point of view of food security and justice, as well as economic reform.

During the 1980s and 1990s, agricultural productivity increased steadily, and per capita production reached a similar level as in developed countries, and food exports increased significantly, and with steady growth in agriculture, rural income increased dramatically. Millions of people were permanently lifted out of poverty. Despite these remarkable achievements, more than 100 million farmers and their families still live in poverty. Rural and urban income gaps remain widespread, and inequality in the rural economy has persisted since the mid-1990s. Urbanization and increasing influence require new demands for food, more land, and more water. As a result, overuse and degradation of resources in major river basins have increased.

12.8.1.2 Passive factors and food security

The challenge of maintaining food security as population and income increase is a research topic and an important issue in China's political agenda. Five anthropological factors that influence the pace of rising food demand are (1) population growth, (2) urbanization, (3) industrialization, (4) changing lifestyle and consumption, and (5) changes in political and economic priorities.

In the first half of the 21st century, China's population is expected to increase to 1.6 billion (Khan *et al.*, 2009). Rural migration to the city, the growth of cities and industry, and changes in consumption patterns that increase with income put additional pressure on land and water resources. Demand for meat and dairy products (which require more water to produce than grain) is on the rise. At the same time, more land will be increasingly needed for transportation, housing, and energy production infrastructure to support the urban population.

12.8.1.3 Land use changes

Changes in land use in China are due to rising demand for food and other economic and political factors. Land sales are an important source of local income in many areas. In 1996, about 67% of China's land was devoted to agriculture. This cultivated area was about 130 million hectares. Between 1978 and 1996, the area fell from 99.4 million to

95.0 million hectares due to structural changes in agriculture, such as the conversion of rice fields to orchards or fish ponds. However, a significant portion of these losses were due to industrial and urban development. Many of the losses occurred in coastal and central areas, while the western region has had little increase in cultivated areas.

The 1990s marked the beginning of the process of agricultural reconstruction, industrialization of villages, urbanization, and economic reforms, which caused the loss of agricultural land. The potential impact of changes in the cultivated area on food production and other ecosystem values differs from the sources of these changes. For example, modifying low-quality land, in general, can offset the loss in productivity. As has happened in southeastern China, crop density and population density are very high, and the amount of arable land per person is very low. According to long-term studies in China, six prominent methods of land degradation are desertification, secondary salinity, agricultural destruction, deforestation, weed degradation, and wetland loss.

Some claim that China's area under cultivation is shrinking. Changes in agricultural and grain-covered areas across China vary with geographical implications. To maintain food production, all losses in the agricultural sector must be offsetby more cultivation in the remaining areas. The long-term sustainability of food production is threatened by soil degradation, pollution, and reduced soil fertility. Additional threats include soil contamination through waste disposal, acid rain from air pollution in cities, and increased flood risk due to urbanization.

12.8.1.4 Food production

China has made significant progress in achieving food security. Since the late 1970s, when China embarked on economic reforms, grain production has increased dramatically. From 1978 to 1997, total output increased by 62%. Corn and wheat production doubled between 1978 and 1997, while rice production increased by about 50%. Land-use changes began in the 1990s, but grain production also reached its highest level. Continued gains in agricultural productivity are needed in China, where more than 300 million workers, or about half of the labor force, depend on agriculture for their livelihoods.

Productivity gains vary across provinces. For example, farm-level data in Jiangsu Province show that the decline in productivity in 1988–1996 was due to a decline in farm production and planting, so understanding food security in China requires regional and provincial data.

The adoption of technology accounted for approximately 40% of the increase in rice productivity between 1980 and 1995, while institutional reforms resulted in a 35% increase in productivity. Wheat, maize, and rice are the most productive. National and international investment in new technology and continued organizational reforms and policies will continue to be important (Khan *et al.*, 2009).

China has pursued a goal of maintaining a high degree of grain self-sufficiency. However, forecasts of Chinese demand and grain supply by 2030 are somewhat different. Per capita consumption of grain across China is expected to increase slowly from its 380 kg in 1995 to a forecasted 400 kg in 2030. Increasing demand for meat in China increases the demand for grain, which is the main feed for livestock, and increasing the production of poultry also uses more feed.

In 2007, China's annual per capita meat consumption (30 kg) was lower than that of South Korea (32 kg), Japan (40 kg), and Malaysia (44 kg), but this per capita was expected to increase, and to compete with Europe and the United States. Pork also accounts for about 75% of Chinese meat consumption, and pork production has less impact on land and water resources than beef production. With rising incomes in China, meat demand may be somewhat diversified, so the demand for grain will increase, and more pressure on land and water resources will increase.

Despite changes in consumption patterns in the 1990s, much of the cost of domestic household products are still in pork (which requires much less water to produce than beef), and possibly the share of fish and poultry in the Chinese diet will have a significant increase. Estimates of the elasticity of food and nutrients from a prosperous region in northern China show that the relative share of grain and protein in calorie intake has declined further.

12.8.1.5 Water resource management

The geographical distribution of water resources in China is uneven. It is estimated that 81% of China's water resources are found in the south, with most of China's fields (64%) in the north. Water availability per hectare of farmland in the north is about 12% of the water available in the south. Similarly, the average supply of groundwater in the south is about four times that of the north. Excessive groundwater harvesting occurs beneath the North China Plain. The consequences of over-pumping include higher pumping costs, excessive energy consumption, saline water infiltration, increased soil salinity, soil erosion, and land decomposition. Excessive pumping of water in the upstream areas completely destroys the river basin. As an example, the Yellow River does not reach the sea for most years.

Water pollution also affects food security. The paper industry, some of which is the largest source of rural pollution in China, accounts for 10% of wastewater discharges and 25% of chemical oxygen demand. Evidence from case studies in heavily polluted cities, such as Chongqing, shows that water pollution reduces GDP by 1.2% in local areas. Losses in agriculture account for the largest share of costs (56%), while losses in the health and industrial sectors account for 20% and 18%, respectively. Many incentives and legal tools have been proposed to reduce water pollution in China.

Sustainable agricultural production and irrigation are vital to securing Chinese food. China has one of the largest irrigation zones in the world (59.3 million ha), which accounts for about half of China's arable land and produces about three-quarters of the grain harvest. The irrigation area increased from 45 million hectares in 1978 to 54.5 hectares in 2004 (Khan *et al.*, 2009). It should be noted that in 2019, the irrigated areas in China reached 69.9 million hectares (Du *et al.*, 2019).

Water management institutions provide financial incentives for water management to encourage water conservation. Inefficient use of water reduces water productivity. The average cereal production per m^3 of water in China is only 0.85 kg or about 50% of productivity in many developed countries, and less than 33% of productivity in Israel (2.32 kg/m^3).

Water transfer projects may be economically feasible, but their social and environmental impacts are controversial. Water transfer projects are based on political reasoning rather than food security concerns. The goal of most of these projects in China is to improve the water environment in the North China basin, which is facing water stress. These engineering measures must be completed with reasonable water prices, investment policies, and legal measures.

12.8.1.6 Summary

China's national development policy holds food security at its heart. But that food security is challenged by several anthropological, social, and political factors, including:

- population growth;
- reputation and industrialization;
- land-use changes and water scarcity;
- revenue growth and nutrition transfer; and
- the turmoil in global energy and food markets.

Sustainable growth in agricultural productivity and sustainable relationships with global food suppliers are both important for ensuring food security. Shortages in domestic food production can affect international food markets. The turmoil in global energy markets can affect food prices and supply costs. Policy protection is essential to safeguard food supply against such forces.

Water resources are essential for agricultural development. China needs to implement policies on the loss of fertile land and must invest in infrastructure such as irrigation, drainage, storage, and transportation. Policies and technical support are also needed to improve water use efficiency and protect water quality in the North China Plain, where groundwater levels are declining. Policies that support the production of genetically modified crops can improve the well-being of China's poor.

Increased water scarcity and symptoms of underground stress are due to complex socio-economic and geophysical factors that require further research. Any future water crisis may be due to a physical shortage of water, but it may also be due to poor water governance. Cross-sectoral policy responses are needed to address the links between water and other development-related sectors, such as population, energy, food, environment, and interactions between them.

12.8.2 The state of water and food resources in Latin America and the Caribbean

The Latin America and Caribbean (LAC) agricultural sector has grown remarkably over the past years, chiefly due to trade liberalization policies which have helped to make the LAC region a significant competitor in the global agricultural market (both concerning food and biofuel production). Furthermore, the region has come to account for an ever-increasing share of the global agricultural market. Accordingly, the LAC region accounted for more than 18.4% of global agricultural trade in 2013, increasing from 11.4% in 1990 (World Bank, 2013). Oilseed, soybean, grain, and animal by-products, albeit to a lesser extent, sustained over half of this export growth.

Latin American water resources face major perils posed by population growth, urban development, land-use changes, and climate change (Jones & Scarpati, 2007). UN estimates have indicated that the population will continue to increase significantly, to the extent that the population in northwestern South America, from Venezuela to Bolivia, is anticipated to grow by a third by 2030. Moreover, other South American countries including Brazil, Argentina and Chile will undergo approximately 20% population growth (World Water Forum, 2012). Meanwhile, Latin America is encountering other developments, such as population shifts from suburbs to cities. As a result, water consumption per capita in urban areas is rising dramatically at the same time. All these occurrences collectively burden local resources, such as the Mexico City aquifer, and lead to issues that include quality degradations seen in groundwaters, aquifer declines, and subsidence. In addition, the increase observed in the number and range of asphalt areas with inadequate drainage provokes overwhelming floods (World Water Forum, 2012).

Climate change triggers an increase in the variability of rainfall and surface runoff, both in terms of time and place, causing an excess or scarcity of water resources and, ultimately, unrelenting disasters. Inevitably, these circumstances lead to changes in electricity generation, agriculture, commerce, and water supply chains. In parallel, some of the practical aftermath of climate change includes the gradual replacement of Amazon rainforests with savanna-like treeless states, changes in agronomic patterns and harvests in the region, heightened vulnerability to floods and drought in Central and South America, additional consequences of El Niño and La Niña cycles, and melting glaciers in the Andes mountains (European Aid, 2009).

LAC countries are in a superior position in terms of blue water availability (surface runoff resources), with only the Caribbean islands of the Dominican Republic and Haiti possessing total actual renewable water resources (TARWR) (km^3/year) less than 3000 m^3/cap/yr. The physical scarcity of water is due to the spatial mismatch between where water is naturally available for use and where it is demanded, all despite the overall water abundance in the area. For example, more than 75% of Mexicans reside in basins, where water consumption amounts to at least twice the quantity of water that is naturally renewed annually (Willaarts *et al.*, 2014).

Green water, also known as soil moisture, presents a paramount role in LAC agriculture and is deemed a key asset in attaining regional and global food and water security. Green water availability (measured per capita of arable lands, in hectares) is, remarkably, adequate for the LAC region (estimated at 0.26 hectares per capita in 2010), and only the Caribbean Islands, Costa Rica, and Colombia hold relatively lower ratios. These green water shortages are mainly compensated for by trade in agricultural areas and do not cause any significant water crises for the countries mentioned above. The most critical threat in terms of food and water security in the LAC region is the inter-annual variability of green water, that is, the high frequency of droughts and floods associated with El Niño and La Niña cycles, and the consequences of this phenomenon on agriculture (Willaarts *et al.*, 2014).

In general, most countries display a positive increase in terms of green water efficiency. Only the Dominican Republic, Cuba, and Belize have encountered drops and declines in agricultural harvests. All in all, these results indicate the gradual separation of agricultural growth from the expansion of the agricultural area, which is a positive sign for increasing food and water security. Nonetheless, irrigation efficiency in the LAC region (39%) is still low in comparison to the global average (56%). Therefore, LAC countries should improve their water use efficiency by:

(1) developing a water accounting system that takes into account the environment;
(2) implementing a transparent and comprehensive accounting system, to be used as an incentive to adopt the best water management practices in agriculture to reduce environmental impact; and
(3) developing effective coordination mechanisms between sector officials and different policies, both at the national level and at the river level, that can ensure that their policies and objectives are mutually consistent and do not undermine each other (Willaarts *et al.*, 2014).

Willaarts *et al.* (2014) highlight the potential consequences of socio-economic change, and the necessity of setting long-term goals to reduce climate change, identifying vulnerable areas and groups, prioritizing research and adaptation, as well as investing to attain water and food security.

12.8.3 Food and water safety considerations on a global scale

The following is a list of the worst countries in the world in terms of food security (Maplecroft, 2011): Afghanistan, Bangladesh, Benin, Burkina Faso, Burundi, Cambodia, Central Africa, Djibouti, Eritrea, Ethiopia, Guinea, Guinea-Bissau, Haiti, Rwanda, Kenya, Liberia, Malawi, Financial, Mozambique, and Zimbabwe. Given that so many countries are in a situation of food insecurity, and in order to improve the situation on a global scale, the United Nations Sustainable Development Goals (SDGs) were developed, to make plans and set goals to face this issue. Special indicators were also considered to assess the situation in different years and at different times.

The 2020 'State of Food Security and Nutrition in the World' report (FAO *et al.*, 2020) complements routine assessments of the world's food and nutrition security with forecasts for 2030, if the trend of the past decade continues. Five years into the 2030 Agenda, it is

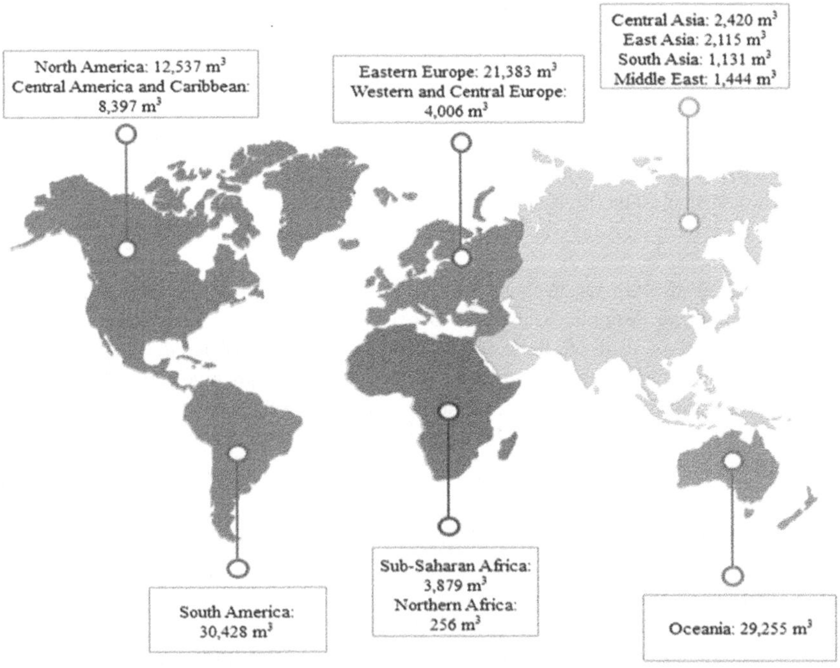

Figure 12.2 Per capita renewable water resources by region.

time to assess progress on whether continued efforts to date will allow countries to achieve the SDG 2 goals (FAO *et al.*, 2020).

The world is not on track to reach SDG 2.1 Zero Hunger by 2030. Combined forecasts that take into account recent trends in population size and composition, total food availability and the degree of food inequality show a PoU index increase of approximately 1%. As a result, the global number of malnourished people in 2030 will exceed 840 million (Galanakis, 2020).

Worldwide, in 2019 the prevalence of stunted or restricted growth in children was 21.3%. Although progress has been made, the slowdown in this stunting prevalence is far less than is needed to meet the World Health Assembly (WHA) target for 2025 and the SDG target for 2030. With the continuation of recent trends, these goals could be achieved in 2035 and 2043, respectively (Galanakis, 2020).

As mentioned above, the structure of water utilization in each region depends on the availability of water and access to resources and technological efficiency, consumption efficiency in the region, and on the distribution of per capita renewable water – which is often is very uneven. Each region or country also faces a heterogeneous distribution of water availability and access within its territory. The availability of water for agricultural purposes is important to meet the food needs of each region, and Figure 12.2 shows the average per capita renewable freshwater resources measured in m^3 per person per year (Kohli & Frenken, 2015). It should be noted that globally, the average consumption of freshwater resources for agriculture for low-, middle-, and high-income countries is 90%, 79%, and 41%, respectively.

Global water demand will grow remarkably over the next twenty years in all three of the industrial, domestic, and agricultural sectors. Although industrial and domestic demand will grow faster than agricultural demand, agriculture will still have the largest share of water demand.

12.9 SUMMARY

Food and water security are interrelated, and this link must always be taken into account when considering their management. Both of these resources are needed for human survival, and various factors, such as population growth, climate change, increasing disease, and the use of biofuels as a solution to reduce greenhouse gases, have led to crises in the field of food and water security. Therefore, it is important to focus more on managing these resources and their consumption with a comprehensive approach that also considers the relationship between them. Achieving global food security requires commitment and investment. To produce more nutrients with less water, innovative technologies are needed to ensure sustainable food production. The following measures are needed to improve product performance:

- implementation of efficient irrigation strategies, reuse of drainage water, development of more precise ways to use fertilizer and water to improve crop protection, reduce post-harvest losses, and create more sustainable livestock and marine production;
- focus on human capacity and institutional framework; agricultural development in underdeveloped countries is largely in the hands of small land owners, most of whom are women. Therefore, it is essential that new organizations focus on responsibility for water regulation, although this focus increases water management responsibilities and increases ownership and user participation;

- improving the value chain of production: throughout harvesting, processing, retailing, consuming, and distribution, efficient water and food recycling strategies should be considered. This can help preserve the needs of the environment.

Overall, under current world conditions, and to ensure sustainable food production under growing water scarcity situations, the following actions need to be taken to empower indigenous communities and revitalize their indigenous knowledge in areas related to conservation and optimal exploitation of water resources:

- invest in producing products that have higher yields per unit of water;
- move towards new irrigation technologies and conservation farming systems;
- increase attention on water rights and water trade;
- make funding available for investment in irrigation;
- increase water supply through the use of uncommon water (reuse of treated wastewater and wastewater).

It should also be noted that the use of virtual water and water footprint is one of the best approaches to establish a quantitative link between water and food resources. In this context, virtual water exchanges within and between countries can be considered as an effective tool to remove environmental constraints by improving water use efficiency and achieving water security in water-poor areas.

REFERENCES

Allan J. A. (2003). Virtual water eliminates water wars? In: A Case Study from the Middle East in Virtual Water Trade. Proceedings of the International Expert Meeting on Virtual Water Trade, A. Y. Hoekstra (ed.), UNESCO-IHE Inst. for Water Educ., Delft, The Netherlands, pp. 137–145.

Bakker K. (2012). Water security: research challenges and opportunities. *Science*, **337**(6097), 914–915.

Boretti A. and Rosa L. (2019). Reassessing the projections of the world water development report. *NPJ Clean Water*, **2**(1), 1–6.

Brauman K. A., Siebert S. and Foley J. A. (2013). Improvements in crop water productivity increase water sustainability and food security a global analysis. *Environmental Research Letters*, **8**(2), 024030.

Cook C. and Bakker K. (2012). Water security: debating an emerging paradigm. *Global Environmental Change*, **22**(1), 94–102.

Cooper R. (2020). Water Security Beyond Covid-19. 4D Helpdesk Report 803. Institute of Development Studies, Brighton, UK.

Du L., Xu L., Li Y., Liu C., Li Z., Wong J. S. and Lei B. (2019). China's agricultural irrigation and water conservancy projects: a policy synthesis and discussion of emerging issues. *Sustainability*, **11**(24), 7027.

EuropeAid. (2009). Climate Change in Latin America. AGRIFOR Consult Group. EuropeAid, Brussels, Belgium.

FAO. (1996). Rome Declaration on World Food Security and World Food Summit Plan of Action: World Food Summit 13–17 November 1996. Food and Agriculture Organization (FAO), Rome, Italy.

FAO. (2009). How to Feed the World in 2050. Paper Prepared for the High Level Expert Forum, October, 12–19, Rome. Food and Agriculture Organization (FAO), Rome, Italy.

FAO, IFAD, UNICEF, WFP and WHO. (2020). In Brief to The State of Food Security and Nutrition in the World 2020. Transforming food systems for affordable healthy diets. Rome, July, 16.

Galanakis C. M. (ed). (2020). Food Security and Nutrition. Academic Press, London.

Godfray H. C. J., Beddington J. R., Crute I. R., Haddad L., Lawrence D., Muir J. F., Pretty J., Robinson S., Thomas S. M. and Toulmin C. (2010). Food security: the challenge of feeding 9 billion people. *Science*, **327**(5967), 812–818.

Graham S., Parkinson C. and Chahine M. (2010). The water cycle. NASA Earth Observatory, https://earthobservatory.nasa.gov/ (accessed 27 May 2021).

Hanjra M. A. and Qureshi M. E. (2010). Global water crisis and future food security in an era of climate change. *Food Policy*, **35**(5), 365–377.

Hoekstra A. Y., Chapagain A. K., Aldaya M. M. and Mekonnen M. M. (2009). Water Footprint Manual. Water Footprint Network, Enschede, The Netherlands.

IWMI. (2007). Water for Food, Water for Life: A Comprehensive Assessment of Water Management in Agriculture. International Water Management Institute (IWMI), Colombo, Sri Lanka.

Jones J. A. A. and Scarpati O. E. (2007). Water resources issues in South America. *GeoJournal*, **70**(4), 227–231.

Kang Y., Khan S. and Ma X. (2009). Climate change impacts on crop yield, crop water productivity and food security–a review. *Progress in Natural Science*, **19**(12), 1665–1674.

Khan S., Hanjra M. A. and Mu J. (2009). Water management and crop production for food security in China: a review. *Agricultural Water Management*, **96**(3), 349–360.

Kohli A. and Frenken K. (2015). Renewable Water Resources Assessment–2015 AQUASTAT Methodology Review. Food and Agricultural Organisation of the United Nations, Rome, Italy, 1–6.

Kuyvenhoven A. (2012). Global Food Policy Report 2011. International Food Policy Research Institute, Washington, DC, USA.

Maplecroft. (2011). Maplecroft Food Security Index and interactive global map. Maplecroft, The Towers, St Stephen's Road, Bath, UK. See: http://maps.maplecroft.com (consulted December 2010; accessed 27 May 2021).

McGuire S. (2015). FAO, IFAD, and WFP. The State of Food Insecurity in the World 2015: Meeting the 2015 International Hunger Targets: Taking Stock of Uneven Progress. FAO, Rome.

Moraru S., Raevskaia I. and Moraru V. (2020). Implementation of the 2030 sustainable development objectives in the first stage. *Journal of Research on Trade, Management and Economic Development*, **14**(2), 26–42.

Pachapur P. K., Pachapur V. L., Brar S. K., Galvez R., Le Bihan Y. and Surampalli R. Y. (2020). Food security and sustainability. In: Sustainability: Fundamentals and Applications, R. Y. Surampalli, T. C. Zhang, M. K. Goyal, S. K. Brar and R. D. Tyagi (eds), John Wiley & Sons, 357–374.

Rockström J. (2001). Green water security for the food makers of tomorrow: windows of opportunity in drought-prone savannahs. *Water Science and Technology*, **43**(4), 71–78.

Rosegrant M. W. and Cline S. A. (2003). Global food security: challenges and policies. *Science*, **302**(5652), 1917–1919.

Turral H., Burke J. and Faurès J. M. (2011). Climate Change, Water and Food Security (No. 36). Food and Agriculture Organization of the United Nations (FAO), Rome, Italy.

United Nations. (2018). The United Nations World Water Development Report 2018. United Nations Educational, Scientific and Cultural Organization (UNESCO), New York, United State.

United Nations. (2020). Policy Brief: The Impact of COVID-19 on Food Security and Nutrition. United Nations, New York, USA.

Vladimirova I., Nguyen N., Schellekens I. and Vassileva I. (2018). Assessment of the Water Productivity Index. Deliverable to Task A4A of the BLUE2 project Study on EU integrated policy assessment for the freshwater and marine environment, on the economic benefits of EU water policy and on the costs of its non- implementation. Report to DG ENV. World Water Assessment, The Netherlands.

Wang S., Hu Y., Yuan R., Feng W., Pan Y. and Yang Y. (2019). Ensuring water security, food security, and clean water in the North China Plain–conflicting strategies. *Current Opinion in Environmental Sustainability*, **40**, 63–71.

White D. M., Gerlach S. C., Loring P., Tidwell A. C. and Chambers M. C. (2007). Food and water security in a changing arctic climate. *Environmental Research Letters*, **2**(4), 045018.

Willaarts B. A., Garrido A. and Llamas M. R. (eds). (2014). Water for Food Security and Well-Being in Latin America and the Caribbean: Social and Environmental Implications for a Globalized Economy. Routledge, London.

World Bank. (2013). World Bank Development Indicators. See: data.worldbank.org/indicator (accessed 17 March 2021).

World Water Forum. (2012). Americas' water agenda: targets, solutions and the paths to improving water resources management. Technical Document. Prepared at 6th World Water Forum, February 2012, Marseille.

Yang H., Reichert P., Abbaspour K. C. and Zehnder A. J. (2003). A water resources threshold and its implications for food security. *Environmental Science & Technology*, **37**(14), 3048–54.

Zhang Y., Zhang J. H., Tian Q., Liu Z. H. and Zhang H. L. (2018). Virtual water trade of agricultural products: a new perspective to explore the Belt and Road. *Science of the Total Environment*, **622**, 988–996.

Chapter 13

Passive defense in water resources

Habib Akbri Alashti[1]*, Omid Bozorg-Haddad*[2*] *and Daniele Bocchiola*[3,4]

[1]*Department of Water Engineering, University of Tabriz, Tabriz, Iran*
[2]*Distinguished Professor, Department of Irrigation & Reclamation Engineering, Faculty of Agricultural Engineering & Technology, College of Agriculture & Natural Resources, University of Tehran, Karaj, Tehran, 31587-77871, Iran. E-mail: OBHaddad@ut.ac.ir*
[3]*Department of ICA, Politecnico di Milano, Milan, Italy*
[4]*EVK2CNR Association, Bergamo, Italy*
**Corresponding author.*

ABSTRACT

Today, both freshwater and wastewater treatment plants are tremendously important for humans and their health. Additionally, water and water systems are sensitive because of their impacts on economic performance. At the same time, economic and population growth, and rising standards of living lead to increased water demand, and freshwater can be scarcely available. The increasing importance of water in modern society has increased the sensitivity and vulnerability of water systems, and the concern about the consequences of any disruption to them, leading to the growing importance of passive defense. Passive defense plays an important role in identifying vulnerability and likely threats, making emergency response plans, and preventing and responding to such threats. Water infrastructures could be physically destroyed, cyber-attacked, or contaminated through the introduction of chemical and biological agents. These threats can have different origins. Some of them may be unintentional and even natural disasters, while others may be due to terrorist attacks, or to regional conflicts and disputes over local water resources. In this chapter, a review of the types of threats pending on water systems is presented, together

 Water Resources: Future Perspectives, Challenges, Concepts and Necessities
Editor: Omid Bozorg-Haddad
doi: 10.2166/9781789062144_0259

with a short history of these threats. The vulnerability of water systems is investigated, and approaches to prevent and respond to them is reported.

Keywords: Biological and chemical warfare, Environmental terrorism, Passive defense, Water and terrorism, Water systems.

13.1 INTRODUCTION

Water is one of the most important substances on earth. All plants and animals must have water to survive. Today, modern life, industry, and the economy are all dependent on water. Human welfare and economic development are increasingly reliant on interconnected water systems and infrastructure to provide freshwater and sanitation. Water systems are also highly vulnerable to both unintentional and deliberate disruption as a result of severe conflict, war, violence, and terrorism.

Water has been used as a political or military target or tool for over 2500 years (Gleick, 2004). However, there have been many water-related accidents in recent history. For example, in 1993, an unintentional outbreak of Cryptosporidium – a waterborne bacteria that can contaminate supplies – in Milwaukee, USA, killed more than a hundred people and caused millions of dollars in material damage (MacKenzie *et al.*, 1994). And attacks on water resources have increased in recent decades, with more than 279 cases of water-related conflict recorded in just ten years (Pacific Institute, 2019a). Also, just as excessive water at the wrong time (such as flooding due to rainfall or a dam break) can cause many deaths and great damage, a lack of water will, for whatever the reason, cause great suffering to any society and negative effects on human health and the economy, creating more political chaos. There is no substitute for water resources and systems. Therefore, water safety and security are fundamental requirements for human comfort and welfare, and threatening them has always been one of the main goals of terrorists who want to threaten human security and international peace.

Carrying out a terrorist attack, such as damage to critical infrastructure, often has multiple goals. Ensuring security is one of the tasks that governments have to employ all possible means to provide. That security may be targeted by terrorists in various sectors, that is, economic, social, political and military sectors, to disrupt the security of society, and to achieve their ominous goals of creating instability and insecurity.

In general, the likelihood of a terrorist attack on water systems may be high (Maiolo & Pantusa, 2018). Unfortunately, water managers and the public often pay too little attention to such threats (Gleick, 2006). A most effective way to deal with such threats is passive defense. Protecting people against threats is one of the most urgent needs in the early stages of urban design, especially in designing sensitive and strategic centers such as water systems. Such an approach aims to provide the most sustainable security against threats with the least effort and inconvenience for the people to be defended. Always, planning and forewarning is the basis for preparing and equipping, and providing adequate protection against possible events. All of these lie in the scope of passive defense.

Another important goal of passive defense in water engineering is to deal with environmental terrorists. While environmental terrorism, that is, the use of resources to serve both as a purpose and as an instrument of armed conflict, has long been used as a form of violence, its magnitude today is unparalleled (Kohler *et al.*, 2019). Recently, deliberate or unintentional disruption has had undeniably a larger effect on human security and on human welfare than in previous decades. Increasing global population and climate change exacerbate these effects, so conservation and monitoring of water systems is increasingly important.

What is less clear is to what extent attacks on water systems can be more effective or easier to accomplish than attacks on other terrorist targets (Gleick, 2006). According to historical analysis, the toll of casualties due to attacks on water systems can be severe, and such attacks may create crises and anarchy in society, and even lead to unexpected reactions from citizens. In general, environmental terrorism is considered more dangerous than conventional civil terrorism because the direct impact of the act of violence is not limited to physical harm, and will, therefore, have long-term, far-reaching consequences for the affected population (Chalecki, 2001).

Water systems can be threatened in many ways. First, there is the physical destruction of water infrastructure due to a terrorist or military attack, or a natural disaster (such as earthquakes and floods). Demolishing a large dam or water treatment system could have terrible impacts. Another threat is the contamination of water, by chemical and biological agents, which can lead to the death and illness of many. Cyber attacks on water systems are also possible, and have attracted much attention recently. Given the scale of the systems involved, and the fact that there is no long-term substitute for water, any disruption to the performance, or control of water systems can cause enormous problems for societies.

Furthermore, climate change due to global warming, increasing population, economic development, and increasing standards of community health, have all increased water demand, which is now considered as a global dilemma (Brzoska & Fröhlich, 2015; Burrows & Kinney, 2016). Gleick (2006) noted the link between water scarcity and national security, and identified water scarcity as the main cause of war and conflict. Therefore, water resources and systems have become more strategic than ever before, and water resources are more likely to be targeted in terrorist conflicts and attacks. Estimates suggest that more than 40% of the world's population will be under water stress conditions by the middle of the 21st century (De Amorim *et al.*, 2018).

In the last decade, much research was focused on the effects of climate change, and of water scarcity on increasing migration and conflict (Brzoska & Fröhlich, 2015; Hsiang, *et al.*, 2013; Scheffran *et al.*, 2012). A most noticeable example is the recent conflict in Syria (Gleick, 2019). Syria experienced several severe droughts before the conflict began in the year 2011, and these droughts, along with population growth and over-ambitious governmental agricultural policies caused water bankruptcy in Syria. As a result, many farmers abandoned their land, and moved to urban areas (Gleick, 2014; Kohler *et al.*, 2019). Many scholars have cited such facts as the main reasons for the conflict in Syria. Researchers have claimed that climate events, such as droughts in areas with high vulnerability could cause migration and even armed conflict (Ide *et al.*, 2014).

Kohler *et al.* (2019) conducted an in-depth empirical study of 100 asylum seekers from 17 countries hosted in Germany, to inform academics and policymakers of the strategic importance of monitoring and preventing threats and violence related to natural resources. These results point to the use of environmental terrorism by governments, opposition groups, or both in Syria, Iraq and Afghanistan. Interviewees discuss how these attacks dramatically reduce or cut off resources such as water or electricity, and pollute water with pathogens, thereby harming their enemies and civilians. Considering that 81% of interviewers reported water or electricity shortages – or both – at their place of birth, the study claims that increasing climate stress on resources is likely to increase the use of environmental terrorism, and thus may cause migration and insecurity of human communities in the regions.

The purpose of this chapter is to present a history of threats to water systems, to highlight the most vulnerable parts of these systems, and discuss the risks that threaten them. The chapter also emphasizes that effective efforts by water managers and planners to minimize risks are beneficial, and can reduce population fears and concerns. Protective measures to prevent and counter threats are also described.

13.2 DEFINITIONS AND CONCEPTS

13.2.1 Passive defense

Passive defense refers to a set of measures that do not require the use of weapons, which may prevent the infliction of financial damage on critical and sensitive military and civilian equipment and facilities, and/or minimize the number of casualties. Among the branches of crisis management, passive defense puts the most emphasis on pre-crisis.

In some literature, these activities have been defined as civilian defense. However, passive defense is not limited to war-related damage, and it is also a tool for mitigating natural threats. Passive defense procedures may include camouflage, concealment, covering, deception, location, scattering, land use planning, and announcements.

13.2.1.1 Camouflage and concealment

Camouflage and concealment are techniques using natural or artificial means to reduce the ability to discover and identify an object, and are used to protect equipment and installations from surveillance and photography. The general concept of camouflage is to match the color and form of installations, equipment and forces to their surroundings. Concealment provides protection against enemy sightings, and camouflage reduces the ability to detect forces, equipment, facilities, and activities.

13.2.1.2 Covering

Covering is the concealing and protecting of equipment, weapons and manpower, against the sighting and shooting of the enemy.

13.2.1.3 Deception

Deception describes any action designed to mislead the enemy from obtaining information and correctly calculating the quantitative and qualitative ability of the other party, or to shed doubt in identifying targets. Deception involves the deviation of the enemy's mind away from true and important goals toward false and trivial ones.

13.2.1.4 Location

Location is one of the principle and most important measures of passive defense. The right location for each facility should be chosen. Facilities built in open, flat areas, such as on flat plains, cannot be kept secret from the enemy's eyes, and vulnerability to threats therefore increases. Creating vital and sensitive installations along highways, main roads, seashores, rivers, and near borders, makes them more easily identifiable and ready to be targeted by the enemy.

13.2.1.5 Scattering

Scattering is the decentralization of deployed forces, equipment, facilities, or activities to reduce their vulnerability to enemy operations, by avoiding being a lumped target.

13.2.1.6 Land use planning

Land use planning is the quantitative and qualitative planning of a position and its location (both at a micro and macro level), taking into account political, military, security, economic, social, and climatic conditions and factors (at the macro level). The most important attributes of land use planning are a comprehensive approach to multiple issues, foresight, and an environmental conservation attitude, all condensed into a development strategy. Proper land use planning is the best tool to improve the quality of life in urban areas.

13.2.1.7 Announcements

Announcements are warnings given to your own forces that an enemy's imminent offensive is approaching. Such warnings should be ready to be issued minutes, hours, days, or even longer before the start of hostilities. Devices used for announcements include, sirens, speakers, messages and warning messages. Radars and visual surveillance are also important tools for surveillance and help make any announcements more accurate and effective.

13.2.2 Sensitive centers

Sensitive centers are centers that, when failing as a whole or in part, can cause a crisis or serious harm to political, leading, control and command, productive and economic, supporting, communication, access, social, and defense systems that are effective in the whole of a country. Artificial water sources fall into this category.

13.2.3 Terrorism

As many scholars have pointed out before, the definition of 'terrorism' is difficult (Martin, 2006; Schmid, 1997; Wardlaw, 1989). Among others, Gleick (2006) stated that one of the challenges of defining terrorism in this context is that there is no exact or standard definition of water systems. However, the US Federal Bureau of Investigation (FBI) has provided a general definition of terrorism as follows: 'The unlawful use of force or violence against persons or property to intimidate or coerce a government, the civilian population, or any segment thereof, in furtherance of political or social objectives' (US Code of Federal Regulations (28 CFR Section 0.85)).

Similarly, U.S. law defines terrorism as 'premeditated, politically motivated violence perpetrated against non-combatant targets by sub-national groups or clandestine agents, usually intended to influence an audience' (U.S. Code, 2006).

Motivation and targets are two important components of terrorist acts. The motive for committing a terrorist act may include a variety of issues – religious, cultural, economic, or of support to political or ideological discourses. Targets usually include governmental and public strategic locations, political figures, events, people, institutions, etc. Thornton (1964) stated that: 'The relatively high efficiency of terrorism derives from its symbolic nature. If terrorists seek a demonstrative action, they will attack targets of maximum symbolic value.'

13.2.3.1 Environmental terrorism

In environmental terrorism, the environment and environmental resources are targeted. Gleick (2006) defined environmental terrorism as: 'The unlawful use of force against environmental resources or systems to harm individuals or deprive populations of environmental benefits in the name of a political or social objective.'

But the definition of environmental terrorism perhaps not so simple. One of the challenging topics faced is the distinction needed between the two topics of environmental warfare and environmental terrorism.

13.2.3.2 Environmental terrorism vs environmental warfare

Some studies have suggested that, in environmental terrorism unlike in environmental warfare, environmental resources are deliberately targeted. In other words, in environmental warfare the environment is not taken hostage, but environmental damage may occur during wars and conflicts (Chalecki, 2001; Schwartz, 1998).

There are two general reasons for expressing the complexity of the difference between environmental warfare and environmental terrorism:

(1) first, military actions taken by governments are usually called wars, whilst military action taken by formally organized groups (such as rebel forces) is called terrorism. But this is not always the reality. Historically, governmental actions have not always been legal. The two most important requirements of a legal war, under the United Nations Charter of 1945 are 'jus ad bellum' (war must be waged for a good reason) and 'jus in bello' (war must be conducted in a just fashion) (Chalecki, 2001). However, many governmental actions have violated these two conditions. On the other hand, many insurgents find governments' actions repressive;

(2) second, much damage to the environment during armed conflict may be due to collateral damage (Gleick, 2006).

13.2.3.3 Environmental terrorism vs eco-terrorism

An important distinction must also be made between environmental terrorism and eco-terrorism, so that the two are not confused. In these two kinds of terrorism, motive and target are different. In eco-terrorism, the motive is to save the environment from destruction and invasion, and social, political or economic resources are targeted. In environmental terrorism, environmental resources are targeted for social, political or economic motives (Chalecki, 2001). To put it simply, in eco-terrorism, environmental advocacy groups use illegal force against people or property to bring public attention to environmental issues (Gleick, 2006).

13.3 BASICS AND LOGIC

In recent decades, due to the global population growth and economic development, there has been a great deal of pressure on resources, including water resources. Water is a scarce resource in many countries, and hence the competition for freshwater has increased significantly. In addition, global warming and climate changes have exacerbated the situation, and created new problems for human society. In many parts of the world, lakes, rivers and aqueducts are drying up, as well as aquifers. Today, tensions related to water are greater than ever, and the situation is predicted to be even worse in the future, when we expect to witness water-related conflicts, wars and terrorist attacks around the world.

Plenty of studies have covered the impacts of climate change on water resources worldwide, and even in the most optimistic scenarios the negative effects are significant (Akbari-Alashti *et al.*, 2018). Some studies have analyzed the extent to which climate events increase water-related violence, conflict, terrorism, and anarchy (Bocchiola *et al.*, 2017; Hsiang *et al.*, 2013). Other studies have shown how conflicts increase societies' vulnerability to climate events (Scheffran *et al.*, 2012). What is certain, is that climate change has raised widespread and growing concerns about the future of human societies. Climate change does not only affect the increase of conflicts and tensions, but also the (extent and nature of the) scarcity of resources, resource vulnerability and the need for inter-basin water transfer (Hsiang *et al.*, 2013), which has increased more than in previous decades. Thus war, conflict, and environmental terrorism can now cause more devastating effects such as deaths, property damage, and anarchy (Chalecki, 2001).

Water is one of the pillars of human life, and an essential resource for economic development and community health. In addition, the attractiveness of water resources depends on their scarcity in a region, their location, vulnerability, and their ability to recover. Where water is scarce, the attractiveness of these water resources increases, increasing their chances of being targeted, and thus requiring more attention and protection (Chalecki, 2001).

Today, hydropower dams play an important and ever-growing role in the supply of energy to communities, and the sensitivity of these centers is very high. Should these important facilities be destroyed or occupied, the economy and welfare of dependent communities would be disrupted (Kohler *et al.*, 2019). The recent experience of the Islamic State of Iraq and the Levant (ISIL) in Syria and Iraq, showed how important the control of water facilities such as dams and also power plants are to the existence of a terrorist group. These experiences show that the threat to major utilities is quite serious and should be taken seriously (Gleick, 2014, 2019).

Various water-related conflicts caused by resource scarcity in the Middle East, India, Pakistan and Afghanistan, as well as inter-state conflicts over water resources such as in Turkey, Syria and Iraq (over the Tigris and Euphrates rivers), or Israel and Jordan (over the Jordan River), show the depth of the global water crisis. Based on these experiences, in 2010 The Human Right to Water and Sanitation (HRWS) was recognized by the UN General Assembly, under the position that the denial of water to civilians constitutes a war crime. In July 2010, UN Resolution 64/292 acknowledged the right to safe, affordable access to freshwater and sanitation, and acknowledged it as human right. Therefore, all countries should strive to achieve and uphold this Resolution (64/292) (United Nations, 2010).

Important hazards that threaten water systems include physical destruction caused by terrorist attacks on essential infrastructure such as pipelines, dams and treatment plants, or chemical and biological attacks on water resources (Gleick, 2006).

The threat to drinking water resources is the most important infrastructural target, especially for military bases (including support bases for logistics and supplies). If these bases become disabled, and are unable to provide services and deliver commodities and ammunition in a timely way, they will quickly affect the performance of the military forces. Researchers from the Air Force and a team of US military engineers have calculated with the help of security technologies that an attack on water supply systems costs between US$0.05–5.00 for each person's death, and that using a simple technology can cause many casualties in a matter of hours (Kroll, 2006).

Sabotage in the water industry is not just about international terrorists and other sabotage groups. Rather, coordinated attacks may also be carried out by water industry employees. The water industry is a vulnerable area, which needs to be considered carefully. A dissatisfied employee, or worker who is familiar with water supply systems, and who knows the vulnerabilities of the water supply system well, may easily perform a sabotage operation. Such dangerous activities, possibly carried out by humiliated/unsatisfied persons can lead to the deaths of many people. In the past, most such incidents reported have been caused by people with a recognizable mental illness, which seemingly makes it necessary to ensure the identification of mentally unstable staff using existing methods and criteria.

Even some simple events, such as a diarrhea epidemic caused by contamination of water resources, may be sufficiently effective to disable a military base, and meet the goals of terrorists.

In general, due to the growing dangers of water resources and passive defense systems, it is becoming increasingly important to identify the most effective ways to prevent and deal with any threat. In Table 13.1 presents a brief summary of recent studies, focusing on definition and assessment of water related risk and conflicts under natural and anthropic threats.

13.4 THE IMPORTANCE OF PASSIVE DEFENSE FOR WATER RESOURCES

Drinking water is an asset and main infrastructure for any country, and is vulnerable to natural disasters (earthquakes, fires, volcanoes, etc.) and terrorist attacks. Access to safe drinking water is one of the vital needs of communities. Part of any water management is the provision of water supply and water distribution networks that, like the circulatory system in a body, enable the continuation of activity and life in society. Due to the importance of water supply facilities, these facilities are sometimes targeted as strategic centers by sabotage operations that can cause serious harm to society, and even cause security crises. An enemy may attack any part of a water facility, destroying or contaminating it, making it very difficult and time consuming to compensate for it, or replace it.

In the destructive strategies introduced by Colonel John Warden for the US Army after the Vietnam War, four steps were introduced to destroy the adversary (Olsen, 2007). A most important task in planning a war is to accurately identify and target the centers of gravity for the invasion, so the invaded country will feel defeat in the first days of the war. In this context, water networks and water reservoirs are defined as a second strategic step, with an indication that cutting off water to people will cutoff the population's resilience capacity. These strategies were used in Kosovo and Iraq.

Many people do not know that water is a good target for terrorists. Attacks on water supply systems could include damage to pipelines or widespread contamination of operating systems. Of course, a terrorist attack on water supply systems usually occurs by contamination of the water distribution network, to cause the deaths of a large number of

Table 13.1 Some recent studies focusing on threats to water systems.

Reference	Topic	Main Output
Gleick (1993)	Water conflicts and international security	Provides a historical discussion of the use of water resources as the objective/means of interstate conflicts, and effects of uneven distribution of water and water stress in over populated countries
Haimes *et al.* (1998)	Analysis of threats against water supply systems	Uses Hierarchical Holographic Modeling (HHM) to provide a foundation for improving the assessment of the risk of willful attacks in the management of civil infrastructure systems
Chalecki (2002)	Environmental terrorism	Discusses the genesis, identification, and tools for mitigation of environmental terrorism
Gleick (2004)	Water conflicts chronology	Provides a chronological depiction of main water conflict, starting from year 3000 BC to present day
Tidwell *et al.* (2005)	Threat assessment of water supply systems	Based on Markov Latent Effect (MLE), proposes an alternative approach to rank the credibility of alternative threat scenarios
Gleick (2006)	Water and terrorism	Reviews history of past attacks on water systems, and vulnerabilities and risks facing modern water systems
Kroll (2006)	Securing of water supply systems	Presents basics of the threats to water infrastructure and steps to prevent such threats
Ezell (2007)	Vulnerability assessment using I-VAM model	Presents vulnerability assessment with the Infrastructure Vulnerability Assessment Model (I-VAM), and its application to a medium-sized clean water system
Gleick (2014)	Effect of water scarcity on the conflict in Syria	Assesses connections between water and conflict in Syria, and sketches future climate-related risks for water systems
USEPA (2005)	Water and wastewater utility vulnerability assessments	Software to assess vulnerabilities to man-made and natural hazards, and evaluate potential improvements
Maiolo and Pantusa (2018)	Vulnerability of drinking water systems	Proposal of a Vulnerability Index for drinking water distribution system against intentional contamination and physical damage
Gleick (2019)	Water as a weapon, and casualty of armed conflict	Focuses on specific instances of violence around water and water systems, in Syria, Iraq, and Yemen
Kohler *et al.* (2019)	Environmental terrorism and climate change	Presents interviews with asylum seekers in Germany, suggesting that increasing climate stress is likely to increase the use of environmental terrorism, migrations and insecurity

people, and does not require a professional terrorist. A terrorist or bioterrorist attack on water supply facilities may kill and affect hundreds, making them sick and leading to critical situations.

Due to their high vulnerability, the thought of a terrorist incident on a water supply facility can be very frightening. For instance, after an attack, anyone who drinks water or consumes water-containing products is likely to become infected and sick. Any human casualty in a water-related terrorist incident would have devastating nationwide psychological effects. Few people would be able to think and act wisely after such an event without feeling threatened. Such attacks undermine community security and weaken public confidence in water industry officials, something which may take months to recover from. For example, in 2009, the Iranian Health Minister stated that the nitrate concentration in drinking water in some parts of Tehran was above the standard, and noted the increased risk to infants. This created a widespread atmosphere of insecurity and anxiety, disturbing people's minds in Tehran. These observed psychological effects indicate that whenever a terrorist incident (e.g., the pollution of metropolitan drinking water) occurs, its effects and complications will not be easily controlled, and will likely lead to a national security crisis for the population.

When terrorists launch an attack, their purpose is not only to create insecurity and intimidation, but also that their attack be seen and well reported. The value of an attack is partly that news of it has the potential to spread internationally, and a relatively cheap, simple technology can meet multiple goals simultaneously.

Because of the vital role of water in human societies, destruction or contamination of water has an effective role in lowering people's morale. As a result, water is an ideal target for a terrorist attack, and the highest level of protection is required for water supply systems.

In fact, deliberate attacks on water supply systems are likely; they have occurred in the past and are still ongoing. Therefore, it is necessary to identify, detect and report any potential threats related to water supply systems, by providing the necessary training to personnel, especially local trusted forces that are indigenous to each region and interested in their national, regional and local interests.

On the role of water, it is generally expected that if a war over water occurred, it would be between countries sharing transboundary catchments. In one of these scenarios, for example, Turkey controls the river sources for both the Tigris and Euphrates, and water tensions, even leading to war, could occur with downstream neighbors, Syria and Iraq. Another example is the use of water by ISIL to achieve its strategic and tactical goals. ISIL occupied the territories adjacent to the Euphrates and Tigris rivers – indeed, the ISIL boundaries are primarily determined by the location of these rivers. As a result, 'water-based tactics', one of the 'most dangerous and cruel tactics', continue to be used in the Mesopotamia region (Veilleux & Dinar, 2018).

13.5 CONFLICTS, TERRORIST AND MILITARY ATTACKS ON WATER FACILITIES

Many water crises have been recorded in history, some due to natural disasters and others due to war or water conflicts. There is also a long history of using water resources as a target, or tool for war and terrorism (Gleick, 1993, 2004). In addition to natural disasters, water systems can be used to harm human societies or used as a pressure lever against other groups. Many cases have been recorded, such as the poisoning of water sources or destruction of dams. This section provides examples of historical water-related events.

According to a report by the Global Terrorism Database (GTD), (https://start.umd.edu/gtd/about/) which collected information on more than 200,000 terrorist incidents between 1970 and 2021, 675 of these incidents in 71 countries were in the water sector. These operations were carried out by 124 terrorist groups, and about 3400 people were killed or sick afterwards.

The first recorded attacks on water systems date back to 4500 years ago. From 2450 to 2400 BC, Urlama, king of Lagash, diverted water to boundary canals, to dry up boundary ditches and deprive the neighboring city-state of Umma of water. His son, Il, later cut off the water supply to Girsu, one of the towns in Umma. In around 600 BC, Solon of Athens besieged Cirrha for a wrong done to the temple of Apollo, and put poisonous hellebore roots (or rye ergot – reports differ) into the local water supply. This reportedly caused the Cirrhaeans to become violently ill, and facilitated the subsequent capture of the city (Eitzen & Takafuji, 1997). This is the earliest example of biowarfare (or bioterrorism) in the world. As shown in Figure 13.1, from the distant past to the present, the number of recorded attacks on water systems has been increasing.

As Figure 13.1 shows, water-related conflicts have increased dramatically over the past four decades. This fact suggests the possibility of increasing water-related conflict and tension in the coming years. Statistics are different in different parts of the world, as shown in Figure 13.2. The Western Asia region has had the most recorded conflicts affecting water systems through history. Most of these conflicts took place in Palestine, Syria, Iraq, and Yemen. South Africa, especially Somalia and Kenya, are other high-tension areas. Other regions of tension include India, Pakistan and Afghanistan (in Southern Asia), Ukraine and the Balkans (in Europe).

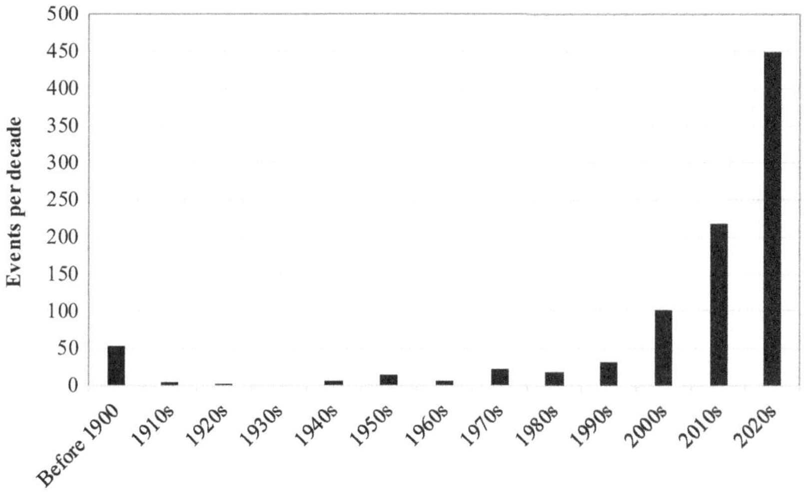

Figure 13.1 Number of water conflict events per decade; data from the water conflict chronology (WorldWater, 2005).

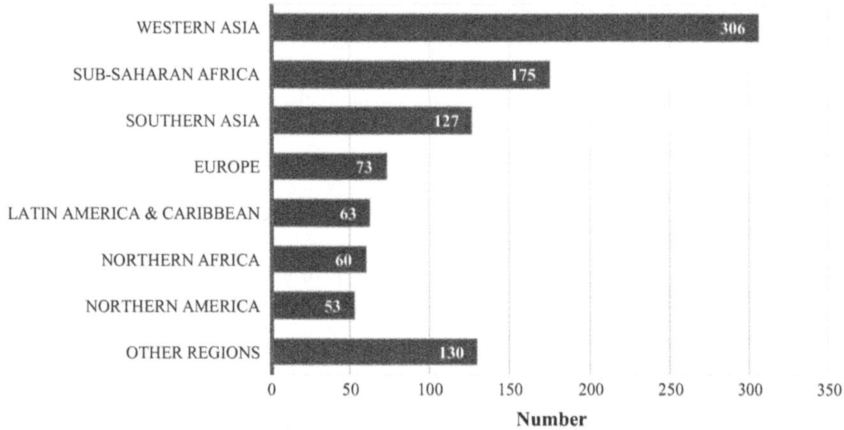

Figure 13.2 Water conflict events in various regions of the world; data from the water conflict chronology (WorldWater, 2005).

One of the motives for violence by individuals and non-state groups in relation to water is perceived inequality. Often, in relation to water development projects or controversial water allocation decisions, the interests of local people – especially in the region of origin of the water – are not taken into consideration, and the government treats protesters with forces and threats that cause deep hatred, possibly leading some day to trouble.

One of the first reported acts to happen in New York City occurred in 1748, in which a ferry house in Brooklyn was set on fire in protest to unfair water allocation. In the 1840s and 1850s, local people in the eastern and central United States, concerned about threats to local health and water supply, attacked a number of small dams and reservoirs.

In another case, between 1907 and 1913, farmers in the Owens Valley, California, repeatedly dynamited the aqueduct system which had been built to divert their water to the growing city of Los Angeles (Reisner, 1993).

The Palestinian region is one of the most eventful regions, when it comes to water system attacks. The first recorded attacks date back to 1965, when diversion pumps were attacked by members of the Palestinian National Liberation Movement (Fatah) (Naff & Matson, 1984). In 2001, in order to expel the Israelis from the settlement of Yitzhar, the Palestinians attacked the pipelines leading to it. In the same year, it was claimed that the Israelis tried to destroy water tanker deliveries, and attacked wastewater treatment plants (ENS, 2001). In 2004, after a Palestinian attack on US diplomatic personnel in the Middle East, the United States canceled plans to build a water development project in the Gaza Strip.

Economic and political issues can add further motivation to water-related conflicts. In 2000, a French analyst, after an accident in France where 5000 liters of sulfuric acid was poured into a branch of the Meuse River, pointed out that 'this is the first time that the environment and public health were taken hostage to exert pressure and this is an unheard of situation until now' (Christian Science Monitor, 2000).

In general, in hostilities, one of the ways in which the enemy chooses to achieve its goals quickly and cheaply is via terrorist acts such as contaminating potable water or damaging basic infrastructure such as pipelines, dams and treatment plants in order to disrupt water supply services and create security crises. The history of putting a chemical or biological agent into local water supplies dates back to World War I. During World War II, the Japanese contaminated China's water resources with Bacillus anthracis, Shigella, Salmonella, Vibrio cholera, and Yersinia pestis. Repetition of such terrorist acts around the world is not unexpected.

Further examples of conflict over water are given below:

- 1972: In Chicago, several members of a neo-Nazi terrorist organization known as the Order of Rising Sun (RISE), whose goal was the creation of a new master race, were arrested while holding detailed plans to contaminate the water supplies of several Midwestern cities, including Chicago and St Louis. They had several biological agents and toxins that had been produced in the laboratory of a local college, including botulism, meningitis, anthrax, and up to 40 kg of typhoid cultures (Kroll, 2006).
- 1973: A German biologist threatened to contaminate water sources through anthrax and botulinum bacilli unless he was paid a ransom of US$8.5 million (Kupperman & Trent, 1979).
- 1977: In North Carolina, a water tank was contaminated with bacterial agents. The safety covers of the tank were removed and the water was contaminated and made unusable, and water supply had to be provided by tanker trucks.
- 1983: Reportedly, Israeli arabs made an unsuccessful attempt to contaminate water resources in the city of Galilee with an unidentified powder (Douglass & Livingstone, 1987).
- 1985: Federal law enforcement authorities found that 'the Covenant, the Sword and the Arm of the Lord' (CSA), a survivalist group in the Ozark Mountains of Arkansas, had acquired a drum containing 30 gallons (c. 114 liters) of potassium cyanide. They tried to contaminate water systems in New York, Chicago, and Washington, believing that this would make the Messiah return more quickly by punishing unrepentant sinners (Monterey Institute for International Studies, 2004).
- 1986–1987: During the Iran–Iraq War, Iraqi forces repeatedly attacked important Iranian dams and power plants, such as the Dez and Abbaspour dams and their power plants.
- 1987: In the Philippines, 19 newly recruited policemen died after receiving and drinking water given to them by anonymous people, and about 140 were hospitalized.
- 1987: Sarin is an extremely toxic synthetic organophosphorus compound. A colorless, odorless liquid, it has been used as a chemical weapon due to its extreme power as a nerve agent. Sarin was used four times against Iranian soldiers at the end of the Iran–Iraq War, helping Iraqi forces to retake control of the al-Faw Peninsula during the Second Battle of al-Faw.
- 1991: A threat was made by anonymous letter to contaminate the water supply of the city of Kelowna, British Columbia, with biological contaminates. The motive was apparently 'associated with the Gulf War'. The security of the water supply was increased in response and no group was identified as the perpetrator.
- 1991: During the Gulf War, Americans targeted Baghdad's water supply systems, such as water facilities and dams on the Euphrates, and Iraqis targeted Kuwaiti freshwater.

- 1993: Saddam Hussein poisoned Shiite water supplies in southern Iraq with chemical pesticides.
- 1999: In Kosovo, water wells were polluted by Serbs. NATO targeted water systems in Yugoslavia and disrupted the water supply system.
- 1999: Engineers discovered an unexploded bomb in a water tank near Pretoria, South Africa. The bomb was strong enough that, if exploded, farmers' water supplies, a military base and a hydrological research center would all have been cut off (Pretoria Dispatch Online, 1999).
- 2000: A simulated terrorist attack on the Lake Nacimiento Dam caused some local panic in central California, until the media were belatedly notified that the situation was merely a disaster preparedness drill (Gaura, 2000).
- 2000: Workers at the Cellatex chemical plant in northern France protested at being deprived of severance/redundancy benefits when the plant was declared bankrupt, and then dumped 5000 liters of sulfuric acid into a tributary of the Meuse River. Whether they were trying to kill or destroy wildlife, people, etc., is unclear (Christian Science Monitor, 2000).
- 2003: Al-Qaida threatened US water systems via a call to a Saudi Arabian magazine. Al-Qaida did not 'rule out… the poisoning of drinking water in American and Western cities' (2003; Waterman, 2003).
- 2003: The main water pipeline in Baghdad. Was sabotaged/bombed. The sabotage was the first such strike against Baghdad's water system. An explosive was fired at the six-foot-wide water main in the northern part of Baghdad, according to the chief engineer for the city's water treatment plants.
- 2003: In Colombia, a bomb blast at the Cali Drinking Water Treatment Plant killed three workers. The workers were members of a trade union involved in intense negotiations over the privatization of the water system.
- 2004: Twelve members of the Indian security forces were killed during a 'counter-insurgency operation in Khanabal in the Anantnag District by an explosive device planted in an underground water pipe' (TNN, 2004).
- 2006: Tamil Tiger rebels cut the water supply to government-held villages in northeastern Sri Lanka.
- 2018: Angry farmers from the agricultural areas in Isfahan province, Iran, sabotaged water supply to neighboring Yazd province, demanding an end to water diversion to Yazd. There were clashes between security forces and enraged farmers. Also, farmers from Isfahan, assembled around the pipes and destroyed the facility. They demanded a halt to the water transfer to Yazd, arguing that sharing their water with the neighboring province has starved their own fields.
- 2019: With the aim of making more space for Israeli settlements, Israel shut off water to 1500 Palestinians to force them out of their homes in Hebron, West Bank.
- 2019: In conflicts in the disputed region of Ukraine, 9 water workers were killed and 26 injured.
- 2019: Due to the ongoing violence in Syria, a water station in Basida, Idlib Governorate, Syria, serving 80,000 people, was damaged (Pacific Institute, 2019b).
- 2019: According to Pakistan, India took threatening action against Pakistan when it released water from a dam on the River Sutlej, causing moderate flooding downstream in Pakistan. Earlier, India had threatened to cut back on water flowing through its rivers to arid Pakistan (Pacific Institute, 2019b).

Examination of these historical examples proves that contamination of water supply around the world is not a new issue and has not been rare. Water and water systems are very attractive as terrorist targets. Recent events in India, Pakistan, the Gulf and the Middle East reaffirm the point that water-related war and conflict are very serious. Water supply systems are a good target, especially for regional terrorists and other groups. Terrorists can make attacks on water using chemical, biological and radiological agents.

13.6 WATER AND WATER SYSTEMS VULNERABILITY

13.6.1 Infrastructure attacks

Damage to water infrastructure, especially water supply itself, hydropower dams, and pipelines can occur as a result of various reasons, including water-related terrorism, war, and natural disasters (floods and earthquakes). Destruction of a large hydroelectric dam on a major river, or of a water treatment plant or a large water supply system could result in a major crisis, with large economic consequences.

A major dam failure could kill many people and cause large damage to other infrastructures and buildings. It will interrupt power generation and negatively affect the performance of other water systems. The disastrous consequences of such destruction are so great and severe that they make dams an attractive terrorist target. Accordingly, this should be considered as a major risk, and basic measures must be taken to counteract such potential occurrences.

Many urban water systems are built with backup and redundancy systems to make them less vulnerable to threats. Others, such as large pipelines, pumping plants or treatment plants, do not have such support systems. The destruction of the large main pipeline entering Yazd, in Iran, in 2018, and the bombing of the major water pipeline entering Baghdad, in Iraq, in 2003 provide examples of the vulnerability of such systems.

There have been many natural disasters that have affected water infrastructure. As one example, the 1975 failure of the Banqiao and Shimantan dams, built on tributaries of the Huang He (Yellow) River in China, resulted in the collapse of dozens of dams downstream, and killed 85,000 people (Yi, 1998).

The weak construction of dams is also a potential hazard to water systems. In 1923, the collapse of the poorly built Gleno hydropower dam in the Italian Alps killed 356 (Pilotti *et al.*, 2011). In 1928, the collapse, due to a flood, of a poorly built dam in Pennsylvania caused extensive damage and killed more than 2000 people. Also in 1928, the failure of the St Francis Dam in San Francisquito Canyon in California killed more than 400 people.

13.6.2 Cyber attacks

With the development of computer technology, a further risk arises for water systems from cyber attacks against valves, pumps, and computer-based equipment for biochemical processing. Cyber attack is a new weapon that is dramatically expanding and necessary measures must be taken to counteract it (Littleton, 1995). Currently, most computer-based control systems have been developed without sufficient attention to security, and need to be re-thought. Most of the supervisory control and data acquisition (SCADA) networks used in water systems to collect data from sensors, and control equipment, may be susceptible to attack and abuse. The real danger is that control of these systems may be gained by hackers, and as a result water supply or quality may be seriously compromised.

These are serious and real dangers. In 2000, in Queensland, Australia, a man gained control of the Maroochy Shire wastewater system and released sewage into parks, rivers and public places, using a computer and radio transmitter (Gellman, 2002). In 2002, US law enforcement and intelligence agencies received evidence that al-Qaida was seeking information on a SCADA network in the USA (McDonnell & Meyer, 2002), reported in numerous wire reports, e.g. MSNBC (IonLife, 2002).

Other reports also indicate the great interest in cyber attacks against US water networks. Therefore, in 1990, the United States issued National Security Decision Directive 42, which states: 'Telecommunications and information processing systems are highly susceptible to interception, unauthorized access and related forms of technical exploitation as well as other dimensions of the foreign intelligence threat. The technology to exploit these electronic systems is widespread and is used extensively by foreign nations and can be employed, as well, by terrorist groups and criminal elements' (National Security Directive, 1990).

In 2011, according to Reuters, 198 reports of cyber attacks or security threats were filed by various US ministries, that is, four times as many as in the 2010. Interestingly, more than 40% of the attacks reported in 2011 targeted the water sector. Many regional water agencies in the United States use a special control system, accessible through the Internet. The system has a flaw that makes it vulnerable to hackers. Reuters reported that agencies were working with the system's developer to fix the flow software update (Finkle, 2011).

13.6.3 Biological and chemical attacks

Given that the introduction of very small amounts of chemical or biological agents into water could threaten the health of millions, people have a right to be concerned that every care is being taken to ensure that contamination will not happen, and that they can be assured of the health of their water intake.

Although it is very difficult to gather the quantities of contaminants necessary to contaminate the water supply network of a large city, attackers could use sabotage operations to contaminate water, usually targeting the most vulnerable, less protected areas. If, during a terrorist attack, water-soluble biological or chemical contaminants are introduced into an urban water source, in the best-case scenario, the contamination will be detected after entering a treatment plant, and it could be removed, and the plant turned off. Citizens can be also warned to use 'boiled water' for consumption. In the worst-case scenario, the contaminant could go undetected and cause illness, panic, and disorder.

Chemical and biological attacks, however, are not very easy to organize. For a chemical or biological weapon to be effective, the following must apply (Gleick, 2006):

- for the intended effect, it must be produced in sufficient quantities.
- it must be viable, dissolvable, stable and transportable in water.
- it must be infectious, viral or toxic enough to be effective (causing illness or death).
- it should have a long persistence so that its effect diminished or disappeared over time. it should also be resistant to water treatment systems.
- it should also be resistant to water treatment systems.

A wide range of chemicals and biological agents could be used in water. These agents each have their strengths and weaknesses for use in water, not described here for shortness.

Some of them have industrial and domestic uses and are found in the market and some are not commercial and can only be found in military warehouses. In any case, these materials may either be produced by sub-national terrorist groups or prepared in other ways (such as stealing from military and laboratory warehouses) (Hickman, 1999; USEPA 2003a).

Because of the difficulty of the production and distribution of military-grade biological and chemical agents, terrorists are more likely to use commercial chemicals to contaminate water sources. However, in 1970 in the US, the radical political group the Weather Underground reportedly attempted to blackmail a homosexual officer at the US Army's bacteriological warfare facility in Fort Detrick, Maryland, in order to get him to supply organisms that could be used to contaminate water resources (Gleick, 2006). However, according to one source, the operation failed when the officer requested several items that were unrelated to his work (Purver, 1995).

The most notable commercial chemicals are pesticides used to kill insects, rodents and plants. These include organophosphate and chlorine-based pesticides. Organophosphates and organochlorine pesticides affect the nervous system. Sodium fluoroacetate, strychnine, and thallium sulfate are also capable of disabling or killing humans if used in proper doses (Hickman, 1999). Several inorganic chemicals such as arsenic and cyanide are also widely available on the market. One of these is sodium cyanide (NaCN), which is stable and highly soluble in water and is relatively abundant on the market (Hickman, 1999).

Studies show that terrorists' use of biological warfare agents in drinking water sources can pose a serious threat to citizens' health. Many biological warfare agents are manufactured for airborne dissemination, but some can be soluble in water and digested and endanger human health. Pathogens (living organisms including bacteria, viruses and protozoa) and toxins (chemicals derived from biological processes) are among the most dangerous (Valcik, 1998).

Many cases have been reported in the open literature about chemical and biological contamination of water systems, with different motives. In Los Angeles, a man was detained by Police and FBI agents while he was preparing to poison the city's water system with biological poison. Another case involved a right-wing, neo-Nazi group known as the 'Order of the Rising Sun', allegedly 'dedicated to creating a new master race'.

In 1992, the Kurdish Workers' Party (PKK), attempted to contaminate water tanks in a Turkish Air Force base in Istanbul, using deadly concentrations of potassium cyanide.

13.6.4 Attacks on and threats to dams in peacetime

Dams are good targets for enemies during war or for terrorist acts. They are also threatened by natural disasters. One of the goals of passive defense is to deal with all such accidents. Below are examples of a number of incidents involving dams:

- Gleno Dam, Italian Alps: This gravity dam was built starting in 1917. Soon after the stone foundation was constructed to close the lower part of the valley, the original design was replaced by a slender reinforced concrete structure of the multiple-arch type. Due to water seepage at the interface between the masonry base and the overlying structure, and the use of poor construction material due to the restriction of World War I, on the morning of 1 December 1923, the Gleno Dam suddenly collapsed just a few days after first filling. Nearly 4.5 million m^3 of water was released. The consequent inundation caused significant destruction along the downstream valley and a death toll of at least 356 lives (Pilotti *et al.*, 2011).

- Vajont Dam, Italy: On 9 October 1963, during the initial filling, a c.180 million m^3 landslide caused a mega-tsunami in the lake, and 50 million m^3 of water overtopped the dam in a wave of c.150 meters, leading to the destruction of several villages and towns, and 1917 deaths. Vajont is considered the most disastrous rockslide ever to have occurred in Europe (Hardenberg, 2011).
- Kremasta Dam, Greece: The Kremasta Dam is an earth-fill embankment dam on the Achelous River in Aetolia-Acarnania, Greece. Shortly after the dam's reservoir, Lake Kremasta, was filled in 1967, a 6.3 Mw earthquake occurred. This earthquake has been attributed to reservoir-induced seismicity. Lake Kremasta is the largest artificial lake in Greece.
- Koyna Dam, India: The dam has withstood many earthquakes, including the devastating 1967 Koynanagar earthquake, resulting in the dam developing some cracks (Hindu Business Line, 2011).
- Banqiao Dam, China: Dam breaks, although rare, can cause great damage. The Banqiao Dam is a dam on the River Ru, Henan province, China. Its failure in 1975 caused as many as 230,000 deaths. The dam was subsequently rebuilt (Human Rights Watch, 1995).
- Aswan Dam, Egypt: An unusual spread of Rift Valley fever (RVF) northward to the Sudan and along the Aswan Dam reservoir to Egypt in 1977–1978 caused hundreds of thousands of animal deaths, as well as sickness in 18,000 humans, and 598 deaths. This outbreak occurred in the wake of major changes in irrigation channels, as a result of the building of the Aswan Dam on the Upper Nile (Cliff *et al.*, 2009).
- Mosul Dam, Iraq: The Mosul Dam was built on the Tigris River in 1984. In February 2016, the United States Embassy in Iraq warned of a 'serious and unprecedented' danger if the dam would collapse, and suggested that plans for evacuation should be made. The cities of Mosul, Tikrit, Samarra, and Baghdad could be at risk in the event of collapse, and up to 1.5 million people could be killed (Borger, 2016).
- Sefīd-Rūd Dam, Iran: The 1990 Manjil–Rudbar earthquake in northern Iran caused significant damage to the body of the dam and to the dam's crest.
- New Orleans Levees (USA): On 29 August 2005, there were over 50 failures of the levees and floodwalls protecting New Orleans, Louisiana, and its suburbs, in the wake of Hurricane Katrina. The levee and floodwall failures resulted in flooding in 80% of New Orleans and caused more damage than the storm (Mittal, 2005).

Clearly, given such evidence, it is clear that a dam break or breach in the wake of a terrorist attack would have catastrophic consequences.

13.7 SECURITY AND PROTECTION MEASURES TO COUNTER THREATS

The chance of an attack on water systems is quite real, and attacks have repeatedly happened in the past. What is important here, is assessing the probability of an attack and estimating its potential consequences. It is also necessary to investigate the vulnerability of water systems, and to take measures to reduce the likelihood of occurrence and the extent of consequences.

Reducing vulnerability requires a series of focused measures, such as (1) training staff and assigning tasks in both times of peace and crisis (2) establishing rapid-response expert teams in times of need; (3) developing and equipping water and water treatment systems; and (4) establishing contamination detection and identification systems.

Establishing a safe and proper system for threat prevention, timely identification, and delay (to create time to reduce potential damages and losses) is essential. Specific guidelines should be set up for dealing with various contamination events (such as, pathogenic microbial agents, chemical pesticides and insecticides, etc.), and should be appropriately communicated to the public (e.g., with posters) to reduce potential casualties and losses. Key people, particularly amongst those who live in the area surrounding water infrastructures, should be selected, and they should be made aware of potential disasters related to water resources and systems, and given dedicated training courses, to help them identify suspicious activity and report it to the relevant authorities.

People around and near water infrastructures are more likely to be harmed than those in remote areas, given that concentrations of contaminants diminish with increasing distance, and mortality and disease rates are therefore higher in areas close to the water sites.

One of the important methods of protecting water systems is to limit access to a few people to such systems. However, given their size (e.g., length of pipelines) and given the public uses of lakes, reservoirs, rivers and land, it is may be complicated to totally deny physical access, such as by locking gates or buildings. Overall, restricting physical access is an important measure but is not sufficient on its own.

Efforts should be made to protect water systems by establishing patrols and posting security guards in sensitive areas, and by restricting access to pumping plants, dams, treatment plants, etc. (Center for Defense Information, 2002). Gleick (2006) noted some recommendations for reducing physical risk to water systems, including:

- periodic control (regular control at appropriate intervals) of facilities, such as treatment plants, reservoirs, dams, storage facilities, pumping plants, intake facilities and control systems. Physical access should be limited to trusted and trained personnel;
- access to water distribution maps and facility plans should be restricted and controlled;
- lighting systems, surveillance cameras, and motion and heat detectors should be installed properly;
- to prevent hacking, SCADA systems should preferably not be connected to the internet, or at least they should be equipped with efficient security measures, such as firewalls, passwords, etc.;
- water treatment chemicals should be regularly checked and inventoried, and stored in safe, restricted areas.

Such security measures, including more extensive monitoring of pipelines and water supplies, and more security at power plants, will be expensive and may mean higher costs for citizens. However, these measures are needed.

Unlike simpler and more traditional weapons used by terrorists in the past, terrorist methods today may be very challenging. As mentioned earlier, attacking a water system could be done through the introduction of a chemical or biological substance. In this case, detection, identification of the nature of the threat, finding the source of the contamination, and understanding how to eliminate and treat it could take a considerable amount of time. Many international efforts have been made to prevent deliberate contamination of drinking water, for example, by protecting water resources and water distribution networks. The Biological and Toxin Weapons Convention (BTWC) prohibits the development, production, stockpiling and use of bacteriological (biological) and toxin weapons (Gleick, 2006).

The safety of water resources in the United States has received special attention since the September 11 attacks in 2001, and also after several episodes of envelopes containing anthrax bacteria. Afterwards, two new institutes, the National Health Systems Resource Center (NHSRC), and Water Sector Decontamination (WSD) of the U.S. Environmental Protection Agency (EPA) were established. In 2002, the US Congress passed the Public Health Security and Bioterrorism Preparedness and Response Act (the Bioterrorism Act), which made safety requirements, including vulnerability assessments and emergency response plans, mandatory (US Congress, 2002).

The implementation of the Bioterrorism Act was emphasized in communities with water systems serving more than 3300 individuals. It was also mandated to assess the protection and vulnerability of water resources and systems, water purification systems, chemical storage sites, disinfection process, electronic control systems, storage reservoirs, water supply networks, wastewater collection and treatment plants, and to report the results to EPA. This type of evaluation, if done correctly, can provide valuable information for planning and reducing vulnerabilities. Eight important and effective factors were highlighted in this emergency plan:

- providing specific information on water supply systems (basic information including population covered, and maps);
- determining the current status of water supply systems and responsibilities (defining the roles and responsibilities of each person);
- determining communication methods: who? what to do? what time? (for each member of the emergency actions team);
- staff safety and security (emergency evacuation plan, staff training, provision of equipment and first aid);
- identification of alternative water sources (water resources near the site and their capacity, identification of capable industries in the preparation of bottled water);
- alternative equipment and preparation of chemical materials (such as pumps and ventilators for emergencies and repairs);
- special protection (to prevent access by others);
- water sampling, analysis, and control (to timely identify potential contaminants);
- According to these requirements, to increase the basic security of water resources, US authorities have devoted considerable funds to several actions, including:
  - o purchase and installation of appropriate equipment to detect violations;
  - o purchase and installation of physical equipment, valves, lighting systems and security surveillance cameras;
  - o ensuring there is no possibility of irregular manipulation of manhole coverings, fire extinguishers and valve chambers;
  - o double locking of doors and entrances;
  - o improvements to electronic, computer and other automated systems and small security systems;

- o participation in training programs and providing guidelines on materials that are related to water security to prepare for countering against terrorist attacks;
- o developing and modifying how to use, store, or transport a variety of chemicals;
- o ensuring proper selection of staff and contract workers.

In addition to the above measures, water treatment has an effective role to counteract against deliberate water pollution. Water treatment processes help to physically remove (e.g., by coagulation and flocculation, sand filtering, and using activated carbon), and inactivate pathogens and protozoan cysts (disinfection). The US army uses reverse osmosis to purify water, which eliminates all forms of pathogens and bioterrorism, such as parasitic cysts, bacteria, viruses and toxins.

Some research has been devoted to determining the fate of *Bacillus anthracis* after water disinfection. Many physical and chemical methods have been found to inactivate *Bacillus anthracis* spores, and other environmental bacteria in water, air and contaminated surfaces such as *Bacillus cereus* and *Bacillus subtilis*.

Chemicals used for disinfection include hypochlorite, hydrogen peroxide, peracetic acid, formaldehyde, glutaraldehyde, sodium hydroxide, ethylene oxide, dioxide and ozone. *Bacillus* spores are also inactivated by UV, gamma and microwave radiation. The efficacy of in-use treatment facilities to remove bioterrorism agents and biotoxins should also be assessed. For example, activated carbon and reverse osmosis units are effective in removing microcystins from drinking water. Cryptosporidium cysts are resistant to disinfectants and are more effectively removed by sand filtering.

Bottled water is also exposed to accidental or intentional contamination by microbes. The most important measures in the drinking water industry are the monitoring and control of all components of the water supply, treatment and distribution systems, and it is not enough to just control treatment plants and water resources. Distribution network vulnerabilities should also be considered for potential attacks, as a main risk for water resources.

There is no feasible available technology for the timely detection of a deliberate or accidental attack on a drinking water supply network. However, the rapid detection and identification of a security breach in a water supply network is critical, and must be pursued as early as possible. If a terrorist attack is identified quickly, the public can be alerted and will be aware of the possible consequences of the attack, and can take preventive and protective measures so that they have enough time to organize health care and self-protection.

13.7.1 Early warning systems (EWS)

Early Warning Systems (EWS) are essential to protect the health of citizens, and should use quick identification technology, so that local or national public health officials and emergency personnel can take immediate response in the event of a crisis. EWS is useful for a wide range of purposes, including regular monitoring of water quality in water systems. Key components of EWS include: (1) contamination detection (2) identification of contaminant characteristics; (3) communication with authorities and citizens; and (4) proper response.

Early detection allows for a proper response. When contamination is suspected, the US EPA recommends field screening tests, including measuring water's radioactivity, pH, chlorine and cyanide, and other necessary tests for the purpose of initial evaluation by water industry officials.

Any EWS must be trustworthy enough to minimize the possibility of any mistake, including assessment of all 'true events' and avoidance of 'false events'. Also, it should be easy to set up and operate, easy to monitor, and fast enough to inform. Continuous monitoring reduces the likelihood of detecting contamination events (Foran & Brosnan, 2000).

EWS tools available include DNA microchip arrays (Betts, 1999a), microrobots (Hewish, 1998), immunological techniques (Betts, 1999b), variety of optical instruments, and molecular probes (Pelley, 1999) including new technologies to detect pathogens in real-time water distribution systems (USEPA, 2005). Standard EWS reduces costs and facilitates repair and replacement, but the development and diffusion of such systems is slow, and most of these technologies are not yet commercially available (Gleick, 2006). However, in recent decades some organizations such as the American Water Works Association (AWWA), and the World Health Organization (WHO) have made efforts to improve these technologies for detection and response. Many seminars have been held to familiarize and train in these technologies, and many publications produced to help in preparation and planning to counter contamination (USEPA, 2003b; WHO, 2003).

13.7.2 Public and governmental responses

Preventive measures, such as creating physical barriers, and EWS are necessary, and yet not sufficient. Both terrorist attacks and accidental threats (such as floods, earthquakes, etc.) on water systems will have a negative impact on society if they occur. Therefore, advanced guidelines, tools, and planning are essential to deal with these catastrophes. Depending on the nature of the attack, the population affected, and the characteristics of the water system involved (Gleick, 2006), progressive measures to counteract threats can include public advisories, temporary shutdown of the system, identification and use of alternative water supplies, chemical and biological treatment and disinfection, additional data collection, continuous monitoring, epidemiologic studies, health cares, etc.

Guidelines and subdivision plans are one of the key elements in responding to threats. They should specify the needs of all groups and individuals involved in the crisis, and be comprehensive enough to bring the level of irregularity and fluster to the minimum. Fortunately, valuable experiences can be called on to help prepare guidelines and compile subdivision plans.

For example, Hurricane Katrina in the United States brought some structural weaknesses to the fore. The US EPA, AWWA, American Society of Civil Engineers, US Federal Emergency Management Agency, the National Infrastructure Protection Center of the FBI, the Emergency Management and Emergency Preparedness Office of the US Health and Human Services, all offer guidelines for water plans, and some effort has been made to develop post-event responses (American Water Works Association, 2006; USEPA, 2003a, b; Waeckerle, 2000). However, it is highly recommended that the 'global' experience used to set up these programs is itself used as a basis on which to modify and improve on for application in other circumstances.

13.8 SUMMARY

Given the increased sensitivity of the water sector for human life today, close attention to water and water systems is essential. There is evidence of a long history of attacks on water resources worldwide, and measures must be taken to protect them, including proper location, proper coverage, early warning systems, etc.

It is crucial to understand that the threat of future attacks is real and the plans for responding to such attacks still often appear to be inadequate. If water resources are captured, destroyed or contaminated, it will have an effective role in abating people's morale. Therefore, passive defense plays an important role in identifying a target's attractiveness and vulnerabilities, and in developing an array of measures to prevent and respond appropriately to these threats. But, given the complex nature of water systems, determining the vulnerability of water resources to a terrorist attack is not easy.

As a result, water is an ideal target for a terrorist attack, and the highest level of protection is required for water supply systems. It is, therefore, necessary to identify, detect and report any threats related to water supply systems, by providing the necessary training to people, especially local trusted personnel, who are indigenous to each region and interested in their national, regional and local concerns.

Among the most important measures of passive defense are the establishment of a risk and vulnerability assessment mechanism for water systems, the development of smart and integrated response strategies, and the implementation of protocols at all levels. Usually, protections systems and facilities already in place are inadequate to identify and eliminate biological and chemical contaminants. It is essential to strengthen these systems, especially where critical infrastructure exists. Also required are the development or strengthening of continuous training of staff, rapid and effective water quality monitoring, creation of effective physical barriers, monitoring, early warning systems, development of effective chemical, and biological treatment systems, and continuous evaluation of their performance.

Most importantly, a widespread and effective public confidence in passive defense and water systems needs to be created. Creating ways to communicate with citizens is important for disseminating information and keeping the trust of the community. Without public confidence and community cooperation, it will be impossible to effectively deal with water threats. In fact, public confidence is the greatest asset during natural disasters and accidents and has great value in post-disaster resilience.

REFERENCES

Akbari-Alashti H., Soncini A., Dinpashoh Y., Fakheri-Fard A., Talatahari S. and Bocchiola D. (2018). Operation of two major reservoirs of Iran under IPCC scenarios during the XXI century. *Hydrological Processes*, **32**(21), 3254–3271.

American Water Works Association. (2006). Water Infrastructure Security Enhancement (WISE) program. See: http://www.awwa.org/science/wise/ (accessed 27 May 2021).

Betts K. S. (1999a). DNA chip technology could revolutionize water testing. *Environmental Science and Technology*, **33**(15), 300A–301A.

Betts K. S. (1999b). Testing the waters for new beach technology. *Environmental Science and Technology*, **33**(16), 353A–354A.

Bocchiola D., Soncini A. and Pelosi M. G. (2017). Effects of hydrological changes upon cooperation in transnational catchments: the case of Syr Darya. *Water International*, **42**, 1–22.

Borger J. (2016). Iraqi PM and US issue warnings over threat of Mosul dam collapse. *The Guardian*, 29 February 2016.

Brzoska M. and Fröhlich C. (2015). Climate change, migration and violent conflict: Vulnerabilities, pathways and adaptation strategies. *Migration and Development*, **5**(2), 190–210.

Burrows K. and Kinney P. (2016). Exploring the climate change, migration and conflict Nexus. *International Journal of Environmental Research and Public Health*, **13**(4), 443–17.

Center for Defense Information (CDI). (2002). Securing US Water Supplies. Center for Defense Information, Washington DC. See: http://www.cdi.org/terrorism/water-pr.cfm (accessed 27 May 2021).

Chalecki E. (2001). A New Vigilance: Identifying and Reducing the Risks of Environmental Terrorism. A Report of the Pacific Institute for Studies in Development, Environment, and Security, Oakland, California. See: http://www.pacinst.org/reports/environment_and_terrorism/ (accessed 27 May 2021).

Chalecki E. (2002). A new vigilance: identifying and reducing the risks of environmental terrorism. *Global Environmental Politics*, **2**(1), 46–64.

Christian Science Monitor. (2000). Ecoterrorism as negotiating tactic. *Christian Science Monitor*, 8.

Cliff A. D., Smallman-Raynor M. R., Haggett P., Stroup D. F. and Thacker S. B. (2009). Infectious Diseases: Emergence and Re-emergence: a Geographical Analysis. Oxford University Press, NY.

De Amorim W. S., Valduga I. B., Pereira Ribeiro J. M., Guazzelli Williamson V., Krauser G. E., Magtoto M. K. and Salgueirinho Osório de Andrade Guerra J. B. (2018). The nexus between water, energy, and food in the context of the global risks: an analysis of the interactions between food, water, and energy security. *Environmental Impact Assessment Review*, **72**(2018), 1–11.

Douglass J. D., Jr. and Livingstone N. C. (1987). America the Vulnerable: the Threat of Chemical and Biological Warfare. Lexington Books, Lexington, Massachusetts.

Ezell B. C. (2007). Infrastructure Vulnerability Assessment Model (I-VAM). *Risk Analysis*, **27**(3), 571–583. doi: 10.1111/j.1539-6924.2007.00907.x. PMID: 17640208

Eitzen E. M. and Takafuji E. T. (1997). Historical overview of biological warfare. In: Textbook of Military Medicine, Medical Aspects of Chemical and Biological Warfare, R. Zajtchuk and R. F. Bellamy (eds), The Office of the Surgeon General, Department of the Army, USA, pp. 415–424.

Environment News Service (ENS). (2001). Environment a weapon in the Israeli–Palestinian conflict. See: http://www.ens-newswire.com/ens/feb2001/2001-02-05-01.asp (accessed 29 March 21).

Finkle J. (2011). U.S. probes cyber attack on water system. *Reuters*. Reuters, London, UK. See: https://www.reuters.com/article/us-cybersecurity-attack-idUSTRE7AH2C320111121 (accessed 27 May 2021).

Foran J. A. and Brosnan T. M. (2000). Early warning systems for hazardous biological agents in potable water. *Environmental Health Perspectives*, **108(10)**; see: http://ehp.niehs.nih.gov/realfiles/docs/2000/108p993-995foran/foran-full.html (accessed 27 May 2021).

Gaura M. A. (2000). Disaster Simulation Too Realistic Media fooled by 'news' of terrorist attack. San Francisco Chronicle, 27 October 2000. See: http://sfgate.com/cgi-bin/

article.cgi?f=/c/a/2000/10/27/MN1002CH.DTL&hw=dam&sn=001&sc=1000 (accessed 27 May 2021).

Gellman B. (2002). Cyber-attacks by Al Qaeda feared. *Washington Post*, 27 June, 2002.

Gleick P. H. (1993). Water and conflict. *International Security*, **18**(1), 79–112 (Summer 1993).

Gleick P. H. (2004). The water conflict chronology. In: The World's Water 2004–2005: The Biennial Report on Freshwater Resources, P. H. Gleick (ed.), Island Press, Covelo, CA, pp. 234–255.

Gleick P. H. (2006). Water and terrorism. *Water Policy*, **8**, 481–503.

Gleick P. H. (2014). Water, drought, climatic events, and conflict in Syria. *Weather, Climate, and Society*, **6**(3), 331–340.

Gleick P. H. (2019). Water as a weapon and casualty of armed conflict: a review of recent water-related violence in Iraq, Syria, and Yemen. *Wiley Interdisciplinary Reviews: Water*, **6**(4), e1351, https://doi.org/10.1002/wat2.1351 (accessed 27 May 2021).

Haimes Y. Y., Matalas, N. C., Lambert J. H. and Jackson B. A. (1998). Reducing vulnerability of water supply systems to attack. *Journal of Infrastructure Systems*, **4**(4).

Hardenberg W. G. V. (2011). Expecting Disaster: The 1963 Landslide of the Vajont Dam. *Environment & Society Portal, Arcadia*, No. 8. Rachel Carson Center for Environment and Society. See: https://doi.org/10.5282/rcc/3401 (accessed 27 May 2021).

Hewish M. (1998). Mini-robots sniff out chemical agents. *Jane's International Defense Review*, **31**(6), 87.

Hickman D. C. (1999). Chemical and Biological Warfare Threat: USAF Water Systems at Risk. Future Warfare Series No. 3. Air University, US Air Force Counterproliferation Center, Maxwell AFB, Alabama, 36. See: http://www.au.af.mil/au/awc/awcgate/cpc-pubs/hickman.htm (accessed 27 May 2021).

Hindu Business Line. (2011). Major project to study earthquake activity at Koyna. *The Hindu Business Line*. See: https://www.thehindubusinessline.com/economy/Major-project-to-study-earthquake-activity-at-Koyna/article20111617.ece (accessed 27 May 2021).

Hsiang M. S., Burke M. and Miguel E. (2013). Quantifying the influence of climate on human conflict. *Science*, **341**(6151), 1235367.

Human Rights Watch. (1995). The Three Gorges Dam in China: forced resettlement, suppression of dissent and labor rights concerns. *Human Rights Watch/Asia*, **7**(1).

Ide T., Schilling J., Jasmin L. S. A., Scheffran J., Ngaruiya G. and Weinzierl T. (2014). On exposure, vulnerability and violence: spatial distribution of risk factors for climate change and violent conflict across Kenya and Uganda. *Political Geography*, **43**, 68–81.

IonLife. (2002). How safe is our water? The threat of terrorism. See: https://www.ionizers.org/water-terrorism.php (accessed 27 May 2021).

Kohler C., Denner C. and Bursztyn M. (2019). Understanding environmental terrorism in times of climate change: implications for asylum seekers in Germany. *Research in Globalization*, **1**. See: https://doi.org/10.1016/j.resglo.2019.100006 (accessed 27 May 2021).

Kroll D. J. (2006). Securing Our Water Supply: Protecting a Vulnerable Resource. PennWell Corp, Tulsa. ISBN 10: 1593700695.

Kupperman R. H. and Trent D. M. (1979). Terrorism: Threat, Reality, Response. Hoover Institution Press, Stanford, CA.

Littleton M. J. (1995). Information Age Terrorism: Toward Cyberterror. Naval Postgraduate School, Monterey, California. See: http://www.fas.org/irp/threat/cyber/docs/npgs/terror.htm#TOC (accessed 27 May 2021).

MacKenzie W. R., Hoxie N. J., Proctor M. E., Gradus M. S., Blair K. A., Peterson D. E., Kazmierczak J. J., Addiss D. G., Fox K. R., Rose J. B. and Davis J. P. (1994). A massive outbreak in Milwaukee of Cryptosporidium infection transmitted through the public water supply. *New England Journal of Medicine*, **331**(3), 161–167.

Maiolo M. and Pantusa D. (2018). Infrastructure Vulerability Index of drinking water systems to terrorist attacks. *Cogent Engineering*, **5**(1), 1456710.

Martin G. (2006). Understanding Terrorism: Challenges, Perspectives, and Issues, 2nd edn. Sage Publications, Thousand Oaks, California.

McDonnell P. J. and Meyer J. (2002). Links to terrorism probed in northwest. *Los Angeles Times*, 13 July 2002. See: https://www.latimes.com/archives/la-xpm-2002-jul-13-na-terror13-story.html (accessed 27 May 2021).

Mittal A. (2005). Lake Pontchartrain and Vicinity Hurricane Protection Project. United States Government Accountability Office (GAO), Washington, DC.

Monterey Institute for International Studies. (2004). A Brief History of Chemical Warfare. Center for Nonproliferation Studies, Monterey, California, http://www.nti.org/h_learnmore/cwtutorial/chapter02_02.html (accessed 27 May 2021).

Naff T. and Matson R. C. (eds). (1984). Water in the Middle East: Conflict or cooperation?. Westview Press, Boulder, Colorado.

National Security Directive 42. (1990). National Policy for the Security of National Security Telecommunications and Information Systems. Unclassified portion. Obtained by Computer Professionals for Social Responsibility under the Freedom of Information Act, April 1992. See: http://www.fas.org/irp/offdocs/nsd/nsd_42.htm (accessed 27 May 2021).

Olsen J. A. (2007). John Warden and the Renaissance of American Air Power. Potomac Books, Washington DC.

Pacific Institute. (2019a). Water Conflict Chronology. Pacific Institute, Oakland, CA. See: https://www.worldwater.org/water-conflict/ (accessed 31 July 2019).

Pacific Institute. (2019b). Pakistan claims India's unexpected release of water from a dam is an act of war. See: https://www.worldwater.org/water-conflict (accessed 27 May 2021).

Pelley J. (1999). Rapid, genetic-based test can identify human viruses in beach water. *Environmental Science and Technology*, **33**(18), 399A.

Pilotti M., Maranzoni A., Tomirotti M. and Valerio G. (2011). 1923 gleno dam break: case study and numerical modeling. *Journal of Hydraulic Engineering*, **137**, 480–492, doi: 10.1061/(ASCE)HY.1943-7900.0000327

Pretoria Dispatch Online. (1999). Dam bomb may be 'aimed at farmers'. See: http://www.dispatch.co.za/1999/07/21/southafrica/RESEVOIR.HTM (accessed 27 May 2021).

Purver R. (1995). Chemical and Biological Terrorism: The Threat According to the Open Literature. Canadian Security Intelligence Service (CSIS), Ottawa, Ontario, Canada. See: http://www.csis-scrs.gc.ca/en/publications/other/c_b_terrorism01.asp (accessed 27 May 2021).

Reisner M. (1993). Cadillac Desert: The American West and its Disappearing Water, 2nd edn. Penguin Books, New York.

Scheffran J., Brzoska M., Link P. M. and Schilling J. (2012). Climate change and violent conflict. *Science*, **336**(6083), 869–871.

Schmid A. (1997). The problems of defining terrorism. In: Encyclopedia of World Terrorism, M. Crenshaw and J. Pimlott (eds), M.E. Sharp, Armonk, New York, **1**, pp. 12–22.

Schwartz D. (1998). Environmental terrorism: analyzing the concept. *Journal of Peace Research*, **35**(4), 483–496.

Tidwell V. C., Cooper J. A. and Silva C. J. (2005). Treatment assessment of water supply systems using Markov latent effects modeling. *Journal of Water Resources Planning and Management*, **131**(3).

Thornton T. P. (1964). Terror as a weapon of political agitation. In: Internal War: Problems and Approaches, H. Eckstein (ed.), Free Press of Glencoe, New York, p. 73.

TNN (Times News Network). (2004). IED was planted in underground pipe. *Times News Network*, December 5, 2004. See: http://timesofindia.indiatimes.com/articleshow/947432.cms (accessed 27 May 2021).

U.S. Code. (2006). Management of Foreign Affairs. Supplement 4, Title 22 – Foreign Relations and Intercourse, Chapter 38 – Department of State, Sec. 2656. Office of the Law Revision Counsel of the United States House of Representatives, Washington, DC. See: https://uscode.house.gov/ (accessed 27 May 2021).

U.S. Congress. (2002). Public Health Security and Bioterrorism Preparedness and Response Act (the Bioterrorism Act). See: https://www.congress.gov/bill/107th-congress/house-bill/3448 (accessed April 2021)

U.S. Environmental Protection Agency (USEPA). (2003a). Response Protocol Toolbox: Planning for and Responding to Drinking Water Contamination Threats and Incidents. US EPA, Washington DC. See: http://www.epa.gov/safewater/watersecurity/pubs/guide_response_module1.pdf (accessed 27 May 2021).

U.S. Environmental Protection Agency (USEPA). (2003b). Large Water System Emergency Response Plan Outline: Guidance to Assist Community Water Systems in Complying with the Public Health Security and Bioterrorism Preparedness and Response Act of 2002. *EPA 810-F-03-007*. US EPA, Office of Water, Office of Ground Water and Drinking Water, http://www.epa.gov/safewater/watersecurity/pubs/erp-long-outline.pdf (accessed 27 May 2021).

U.S. Environmental Protection Agency (USEPA). (2005). The Monitor: The Newsletter of the ETV Advanced Monitoring Systems (AMS) Center, **8**(6). http://www.epa.gov/etv/pdfs/newletters/monitor/01_mon_nov05.pdf (accessed 27 May 2021).

United Nations. (2010). The human right to water and sanitation. *Resolution 64/292*. United Nations General Assembly, New York. http://www.un.org/waterforlifedecade/human_right_to_water.shtml (accessed 31 July 2019).

Valcik J. E. (1998). Biological Warfare Agents as Potable Water Threats. *Medical Issues Information Paper No. IP-31-017*. US Army Center for Health Promotion and Preventative Medicine, Aberdeen Proving Ground, Maryland.

Veilleux J. and Dinar S. (2018). New Global Analysis Finds Water-Related Terrorism Is On the Rise. New Security Beat (Blog of the Wilson Center's Environmental Change and Security Program). See: https://www.newsecuritybeat.org/2018/05/global-analysis-finds-water-related-terrorism-rise/ (accessed April 2021)

Waeckerle J. F. (2000). Domestic preparedness for events involving weapons of mass destruction. *Journal of the American Medical Association*, **283**(2), 252–254.

Wardlaw G. (1989). Political terrorism, Theory, Tactics and Counter-measures, 2nd edn. Cambridge University Press, Cambridge.

Waterman S. (2003). Al-Qaida Threat to US Water Supply. United Press International (UPI), Washington, D.C., and Boca Raton, Fla.

World Health Organization. (2003). Public Health Response to Biological and Chemical Weapons: WHO Guidance, 2nd edn. (Draft, March 2003). World Health Organization, Geneva. See: http://www.who.int/csr/delibepidemics/biochemguide/en/index.html (accessed 27 May 2021).

WorldWater. (2005). Water conflict chronology. See: https://www.worldwater.org/conflict/list/ (accessed April 2021).

Yi S. (1998). The world's most catastrophic dam failures: the August 1975 collapse of the Banqiao and Shimantan Dams. In: The River Dragon Has Come, Q. Dai (ed.), M.E. Sharpe, New York.

Chapter 14

Forensic engineering

Manizhe Zarei[1], Omid Bozorg-Haddad[2] and Vijay P. Singh[3]*

[1]Water Engineering M.Sc. Student, Department of Irrigation & Reclamation Engineering, Faculty of Agriculture Engineering & Technology, College of Agriculture & Natural Resources, University of Tehran, Karaj, Alborz, Iran.
E-mail: Zarei.manizhe@alumni.ut.ac.ir
[2]Distinguished Professor, Department of Irrigation & Reclamation Engineering, Faculty of Agricultural Engineering & Technology, College of Agriculture & Natural Resources, University of Tehran, Karaj, Tehran, 31587-77871, Iran.
E-mail: OBHaddad@ut.ac.ir
[3]Distinguished Professor, Regents Professor, Caroline & William N. Lehrer Distinguished Chair in Water Engineering, Department of Biological and Agricultural Engineering & Zachry Department of Civil & Environmental Engineering, Texas A&M University, 321 Scoates Hall, 2117 TAMU, College Station, Texas 77843-2117, USA.
E-mail: VSingh@tamu.edu
**Corresponding author.*

ABSTRACT

Throughout history, natural events such as floods, droughts, fires, lightning, and storms have caused significant losses of life and property. To mitigate the hazardous consequences of such events, or 'failures' (as they are referred to), a number of questions can be asked, such as: 'What are the causes of these events?'; 'What natural factors cause these events?'; 'What is the human role in the occurrence of these events?'; 'Who is to blame for such events?'; and 'What actions should be taken to prevent such events from happening?' The forensic engineering approach allows us to answer these questions. Forensic engineering, a term developed in recent years, allows us to identify the causes of events by looking back and

 Water Resources: Future Perspectives, Challenges, Concepts and Necessities
Editor: Omid Bozorg-Haddad
doi: 10.2166/9781789062137_0287

analyzing the relationship between an event's causes and their consequences; it is a useful tool for determining the natural or human causes of events that lead to disasters. Forensic hydrology is a branch of forensic engineering and applies directly to floods and droughts but is not limited to these events. Forensic hydrology is also used for the historical assessment and analysis of events such as water pollution, drying of lakes and rivers, the drying up (or significant reduction in the water table) of wells, and the infiltration of saline water into freshwater. Forensic hydrology analyzes event evidence and data from a variety of perspectives. Examining the origins and mechanisms of such events to find their causes can lead to better water management, allocation and improved use, and can also help to prevent or minimize severe damage. This chapter provides an introduction to forensic engineering and describes the processes which should be followed to evaluate hazardous events.

Keywords: Forensic engineering, Forensic hydrology, Failures, Natural and human causes, Droughts, Floods, Evidence.

14.1 INTRODUCTION

In recent years, many natural events have caused devastating losses. In 2018, storms hit Asia and hurricane Florence hit the east coast of the United States causing widespread damage, flooding and landslides; Scandinavia, Spain, Portugal, Britain, North America, and South Africa all experienced severe wildfires and forest fires (AghaKouchak *et al.*, 2018). In December 2019, Australia experienced severe wildfires that killed at least 34 people, while 400 people died from smoke inhalation. In January 2020, the Indonesian capital Jakarta and its neighboring areas were damaged by devastating floods. That same month, the Taal volcano erupted in Luzon and many earthquakes shook Turkey, China, Iran, Russia, the Philippines, and India. What were the causes of these events? What role do natural and human factors have in the occurrence of these events? Is there a connection between them? Many other questions following the occurrence of such events highlight the importance of forensic engineering in finding answers to these questions. Forensic engineering, developed in recent decades, is the application of engineering principles to answer questions about the occurrence of an event, and find the source and causes of it. Forensic engineers use knowledge, skills, methods, and principles of engineering to try to answer questions about events. These questions are:

- What exactly happened? (Definition of event)
- What is the magnitude or severity of the event?
- Why did the event happen?
- What can be done to prevent the event from happening again?

Looking back and evaluating cause–effect relationships, a forensic engineer can determine the contribution of each of the natural and human factors to the formation and occurrence of an event. They must provide fact-based opinions, regardless of the outcome. Lawyers are advocates, but forensic engineers are advocates of views. Forensic engineers need considerable training and experience, beyond mere academic training. They need a wealth of information in various fields and disciplines to provide forensic reasoning in practical fields, which is beyond the education offered by a university. The recognition of a forensic engineer by the courts requires more than formal training (Noon, 2001). A forensic engineer is someone who searches, investigates, and gathers evidence, and can provide conclusive proof of what needs to be clarified. Sometimes the evidence is not easily gathered, and a more in-depth assessment is needed to determine the true causes of an event (Noon, 2001).

Forensic engineering is a scientific method for investigating events that do not have direct data available (Bronstert *et al.*, 2018) and has become an interdisciplinary field by also considering other specialties such as meteorology, sociology, geology, and economics. Forensic engineers require simultaneous expertise in several disciplines. For example, suppose the event being investigated is a flood. In order for a forensic engineer to find the causes of the flood, including natural and human causes, they must knowledge of several different fields of meteorology, hydrology, groundwater, geomorphology, sediment transport, and risk analysis (Bronstert *et al.*, 2018). In addition to quantitative data, qualitative data can also be used, such as flood peak and maximum water elevation marks (Ramirez & Herrera, 2016), and the use of separate and independent data can reduce the uncertainty of analysis and increase accuracy. Sometimes the experience and expertise of a forensic engineer are not enough to find the causes of event, and they should seek help, where possible, from other forensic engineers with different experience to understand the root causes of the failure.

Forensic hydrology has emerged from forensic engineering as a specialized field in recent years. Forensic hydrology identifies the causes and mechanisms of hydrological events, examining events such as floods, droughts, sediment transport, infiltration of saline water into freshwater, well drying, etc., focusing on hydrological events by collecting and analyzing various types of data, information, and evidence. However, the information and evidence obtained may or may not be obtained directly, i.e. they may not be related to the place and time of the incident. Frequently asked questions by a forensic engineer in the field of hydrology relate to: (1) identifying potential sources of water pollution; (2) identifying the sources of water, such as local rain, infiltrating surface water, sewage leakage system, seawater and halite leaching; (3) determining the characteristics of a porous environment, such as permeability, hydraulic resistance, and porosity; and (4) detection of geochemical reactions.

Forensic hydrology can be called disaster forensic engineering (Bronstert *et al.*, 2018), where many different mechanisms and factors, including natural disasters and catastrophes, together with human factors, create a disaster with financial and human losses. For example, in the case of material damage and the loss of human life due to floods, the root causes, mechanisms, and various natural and human factors are analyzed, which clarifies the causes of an event that, in particular, will help prevent the recurrence of floods in the future. If possible, forensic hydrology analysis should be done close to the time of an event, although it may also be performed long after the event has occurred (Bronstert *et al.*, 2018).

14.2 FORENSIC HYDROLOGY LITERATURE REVIEW

The term forensic hydrology was first introduced to the scientific community in 2007 by the Journal *Southwest Hydrology*, which used forensic hydrology as a useful tool to determine the occurrence of an event, its natural or human factors, and evaluating ways to prevent it from occurring (Witte *et al.* 2019). In recent years, a number of studies following the concept of forensic hydrology have identified the causes of events.

14.2.1 Floods

From a climatic, hydraulic, and hydrological perspective, Loaiciga (2001) assessed the causes of flood damage that occurred on 10 March 1995 in the San Luis Obispo Creek, in a floodplain in California, USA. He showed that not only the rainfall and humidity of the basin had led to floods and damage, but also the development of floodplains and other measures, such as agriculture and park construction. Analyzing the historical data, he realized that the development of the floodplain had increased its hydraulic roughness and, as a result, had led to an increase in the flood level and consequent damage. Also, the floodplain development had increased the flood return period to 50 years, while the rainfall return period was only ten years.

Aparicio *et al.* (2009) investigated the causes of floods in Tabasco, Mexico, from 28–30 October 2007, from the perspectives of hydrology, meteorology, dam operation and management, as well as the flood control system and flood alert system. They concluded that the possible causes of floods could be divided into two groups: (1) immediate causes, such as exceptional rainfall; and (2) causes with delays, such as the non-observance of the frontage of rivers, with improper construction of settlements, the lack of hydraulic control of infrastructure, the lack of accurate and timely forecasting of meteorology and hydrology, lack of timely use of warning system floods, lack of integrated flood management programs (such as the conversion of forests into agricultural lands and rangelands), lack of soil protection, and inadequate land use control. They also provided guidelines and strategies for minimizing the financial and human damage caused by future floods, including adopting a modern and integrated management approach, relocating populations from critical areas with a 100-year return period, soil conservation and watershed management, social participation, education and dissemination of information among the population and stakeholders, urban growth planning to reduce the vulnerability of the urban area, further studies on the relationship between deforestation and rainfall-runoff and sedimentation, and updating of plans of flood risk zoning.

Van den Honert and Mcaneney (2011) assessed the causes, effects, and some consequences of the 2011 flood in Brisbane. They found that between December 2010 and January 2011, a strong La Nina event occurred with heavy rainfall and widespread flooding in eastern Australia. The flood was named the 'Dam Release Flood' by hydrologists with the Australian Insurance Council. The leading cause of flood was the release of water from the Wivenhoe Dam, where the dam operators acted only following the dam operation guidelines. Their model did not take into account the precipitation forecast when determining the water level behind the dam and, as a result, caused a sub-optimal strategy.

Bronstert *et al.* (2018) used the forensic hydrology approach to assess the natural causes of flash flood that occurred in southern Germany at the end of May/early June 2016. They analyzed meteorological, hydrological, geomorphological, sediment transport and damage data, and risk analysis data, to examine the contribution of meteorological and hydrological events affecting the occurrence of flash floods. They concluded that scarce rainfall in combination with local basin characteristics led to the production of very severe surface runoff with severe geomorphological hazards, such as debris flow, which resulted in massive damage in the small village of Braunsbach. They found that the flash flood, with a return period of more than 100 years, was a record-breaking disaster for all precipitation, peak discharge, and sediment transport.

Ho *et al.* (2020) assessed the root causes of floods in Thailand in 2011. By analyzing hydrological data from 23 measuring stations along the Chao Phraya River and other remote sensing data, they separated the share of river overflow volume from local rainfall affecting the flood event. They found that local rainfall had a larger share of responsibility for floods. In fact, of the total estimated 77.6 km^3 of flood, 73.7% was related to rainfall, and only 26.3% was related to river discharge, so more accurate meteorological predictors and upstream dam management would have had little effect on reducing the flood volume.

14.2.2 Groundwater level decline

Witte *et al.* (2019) evaluated the human factors affecting groundwater recharge during the period 1950–2010 in the Netherlands, which had a more significant decline than that shown by hydrological model simulations. To answer the question 'What human factors caused the groundwater table to drop by about 0.2 to 0.3 m over 60 years?', they analyzed historical data, showing that increasing the yield of agricultural products in this period had led to an increase in water consumption and evapotranspiration, which led to a decrease in groundwater recharge. Further, land-use change, urban development, and deforestation had reduced groundwater recharge in this period, because in urban areas a large part of rainfall-storm flow had gone to the urban sewage sector.

14.2.3 Drying up of lakes

Cong *et al.* (2009) assessed the causes for the Yellow River drying up during the 1990s in China. They simulated natural runoff based on observational data and results of a simulation model, combining historical meteorological data and landscape geographic information with a distributed hydrological model. They quantitatively analyzed hydrological trends to assess the effect of climate change and human activities. Results showed that the leading cause of intensification of lake drying was an arid climate that had existed since the 1990s and the main reason for the improvement of the river drying in the 21st century was the improvement of water resources management since 2000.

Ramezani Mehrian *et al.* (2016) evaluated the leading causes of the environmental crisis of Lake Urmia, Iran. Using the technique of remote sensing and analysis of climatic data time series, they tried to achieve a clear picture of the processes and trends. Results showed that the expansion of irrigation in the basin in the period 1999–2014, regardless of water resources limitations, had led to this environmental disaster. This suggests that climatic parameters, such as rainfall and temperature, cannot be the leading causes of declining water levels of Lake Urmia.

Alizade Govarchin Ghale *et al.* (2018) evaluated the causes of drying of Lake Urmia, Iran, and the effect of drought intensity on it in the period 1985–2010, and considered a novel hypothesis to quantify the effects of human activities and climate on reducing the lake's volume. Results showed that significant changes in influential variables involved in reducing the level of Lake Urmia had been observed since 1998, and they noticed a significant leakage from Lake Urmia to groundwater. They concluded that the effect of human activities on the lake was much more important than climatic factors, being 80 and 20% effective in drying the lake, respectively. Further, they expressed recommendations for the recovery of Urmia Lake: the first step being a revision of surface water management, management of dams and groundwater resources; and the second step being the classification of growing agricultural products in the region in terms of water consumption and cost-effective irrigation method, with the need to educate local people about this classification.

Alborzi *et al.* (2018) investigated the role of meteorological drought effects along with unsustainable management of water resources in the environmental disaster of Lake Urmia, Iran. They analyzed a set of hydrological, water demand, water abstraction, as well as water management infrastructure data, i.e. reservoir capacity, and management policies. Further, they quantitatively assessed water resources and found that a critical point had been reached in Lake Urmia in the early 2000s. Moreover, during a relatively normal water period immediately after the drought in the period 1998–2002, the lake's surface water level did not reach the designated ecological threshold, i.e. 1274 m above sea level. This occurred due to the increased hydrological capacity of the basin, and severe climatic stresses had accompanied anthropogenic drought growth. They also proposed a dynamic environmental flow program for various dry, wet and near-normal climatic conditions in combination with water harvesting scenarios to aid the recovery of Lake Urmia and manage the failure. Further, they estimated the expected time needed for the recovery of the environment of Lake Urmia for each environmental flow scenario.

Khazaei *et al.* (2019) evaluated the causes of the 8 m decrease in the water level of Lake Urmia, in Iran since 1995. Analyzing climatic (rainfall and temperature), hydrological (soil moisture and water level), and vegetation (agricultural products and other vegetation) data during the period 1981–2015 and using classical statistical methods, they assessed the role of climate change and human activities in the drying of Lake Urmia. They showed that changes in temperature, precipitation, and soil moisture could not justify the decrease in the lake water level until 2000. However, an increase in agricultural vegetation in the basin had a good correlation with the decrease of the lake water level, which indicated that the human activity of increasing agricultural vegetation and the expansion of irrigation had been able to play an essential role in the drying of the lake. Increased irrigation had increased the evapotranspiration of the agricultural vegetation, which had led to more water diversion and storage changes in the basin.

14.3 DEFINITIONS AND TERMS

14.3.1 Failure events

Failure events are the result of natural or human factors, or both, or the occurrence of compound failures across multiple spatial and temporal scales, which cause damage in the form of loss of life, or cause damage to health, the environment, or to financial and economic aspects (Mendoza & Schwarze, 2019).

14.3.2 Exposure

Exposure places people, goods, and systems in areas that leaves them vulnerable to possible damage or danger by failure events (Mendoza & Schwarze, 2019).

14.3.3 Vulnerability

Vulnerability is the predisposition of a system to be negatively affected, which arises from the sensitivity of the system and its components to damage, combined with a lack of short-term coping capacity and long-term adaptation capacity (IPCC, 2014).

14.3.4 Failure risk

Failure risk is the possibility of drastic changes in the normal functioning of communities in a given period due to the occurrence of failures in interaction with vulnerable aspects of society. It leads to negative impacts on a society, economy, or environment that, to meet the vital needs of human beings, requires an immediate response and may require external support for recovery. The potential for negative impacts and consequences is not only due to events (drought, floods, etc.) but also due to the interaction of events, exposure, and vulnerability (IPCC, 2012, 2014).

14.3.5 Cascading and compound events

One event may be the cause of another event, even if they occur across temporal and spatial scales. Cascading and compound events are a series of unfortunate events which, like dominoes falling, occur one after the other, with each one being the cause of another. Unprecedented severe fires, for example, may be caused by global warming and indicate a link between extreme events. Many fires occur following long periods of drought. One event can provide the chance or opportunity for other future devastating events (AghaKouchak *et al.*, 2018).

14.3.6 Resilience

Resilience is the ability of a system and its components to predict, adapt or recover promptly against the impacts of failures, with efficient methods including reliably maintaining, recovering or improving basic structures and functions (IPCC, 2012).

14.3.7 Reverse engineering

Initially, a forensic engineer only knows the final result of a process or series of events, i.e. it is clear that a failure has occurred, such as damage to crops. Forensic engineers start from this endpoint and work in reverse, gathering evidence and information and trying to understand by asking what, who, where, when, why, and how the failure happened. A forensic engineer uses reverse engineering to reach the starting point, i.e. the causes of failure, which is why forensic engineering is also referred to as reverse engineering (Noon, 2001).

14.3.8 Main and root causes

Main and root causes are not identical. To clarify the meaning and difference between these two terms, suppose an event is a flood that has caused significant damage. In this case, the main cause of the damage is the flood itself, but the flood is not the root cause. The root causes of a flood can be a number of things, including intensive rainfall, lack of timely flood prediction, and improper operation of a dam. This example illustrates the importance of finding the root cause.

14.3.9 Failure analysis and root analysis

Failure analysis and root analysis are often used interchangeably and synonymously in forensic engineering. However, sometimes there are differences in the meaning of these terms. Forensic engineering is called failure analysis if it is performed according to engineering methods and determines how a particular part of a system fails. However, root analysis determines how an event fails a whole system, not part of the system, and focuses more on the managerial aspects of the failure and discusses how to improve management methods and techniques to prevent the failure from recurring. In general, forensic engineering is called failure analysis if it is in a specific domain, and root analysis if it is in a general domain (Noon, 2001).

14.3.10 Investigation pyramid

The structure of scientific investigation and analysis of failure can be likened to a pyramid. In this way, the bottom of the pyramid, which is wider, is the place of great provable facts and evidence of failure. These facts are the basis of failure analysis, and forensic engineers analyze the natural and human causes of failure. Ultimately, facts, evidence, and analysis lead forensic engineers to achieve results – which form the top of the pyramid (see Figure 14.1). Results should be directly the result of facts and analysis, and not based on other results or hypotheses. If facts are examined in an integrated manner, the results will be evident (Noon, 2001).

14.3.11 Event reconstruction

When an event is described, it is said that the event has been reconstructed, which is why forensic engineers are called event reconstruction experts (Noon, 2001). Forensic engineering reconstructs an event to determine what happened, and establish what factors (and who) were the leading causes of the event.

14.3.12 Statement interview

Eyewitnesses of an event are valuable resources for forensic engineering and should be skillfully interviewed about how the failure occurred. Witnesses may combine their opinions and conclusions with what happened, or witnesses' descriptions of a failure may take on an aspect of their tastes, culture, or even education, or they may describe the failure a certain way due to their hostility to individuals, or for the benefit of friends and family involved in the accident. For example, in the case of well water contamination, the

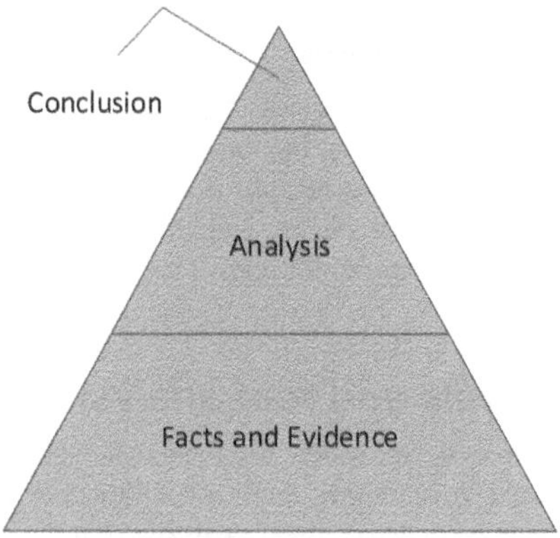

Figure 14.1 Investigation pyramid (Noon, 2001).

well owner may blame a person or persons for the incident out of hostility. This is why the method of interviewing eyewitnesses about a failure is crucial. Interview questions for witnesses should be both precise enough to reveal the truth and to help separate what happened from their conclusions. Moreover, they should allow the witnesses to review their memory and define the failure in detail, rather than experts suggest answers to witnesses. (Noon, 2001).

14.3.13 Freeze interview

Interviews with witnesses are usually recorded on tape and immediately transcribed in the presence of witnesses as a statement; this is called freeze. The interview should be kept frozen because time and subsequent conversations with others can lead to changes in a witness's recall (Noon, 2001).

14.4 BASICS AND LOGIC

Forensic engineers apply logic. Logic and clear facts provide principles and methods that affect a particular case. At first, the evidence can appear to be like scattered pieces of a puzzle. To find the cause and how failure occurs, the different pieces have to be matched. When pieces are matched, part of the image emerges; eventually, when all the pieces are put together, the puzzle is solved and the cause of the event is revealed (Noon, 2001).

A general principle is that failure is not due to one factor, but to a chain of factors (Noon, 2001), each of which alone may not play a role in shaping the failure, but they create it together. There is a difference between the way that forensic engineers and other engineers apply this principle. Engineers evaluate the probability of the occurrence of single events, without considering the relationship between events. By contrast, forensic engineers take a holistic view of events and consider the interaction of events in their risk analysis. If events are assessed independently, their origins and causes cannot be understood, and proper planning cannot be undertaken to prevent future disasters. By answering the question 'What factors can increase the likelihood of the occurrence of an event?', forensic engineers consider these factors and events to prevent failure.

Another point that forensic engineers consider in their event analysis is that the conclusion should be based directly on facts, not on other conclusions or hypotheses. If the facts are examined logically and systematically, the complete results will be evident (Noon, 2001). As an example, we can consider a new well, drilled beside a primary well which itself dried up shortly after the new well was drilled. Sometimes, a newly drilled well dries up a primary well and sometimes it does not. Therefore, based on a purely statistical analysis, it is possible that the newly drilled well has dried up the primary well. However, the fact that the primary well dried up after the operation of the new well may be the result of another accident or event. If an expert concludes that the primary well dried up based on the start of operation of the new well, they have drawn that conclusion based on the hypothesis; instead, they must gather more evidence to be able to confirm or reject this hypothesis.

14.5 IMPORTANCE AND NECESSITY OF THE SUBJECT

Finding the causes of a failure is essential in reducing the risk its recurrence. Further, following the occurrence of an event, there is always a tendency to determine that something or someone else is to blame. Investigating the causes of failure and seeking information by forensic engineers requires the examination of the responsibilities of all parties to the failure, and ultimately the forensic engineer can determine the amount of damages that parties involved must pay to compensate for the failure. For example, suppose a river on the border of two countries is dried, this river may be the only source of water needed by both countries or just one of them. If human causes play an influential role in drying up the river, then one or both of the countries are to blame for this catastrophe. This example helps make apparent the importance of determining the causes and the culprit in drying up of the river, and determining the damages that the culprit must pay to the other in compensation.

Forensic engineers deal with a wide range of legislators and lawyers.. They must be familiar with legal proceedings, where their testimony and opinion is often required as an expert witness, to help determine the culprit responsible for a failure. The need for a forensic engineer to appear in legal hearings is required for some or all of the following:

- to quantify the extent and severity of the event;
- to provide eyewitnesses to the failure;
- to determine the root causes of failure and its culprits, according to the responsibilities of the parties involved; and
- to quantify courses of action to compensate for failure and prevent a recurrence.

Legal proceedings need this information to reach a verdict and allocate responsibility to guilty parties for the failure, and all of this information comes from the judge's and the court's questions, and from the testimony of forensic engineers and other witnesses.

In a growing and increasingly urbanized world, water management and distribution are necessary and sometimes challenging tasks, particularly when flood and drought occur, whether due to human or natural causes, or to a combination of the two – which can exacerbate their effects (Berghuijs *et al.*, 2019; Namias, 1995; Raymond *et al.*, 2020). In these two sections, the definitions of drought, flood and their causes are described, then the necessary role of forensic engineering in their management and prevention is described.

14.5.1 Drought

Drought is generally referred to as water scarcity when compared to normal conditions (Hagenlocher *et al.*, 2019; Van Loon *et al.*, 2016). Droughts begin slowly, with direct and indirect effects on humans and natural systems, including water resources, water quality, agriculture, food security, health, energy, and the economy (Hagenlocher *et al.*, 2019; Schwalm *et al.*, 2017). Generally, they are classified into four categories – meteorological, hydrological, agricultural, and socio-economic droughts – and these droughts occur in a sequence. Climate change causes a precipitation deficit. Meteorological droughts occur when rainfall for any given period (monthly or seasonal) is less than its long-term average. Continuing long-term precipitation deficit and high potential evapotranspiration leads to agricultural drought in which the amount of moisture in plant roots decreases to such an extent that it causes wilting and eventually reduction of agricultural products. The previous two factors combined with the pressure of anthropogenic demands lead to falling water levels of rivers, reservoirs, lakes, and groundwater below their long-term averages. Socio-economic drought is associated with an inadequate supply of goods due to meteorological, hydrological, and agricultural droughts (Hagenlocher *et al.*, 2019; Van Loon *et al.*, 2016; Wilhite, 2000; Zargar *et al.*, 2011).

In the glossary of the IPCC SREX report (IPCC, 2012) drought is defined as 'an abnormal long-term period of dry weather that leads to severe hydrological imbalances. Drought is a relative term, so any discussion of a lack of rainfall should refer to the specific rainfall activity. Megadrought is a long-term, pervasive drought that lasts much longer than usual, usually lasting a decade or more' (Seneviratne *et al.*, 2012).

The leading causes of drought sequences (sequence of meteorological drought, agricultural drought, hydrological drought and socio-economic drought) are deficit rainfall and/or increased evapotranspiration, although studies have suggested deficit rainfall as a more prominent cause. The effects of evapotranspiration have been explicitly or indirectly considered in several drought indicators (Seneviratne *et al.*, 2012). In the context of climatic projections, analyses consider changes in simulated agricultural drought to be more influenced by changes in rainfall, with increased evapotranspiration higher than vapor pressure, and available radiation moderating some changes (Burke & Brown, 2008; Orlowsky & Seneviratne, 2012; Sheffield & Wood, 2008; Seneviratne *et al.*, 2012). Other causes of drought sequencing are: declining soil moisture, rising temperature, land-use change, water management without proper policies, increased urbanization, wetland development, and dense forests – all of which can exacerbate the effects of droughts. Figure 14.2 shows the causes of drought sequences.

Agriculture is one of the sectors that suffer the most from drought. The agricultural sector includes a multitude of heterogeneous agricultural systems and a complex network of food supply, needs, and trade that reflect the systems' high risk (Jahn *et al.*, 2018; Raymond

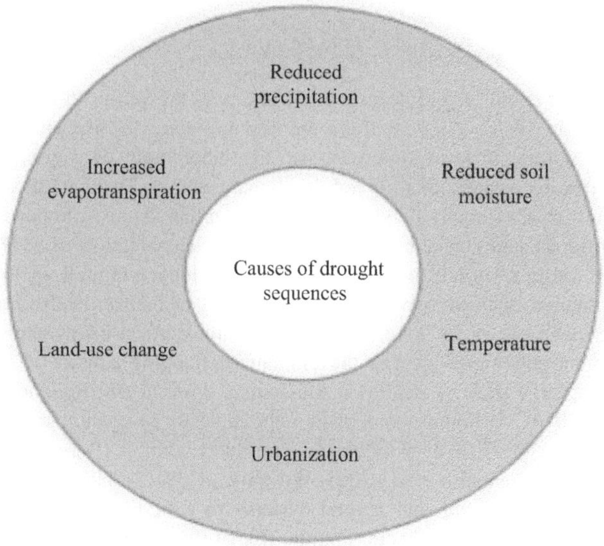

Figure 14.2 Causes of drought sequences.

et al., 2020). Drought can threaten the food security of a region and the world. Heatwaves and fires accompany drought, and crops are vulnerable to hot and dry conditions. The harmful effects of heatwaves during crop growth are considered a significant threat. During the period of lack of soil moisture, the earth's surface becomes warmer, because the existing energy, instead of evaporating and becoming the latent heat of evaporation, remains on the earth's surface and is called sensible heat (Chiang *et al.*, 2018; Fischer *et al.*, 2007; Seneviratne *et al.*, 2006). Drought causes more water to be harvested by farmers, and reduces basal flow in rivers and streams. Reduced stream flow and snow melting in dry periods reduces groundwater and surface water flow in a cycle that endangers local habitats (AghaKouchak *et al.*, 2015; Barnett *et al.*, 2008; Power *et al.*, 2015). Rising temperatures when water is scarce increase the energy demand to generate cooling power, and increase the tension between human priorities and nature (AghaKouchak *et al.*, 2015).

Learning lessons from past droughts, forensic engineering predicts the potential effects on water resources, agriculture, food security, environment, and economy, and – according to the causes of the drought – provides strategies, such as water conservation and storage, more comprehensive irrigation, wastewater reuse, saline desalination, optimal cultivation patterns, and virtual water imports, as well as providing a framework for long-term drought monitoring. By considering plans for managing lands, rivers, and wetlands, and setting long-term plans that cover all sectors, forensic engineers help manage drought and increase system resilience. Forensic engineers focus on increasing drought management needs and reducing water storage over time by updating drought management strategies. Meeting environmental needs in drought conditions is an example of the success of adaptation programs (AghaKouchak *et al.*, 2015). A drought management plan should go beyond simply meeting immediate needs and should also target local water development, allocation, and adaptation programs.

14.5.2 Floods

Floods are the most common and devastating disasters in the world. In the glossary of the IPCC SREX report (IPCC, 2012), a flood are said to occur: 'as a result of overflowing water from the normal flow of waterways or any other body of water, or due to the accumulation of water in areas that are not usually submerged. There are different types of floods, including river (fluvial) floods, flash floods, urban floods, pluvial floods, sewer floods, coastal floods, and glacial lake outburst floods' (Seneviratne *et al.*, 2012).

The leading causes of floods are intensive and/or prolonged rainfall, melting snow and ice, or a combination of these two causes, as well as dam breaks, landslides, and severe local storms (Seneviratne *et al.*, 2012). Rising temperatures affect the dynamics of snow in cold regions, thus potentially altering the amount of flooding caused by melting snow (Berghuijs *et al.*, 2019; Hall *et al.*, 2014; Musselman *et al.*, 2018; Regonda *et al.*, 2005; Vormoor *et al.*, 2016). At higher temperatures, the intensity of precipitation increases and potentially increases the risk of flooding due to heavy rainfall (Berghuijs *et al.*, 2019; Kendon *et al.*, 2014; Trenberth *et al.*, 2003; Westra *et al.*, 2013). Human systems can also interact with the environment and natural systems to lead to the formation of severe hydrological events, significantly increasing the risk of floods caused by precipitation –for example, urban drainage systems modulate the number of surface floods and water quality at discharge points due to the correlation of sewage overflow and intensive rainfall

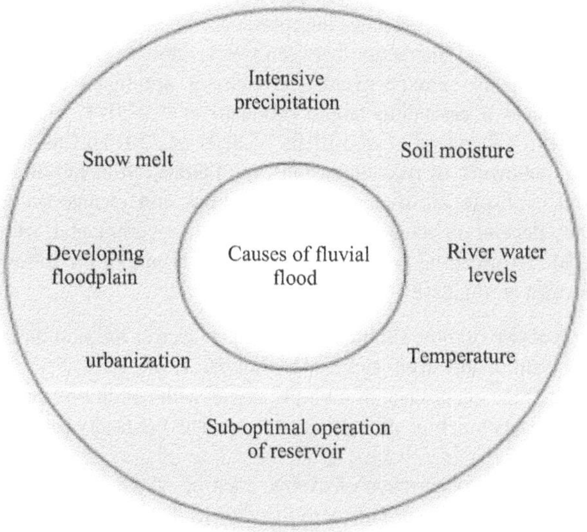

Figure 14.3 Causes of fluvial floods.

(Raymond *et al.*, 2020). Other causes of flooding include previous soil moisture, urbanization, flood plain development, and improper operation of dam reservoirs. Figure 14.3 shows the causes of river floods.

Forensic engineering evaluates natural causes and culprits and leads to the management, optimal operation of dam reservoirs, and clarification of responsibilities. According to the cause of a flood, appropriate solutions and courses of action are proposed such as timely prediction of meteorology and hydrology, soil protection, land use control, urban development planning, respect of river frontages, stream improvement, construction of flood control reservoirs, dam diversion structures, flood repellent structures, and the use of timely flood warning systems; in this way, forensic engineering can help reduce or eliminate casualties, and offer a framework for integrated flood management.

14.5.3 Cascading failures

Although individual events can have severe effects, the nature of interconnected natural factors shows that cascading failures, especially with increasing temperature, can have more destructive effects. Floods, droughts, heatwaves, and fires are often the result of the interaction of several processes at different spatial and temporal scales (Zscheischler *et al.*, 2018). The combination of these processes (natural and human factors, and failures) can lead to severe and destructive effects called cascading failures, and in this way a chain of small events can cause disasters. For example, drought and heatwaves often occur together. Drought leads to the drying of soil, which prevents the release of solar energy as evaporation and warms the earth's surface. Dry and hot conditions increase the likelihood of fires that cause the destruction of soil, and provides the conditions for future floods and landslides. At the same time, temperature changes, fires, and floods affect dams and reservoirs (AghaKouchak *et al.*, 2018), where the importance of managing water resources to prevent more destructive disasters comes into play.

The interdependence and interaction between two or more event factors, each of which alone may not lead to an extreme event, can cause a significant extreme failure, in a phenomenon known as a cascading failure (Leonard *et al.*, 2014; Mehran *et al.*, 2017; Sadegh *et al.*, 2018; Vahedifard *et al.*, 2016; Wahl *et al.*, 2015). Cascading failures are embedded in the framework of risk dimensions, i.e. failures, vulnerabilities and exposure. Vulnerability and exposure are dynamic vulnerabilities and change on a temporal and spatial scale. They depend on economic, social, cultural, governmental, and environmental factors (IPCC, 2012). Cascading failures may occur due to one of the following conditions (IPCC, 2012; Sadegh *et al.*, 2018):

(1) as a result of two or more simultaneous failures (such as the simultaneous occurrence of heavy rainfall and storm surges) (Moftakhari *et al.*, 2017);
(2) as a result of the occurrence of multiple events with conditions that strengthen each of their impacts (such as droughts and heatwaves) (Mazdiyasni & AghaKouchak, 2015);
(3) as a result of the occurrence of events, each of which alone may not lead to an extreme event, but the sum of them together leads to an extreme event (such as a moderate coastal flood occurring during an above average tide) (Moftakhari *et al.*, 2015).

Traditional risk analysis methods typically consider only one event, which leads to an inaccurate estimation of the causes of failure and underrates risk, because the event factors often interact and are spatially and temporally interdependent. Forensic engineers evaluate the probability of failures by considering their relationships and cascading failure scenarios. Ignoring the relationships of failures causes their leading causes and effects to be ignored. Forensic engineers use a new framework to analyze events based on the cause–effect relationship between them, and make joint assessments of natural and human factors and failures occurring at different spatial and temporal scales to improve risk management in response to events.

14.6 FORENSIC ENGINEERING GOALS

Forensic engineers use a monitoring network to track failures to find out the cause of each event, and develop solutions to prevent failures. Minor events, which are not typically considered major failures, can have profound implications for future disasters. Failure evaluation models should be able to consider failure factors in a multifaceted way. By identifying the vulnerable infrastructure of a system, the goal of forensic engineers is to improve the resilience of that system and emergency response to events, by evaluating strategies, reducing the likelihood of failures and system vulnerability to them, to minimize financial and human losses.

14.7 METHODOLOGY

In forensic engineering, the historical evaluation and analysis of an event is done in the form of a root analysis, with the aim of determining the root causes of failure, including natural and human causes. This section explains the path that forensic engineers must take to reach this goal.

14.7.1 Collecting quantitative and qualitative historical data about the occurrence of events

A forensic event investigation requires different categories of data, including data on natural and human factors, damage, and preventive measures. The data on natural and human failure factors include exposure data (failure-prone regional data) and vulnerability data for the area subject to failure. Damage data include data of various damages related to failure, and preventive measures data include data on measures taken to reduce the risk of failure in the preventive phase (Mendoza & Schwarze, 2019).

Forensic engineers may not always have access to data; sometimes, historical data may not be available or may be scattered if it is present. Estimating missing values using data generation techniques can reduce the accuracy of the final result and increase its uncertainty (Ramirez & Herrera, 2016). The use of GIS and remote sensing data can be helpful sources in these cases (Johnson, 2008; Kumar, 2002).

14.7.2 Interviews with eyewitnesses

Reconstruction of the event by eyewitnesses of any failure is essential. Evidence from interviews with witnesses can provide valuable and vital information for forensic engineers to analyze failures. Forensic engineers should always consider interviewing failure witnesses.

14.7.3 Assessing conditions, exposure and vulnerability by analyzing historical data

In this step, forensic engineers describe the characteristics of the system and aspects of its exposure and vulnerability, and evaluate the current situation against the acceptable situation in order to identify the factors that affect the system in a way which leads to unacceptable performance. An accurate definition of failure to achieve a goal, determining the causes of failure, and providing solutions to prevent the recurrence of that failure, are of particular importance.

Forensic engineers must consider both vulnerabilities and opportunities. Vulnerabilities include developments that prevent the system from achieving its goals, whilst opportunities include improvements that can be made to assist the system in achieving them (Haasnoot *et al.*, 2019). Initially, only the result is known. Forensic engineers work to determine under what, who, how and why the failure occurred (Noon, 2001).

14.7.4 Using engineering knowledge to present different scenarios, to link different facts, data and evidence

In this step, forensic engineers evaluate natural factors, functions, and actions, and define static scenarios. Such scenarios describe the endpoint (failure) in contrast to time-dependent scenarios that describe changes over time (Haasnoot *et al.*, 2019). The presentation of these scenarios requires a precise definition of the system, the endpoint (failure) and expertise in several disciplines, because failure often does not occur due to one factor. Instead, a set of factors interacting with each other often causes failure occurrence, or a sequence of failures causes another failure to occur. These factors and failures may each be related to specialized knowledge. Therefore, forensic engineers need to have mastered several related disciplines to examine and evaluate each factor and its impact on other factors. Sometimes the participation and experience of several engineers is needed to evaluate failure and the factors that cause it.

Scenarios are hypotheses with uncertainties about a set of factors or failures in the past that have led to unacceptable system performance that requires historical data analysis. Statistical methods often analyze historical data. In this step, forensic engineers must discover what story is behind the data and use their knowledge, expertise, and experience to verify the scenarios. Forensic engineers must take a holistic approach to statistical analysis of data to determine the causes of failure. Each scenario considers a chain of causes or failures – in other words, each cause changes other factors, and changes in these factors lead to changes in other factors, and it is this chain of factors that eventually causes failure. Analyzing data, forensic engineers can understand the accuracy of the scenarios and find the root causes of failure. When determining the causes of failure, efficient use of cost and time resources is essential.

14.7.5 Evaluating and proposing solutions to manage or prevent recurrence of failure

Forensic engineers evaluate strategies to manage a failure or prevent its recurrence, depending on the causes of the failure. The goal is to provide conditions that lead to the acceptable performance of the system or prevent its unacceptable performance. To achieve this goal, it is crucial to define the current state of the system, and to assess exposure and vulnerability in areas prone to failure. Also, evaluating past failures, their causes, and the effective and ineffective risk management methods that have been used can provide forensic engineers with valuable information to produce solutions and prevent recurrences of failures.

For example, if the cause of flood is an incorrect prediction by a meteorological agency, the goal of the forensic engineer is to develop a method that leads to timely prediction of dam reservoir discharge and effective measures to manage water resources, and thus prevent floods in the future. In terms of controlling and managing significant groundwater decline, forensic engineers evaluate strategies, such as composite exploitation of surface water and groundwater, virtual water imports, reducing crop cultivation area, cultivating crops with fewer water requirements, stopping or restricting urban growth, and overseeing strict practices and regulations to prevent unauthorized abstraction of groundwater. Figure 14.4 shows the methodology of forensic engineering.

14.8 FORENSIC ENGINEERING APPROACHES

The need for failure analysis has led to many advances in forensic disaster approaches, including statistical approaches (such as copula and complex network approaches) and socio-physical approaches (such as dynamic adaptive policy pathways) that are described in this section). Table 14.1 shows a number of forensic disaster approaches.

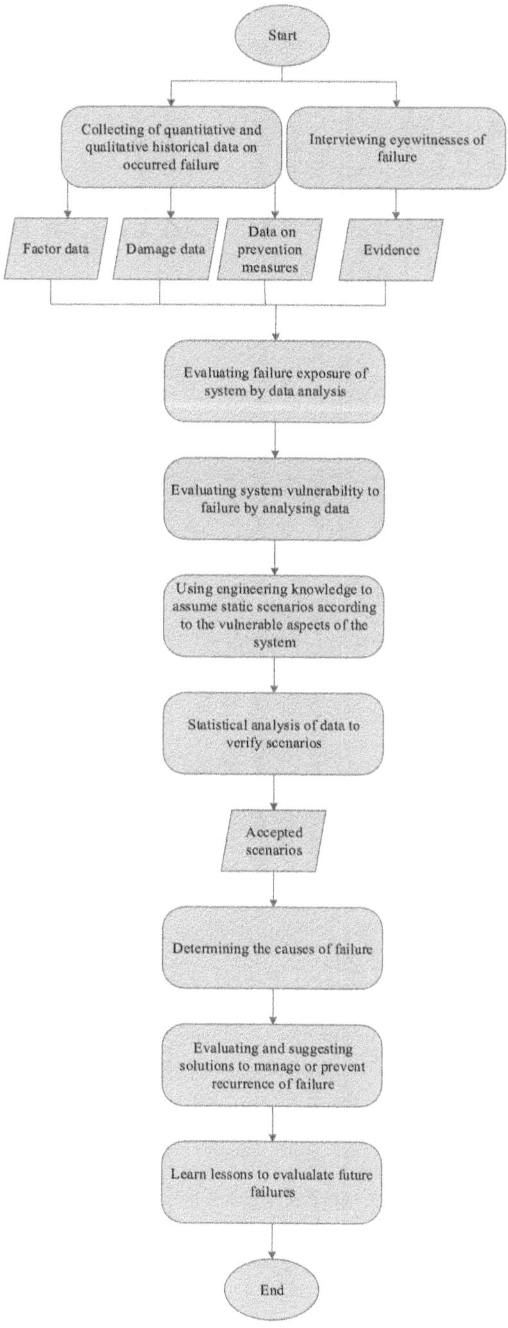

Figure 14.4 Forensic engineering methodology flowchart.

Table 14.1. Forensic disaster approaches (Mendoza & Schwarze, 2019).

Approach	Year	Objective(s)	Characteristics	Methodology
Forensic investigation of disasters, by Integrated Research on Disaster Risk (IRDR)	2010–2016	Identifying root causes so that they can be evaluated and addressed by policy and practice.	Structured around the disaster risk reduction management cycle (risk reduction and preparedness, response, recovery).	Desk review, fieldwork, semi-structural and informal interviews.
Post-event review capability, by Zurich Insurance Group and ISET-International	2015	Uncovering root causes, successes and failures in the management of disaster risk; identifying critical gaps and actionable opportunities for disaster risk reduction.	Structured around the disaster risk reduction management cycle (risk reduction and preparedness, response, recovery).	Desk review, fieldwork, semi-structural and informal interviews.
Disaster root causes framework and tool, by the German Committee for Disaster Reduction (DKKV)	2012	Identifying root causes and defining areas for interventions to address them.	Analysis of causal linkages between impacts and root causes, integrating the disaster risk reduction cycle.	Desk review, expert interviews.
Forensic disaster analysis in near real-time, by the Center for Disaster Management and Risk Reduction Technology (CEDIM)	2012	Improving understanding of how natural hazards become disasters, providing results in the immediate aftermath of a disaster.	Focus on near real-time disaster analyses by interdisciplinary teams, complements forensic investigation of disasters.	New near real-time methodologies, interdisciplinary approach.

14.8.1 The copula approach

Failures are typically multivariate. For example, the definition of drought based on one variable such as rainfall or soil moisture or runoff may not be sufficient for high-reliability risk analysis and decision-making (Hao & AghaKouchak, 2013). The copula approach is used to model the dependence of variables of a failure, or the dependence of causes of a failure. Copulas are mathematical functions that link or couple two or more time-independent variables (Sadegh *et al.*, 2017); in other words, copulas join a multivariate distribution function to its one-dimensional marginal distribution functions (Nelsen, 2003). The concept of a copula is an approach taken from a set of multivariate analysis methods, and is widely used to model the dependence of two or more random variables. Copulas can be used in hydrology and climate change (Zhang & Singh, 2019) and in many other fields including integrated risk management, risk analysis, flood and drought monitoring, etc. (Sadegh *et al.*, 2017). Hao and AghaKouchak (2013) developed a standard multivariate index with a copula approach that joins the standard precipitation index (SPI) to the standard soil moisture index (SSI) to assess drought. Renard and Lang (2007) used a Gaussian copula to assess the risk of hydrological events.

The strengths of the copula approach are that it is common and well-developed, whilst its weaknesses are that it does not consider the cause and effect relationships in failure modelling. Also, if the number of data is limited, fitting can be difficult (Raymond *et al.*, 2016). Interested readers can refer to Nelsen (2007) and Zhang and Singh (2019) for more information about the copula approach.

14.8.2 The complex network approach

Before 2010, researchers studied existing processes in one network (Valdez *et al.*, 2020), while the real world is like a complex network that includes a collection of different networks. Within a network, connecting links connect the nodes, while dependency links connect the nodes of different networks (Valdez *et al.*, 2020). Failure of a fraction of the nodes of one network may cause the failure of other networks because the malfunction of the nodes of the first network will cause malfunction of other networks. These cascading failures will result from failure of only a fraction of nodes in one network and cause the disorder of the function of other networks (Buldyrev *et al.*, 2010). For example, California has been experiencing drought since 2012, and population growth and increased agriculture almost doubling water use since the 1950s has exacerbated the drought. Twelve million trees have been lost. Known as the United States' salad bowl for its fruit and vegetable production, California suffered $2.2 million damages in 2014 following that year's drought, and 17,000 jobs were lost, especially in poor rural areas (Adams *et al.*, 2009; AghaKouchak *et al.*, 2015; Herring *et al.*, 2014; Howitt *et al.*, 2014; Shukla *et al.*, 2015). Cascading failures are a devastating process that spreads across a complex real-world network and can affect all infrastructures, natural systems, and social cohesion. One of the key features that make a complex network vulnerable to failures is the interdependence between its components, which exposes networks to a significant risk of destabilizing failures. Modelling cascading failures using the complex network method can be an approach to strengthen a real-world network and reduce or stop failures (Valdez *et al.*, 2020).

Climatic observations are often correlated over long distances, and extreme events and failures (such as heatwaves and floods) are often assumed to be related to such 'teleconnections' (Boers *et al.*, 2019; Hong *et al.*, 2011; Lau and Kim, 2012); the term teleconnections is often used in atmospheric science to describe climatic links between geographical areas (Nigam, 2003). For example, Lau and Kim (2012) showed that in 2010 heatwaves and fires in Russia were associated with the floods in Pakistan. Extreme events and failures can be predicted using the complex network approach (Boers *et al.*, 2014, 2015). Boers *et al.* (2019) demonstrated a global pattern of intensive rainfall events with a complex network approach. The key idea in this approach is to identify time series measured at different locations with different network nodes, and find strong symmetries of extreme events in the time series, with network links connecting the respective nodes (Boers *et al.*, 2015).

One of the advantages of the complex network approach is the identification of dynamic delays between events in interactions in time series (Boers *et al.*, 2015). However, it is a method which requires heavy calculations, and its interpretation requires in-depth knowledge of the system (Raymond *et al.*, 2020). Interested readers can refer to Boers *et al.* (2015) for more information.

14.8.3 The dynamic adaptive policy pathways approach

Forensic engineers are faced with many uncertainties about future variable conditions in order to provide solutions to prevent recurrence of failure. For example, in providing solutions to manage the operation of dam reservoirs to prevent the occurrence of floods, they are faced with uncertainty about the volume of inflow to the reservoirs in coming months. If an operator drains a volume of water from a dam reservoir every month, regardless of meeting needs in low water seasons, the risk of failure of the reservoir will increase. To provide solutions, forensic engineers must be confident that they will be able to cope with changing conditions in the future; that is, they will not fail to be able to achieve their goals under changing conditions.

The dynamic adaptive policy pathways approach examines a series of alternative decisions (adaptive paths) for different situations in the future. It recognizes that decisions made over time interact dynamically with a system, and clarifies their dependence. This approach opens up a decision-making space and helps break policy paralysis about future uncertainty. This approach also leads to the design of a dynamic adaptive strategy, including initial actions and long-term options, and identifies when it is necessary to implement long-term options and review decisions (Haasnoot *et al.*, 2019). By simulating the future, this approach determines short-term actions and sets a framework for guiding future actions (Haasnoot *et al.*, 2013).

Haasnoot *et al.* (2013) called the combination of the two complementary approaches of adaptive policymaking and adaptation pathways the dynamic adaptive policy pathways approach (see Figure 14.5). Adaptive policymaking is a theoretical approach that describes the planning process of various actions and monitors when an adaptive approach is needed. Adaptation pathways are an analytical approach to explore and sequence a set of possible actions based on alternative external developments over time. The dynamic adaptive policy pathways approach includes transient scenarios that reflect types of uncertainties and their development over time, and different types of measures to address vulnerabilities and opportunities. For example, Haasnoot *et al.* (2012) examined adaptation

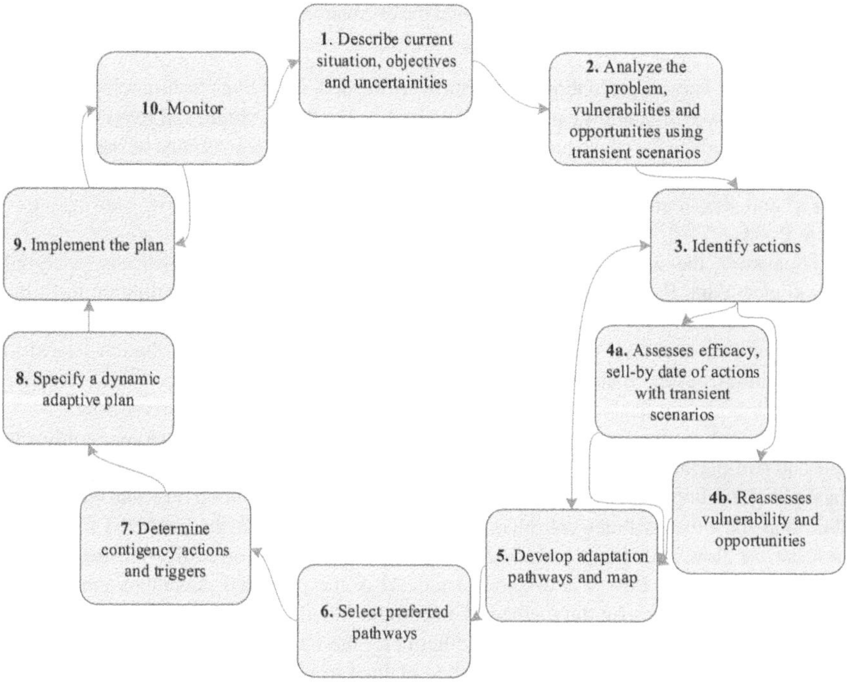

Figure 14.5 Development of a dynamic adaptive policy pathways approach (Haasnoot *et al.*, 2013).

pathways to achieve sustainable management of water resources under changing and uncertain conditions in the future. They developed a method for adapting pathways, including the dynamics of natural variability and identified the interaction between a water system and the community.

One of the advantages of this approach is that it allows forensic engineers to make decisions and provide solutions to prevent failures despite future uncertainties, although it may require a large number of hypotheses about the future (Raymond *et al.*, 2020).

14.9 EXAMPLE OF FORENSIC ENGINEERING METHODOLOGY FOR EVALUATING FLOOD EVENT

In 2002, the River Vltava flooded in Prague, in the Czech Republic, causing unprecedented damage to houses, hotels, schools, churches and offices. Holický and Sýkora (2009) answered the following questions with a judicial engineering approach to provide a model for future floods:

- Was the flood an exceptional and unpredictable event?
- What are the failure causes of buildings?
- What is the return period of the flood according to the measured water level?

- Do recorded flood measurements affect the estimation of the values of structural design characteristics?

After flooding, various precautionary measures are needed, including creating safety barriers, managing the river by building polders and correcting the depth, width, and roughness of the river canal to reduce the consequences of future floods. These decisions must be based on cost optimization and taking into account the possible social and economic consequences of floods, and this requires a suitable theoretical model to predict the flow and extent of future floods.

To answer the above questions, Holický and Sýkora (2009) statistically analyzed hydrological data. First, they identified the most appropriate possible distribution to fit the maximum annual flow data of the Vltava River, using Kolmogorov–Smirnov and Chi-square statistical tests and the experience of fitting the possible distribution related to other samples. They found that two-parameter lognormal distribution was a suitable theoretical model for analyzing data and predicting the maximum annual river flow.

The researchers then predicted maximum annual river flow and design values according to data and two-parameter lognormal distribution, using the method of moments and method of maximum likelihood for two cases: one considering the maximum flood flow that occurred in 2002 and the other without considering it. They concluded that the predicted maximum annual river flow and design values given the maximum flow of 2002 were higher than when that maximum flow was not considered. Also, the predicted maximum annual river flow and design values by the method of maximum likelihood were more than the values predicted by the method of moments. Furthermore, the recommended partial safety factor of 1.5 was significantly less than the amount obtained from the data.

The values of the return period related to the maximum flood flow which occurred in 2002 using the method of moments and the method of maximum likelihood estimation were almost 350 and 210 years, respectively, which exceeded the standard return period of 100 years.

Based on the evaluation of the cases obtained from statistical analysis of data, they concluded that the main cause of the destruction of buildings due to floods is the lack of structural robustness and that the causes of destruction were divided into two parts: geotechnical and structural. Geotechnical causes include insufficient foundation (such as depth and width), underground sediment transfer, human-made caves, and increased ground pressure due to rising groundwater. Structural causes include insufficient structural strength such as not using ring beams in the construction of buildings, insufficient construction materials used in the buildings, and changes in material properties due to moisture, volume, and the intensity of water.

Holický and Sýkora (2009) also made suggestions to prevent the destruction of buildings due to floods:

- before estimating of the values of design characteristics of flooded structures, care should be analyzed for the flow data, as the predicted maximum flow may be significantly different from the standards.
- robustness aspects should be considered in the design phase of building structures to prevent potential flood damage. Adequate strength of structures is achieved due to an adequate tie system, increased strength of critical beams, secondary protection of critical beams, and the use of indestructible structural details.

14.10 SUMMARY

In recent decades, the number of disaster occurrences has increased significantly, raising the need to determine the causes of disasters and to analyze the relationship between failures that occur at different spatial and temporal scales. When exposing systems to fail and be vulnerable, disasters can happen. Finding the causes of disasters is essential to reduce the risk of them happening in the first place, and to prevent their recurrence. Forensic engineering allows us to find the roots, mechanisms, and causes of disasters by gathering evidence, information and analyzing data, as well as analyzing the cause–effect relationship between factors and failures, and analyzing the interaction of the risk dimensions: failure, exposure, and vulnerability. The purpose of forensic engineering is to determine or clarify the causes of failure of a particular system, and to prevent the unacceptable performance of that system in the future. Forensic engineers should see the occurrence of failures as a valuable opportunity to learn lessons for effective prevention of future disasters, and their policies should be based on the magnitude of disaster risk. It is hoped that this chapter will provide readers with an overview of forensic engineering and enable them to take steps to reduce disaster risk and enhance their integrated management skills.

REFERENCES

Adams H. D., Guardiola M., Barron-Gafford G., Villegas J., Breshears D., Zou C., Troch P. and Huxman T. (2009). Temperature sensitivity of drought-induced tree mortality portends increased regional die-off under global-change-type drought. *Proceedings of the National Academy of Sciences of the United States*, **106**, 7063–7066.

AghaKouchak A., Feldman D., Hoerling M., Huxman T. and Lund J. (2015). Water and climate: recognize anthropogenic drought. *Nature*, **524**(7566), 409–411.

AghaKouchak A., S. Huning L., Mazdiyasni O., Malakpour I., Sadegh M., Vahedifard F. and Moftakhari H. (2018). How do natural hazards cascade to cause disasters? *Nature*, **561** (7724), 458–460.

Alborzi A., Mirchi A., Moftakhari H., Mallakpour I., Alian S., Nazemi A., Hassanzadeh E., Mazdiyansi O., Ashraf S., Madani K., Norouzi H., Azarderakhsh M., Mehran A., Sadegh M., Castelletti A. and AghaKouchak A. (2018). Climate-informed environmental inflows to rivive a drying lake facing meteorological and anthropogenic droughts. *Environmental Research Letters*, **13**(8), 1–21.

Alizade Govarchin Ghale Y., Altunkaynak A. and Unal A. (2018). Investigation anthropogenic impacts and climate factors on drying up of Urmia lake using water budget and drought analysis. *Water Resource Management*, **32**(1), 325–337, doi: 10. 1007/s11269-017-1812-5.

Aparicio J., Martinez-Austria P. F., Guitron A. and Ramirez A. I. (2009). Floods in Tabasco, Mexico: a diagnosis and proposal for courses of action. *Journal of Flood Risk Management*, **2**(2), 132–138.

Barnett T. P., Pierce D. W., Hidalgo H. G., Bonfils C., Santer B. D., Das T., Bala G., Wood A. W., Nozawa T., Mirin A. A., Cayan D. R. and Dettinger M. D. (2008). Human-induced changes in the hydrology of the western United States. *Science*, **319** (5866), 1080–1083. doi: 10.1126/science.1152538.

Berghuijs W. R., Harrigan S., Molnar P., Slater L. J. and Kirchner J. W. (2019). Relative importance of different food-generating mechanisms across Europe. *Water Resources Research*, **55**(6), 4582–4593.

Boers N., Bookhagen B., Barbosa H. M. J., Marwan N., Kurths J. and Marengo J. (2014). Prediction of extreme floods in the eastern central Andes based on a complex network approach. *Nature Communications*, **5**(1), 5199. doi: 10.1038/ncomms6199.

Boers N., Rheinwalt A., Bookhagen B., Marwan N. and Kurths J. (2015). A complex network approach to investigate the spatiotemporal co-variability of extreme rainfall. In: Machine Learning and Data Mining Approaches to Climate Science, V. Lakshmasnan, E. Gilleland, A. McGovern and M. Tingley (eds), Springer International Publishing, Switzerland, doi: 10.1007/978-3-319-17220-0_15.

Boers N., Goswami B., Rheinwalt A., Hoskins B. and Kurths J. (2019). Complex networks reveal global pattern of extreme-rainfall teleconnections. *Nature*, **566**(7744), 373–377.

Bronstert A., Agarwal A., Boessenkool B., Crisologo I., Fischer M., Heistermann M., Kohn-Riech L., Lopez-Tarazon J. A., Maron T., Ozturk U., Reinhardt-Imjela C. and Wendi D. (2018). Forensic hydro-meteorological analysis of an extreme flash flood: the 2016-05-29 event in Braunsbach, SW Germany. *Science of the Total Environment*, **630**, 977–991.

Buldyrev S., Parshani R., Paul G., Stanley H. E. and Havlin S. (2010). Catastrophic cascade of failures in interdependent networks. *Nature*, **464**(7291), 1025–1028, doi: 10.1038/nature08932.

Burke E. J. and Brown S. J. (2008). Evaluating uncertainties in the projection of future drought. *Journal of Hydrometeorology*, **9**(2), 292–299.

Chiang F., Mazdiyasni O. and AghaKouchak A. (2018). Amplified warming of droughts in southern United States in observations and model simulations. *Science Advances*, **4**(8), 1–6.

Cong Z., Yang D., Gao B., Yang H. and Hu H. (2009). Hydrological trend analysis in the Yellow River basin using a distributed hydrological model. *Water Resources Research*, **45**(13), 1–13.

Field C. B., Barros V., Stocker T. F., Qin D., Dokken D. J., Ebi K. L., Mastrandrea M. D., Mach K. J., Plattner G.-K., Allen S. K., Tignor M. and Midgley P. M. (eds) (2012). IPCC 2012: Managing the Risks of Extreme Events and Disasters to Advance Climate Change Adaptation. A Special Report of Working Groups I and II of the IPCC. Cambridge University Press, Cambridge.

Fischer E. M., Seneviratne S. I., Lüthi D. and Schär C. (2007). Contribution of land-atmosphere coupling to recent European summer heat waves. *Geophysical Research Letters*, **34**(6), L06707.

Haasnoot M., Middelkoop H., Offermans A., Beek E. and Deursen W. (2012). Exploring pathways for sustainable water management in river deltas in a changing environment. *Climatic Change*, **115**(3–4), 795–819, doi: 10.1007/s10584-012-0444-2.

Haasnoot M., Kwakkel J., Walker W. and Maat J. (2013). Dynamic adaptive policy pathways: a method for crafting robust decisions for a deeply uncertain world. *Global Environmental Change*, **23**(2), 485–498.

Haasnoot M., Warren A. and Kwakkel J. (2019). Dynamic adaptive policy pathways (DAPP). In: Decision Making under Deep Uncertainty, V. Marchau, W. Walker, P. Bloemen and S. Popper (eds), Springer, Cham, Switzerland, doi: 10.1007/978-3-030-05252-2_4.

Hagenlocher M., Meza I., Anderson C., Min A., Renaud F., Walz Y., Siebert S. and Sebesvari Z. (2019). Drought vulnerability and risk assessments: state of the art, persistent gaps, and research agenda. *Environmental Research Letters*, **14**(8), 083002.

Hall J., Arheimer B., Borga M., Brázdil R., Claps P., Kiss A., Kjeldsen T. R., Kriaučiūnienė J., Kundzewicz Z. W., Lang M., Llasat M. C., Macdonald N., McIntyre N., Mediero L., Merz B., Merz R., Molnar P., Montanari A., Neuhold C., Parajka J., Perdigão R. A. P., Plavcová L., Rogger M., Salinas J. L., Sauquet E., Schär C., Szolgay J., Viglione A. and Blöschl G. (2014). Understanding flood regime changes in Europe: A state-of-the-art assessment. *Hydrology and Earth System Sciences*, **18**, 2735–2772, doi: 10.5194/hess-18-2735-2014.

Hao Z. and AghaKouchak A. (2013). Multivariate standardized drought index: a parametric multi-index model. *Advances in Water Resources*, **57**(2013), 12–18.

Herring S. C., Hoerling M. P., Peterson T. C. and Stott P. A. (eds) (2014). Explaining extreme events of 2013 from a climate perspective. *Bulletin of the American Meteorological Society* **95**, S1–S96.

Ho H. L., Park E., Chitwatkulsiri D., Lim J., Yun S.-H., Maneechot L. and Phuong D. V. (2020). Local rainfall or river overflow? Re-evaluating the cause of the great 2011 Thailand flood. *Journal of Hydrology*, **589**(1–2), 1–32, doi: 10.1016/j.jhydrol.2020.125368.

Holický M. and Sýkora M. (2009). Forensic investigation of fluvial flood damage in Czech Republic. *Proceedings of the Institution of Civil Engineers – Civil Engineering*, **162**(5), 33–37.

Hong C. C., Hsu H. H., Lin N. H. and Chiu H. (2011). Roles of European blocking and tropical–extratropical interaction in the 2010 Pakistan flooding. *Geophysical Research Letters*, **38**(13), L13806.

Howitt R. E., Medellin-Azuara J., MacEwan D., Lund J. R. and Sumner D. A. (2014). Economic Analysis of the 2014 Drought for California Agriculture. University of California, Davis.

IPCC (2012). Managing the risks of extreme events and disasters to advance climate change adaptation. In: A Special Report of Working Groups I and II of the Intergovernmental Panel on Climate Change, C. B. Field, V. Barros, T. F. Stocker, D. Qin, D. J. Dokken, K. L. Ebi, M. D. Mastrandrea, K. J. Mach, G.-K. Plattner, S. K. Allen, M. Tignor and P. M. Midgley (eds), Cambridge University Press, Cambridge, UK, and New York, NY, USA, pp. 109–230.

IPCC (2014). Climate change 2014: impacts, adaptation, and vulnerability. Part B: regional aspects. In: Contribution of Working Group II to the Fifth Assessment Report of the Intergovernmental Panel on Climate Change, V. R. Barros, C. B. Field, D. J. Dokken, M. D. Mastrandrea, K. J. Mach, T. E. Bilir, M. Chatterjee, K. L. Ebi, Y. O. Estrada, R. C. Genova, B. Girma, E. S. Kissel, A. N. Levy, S. MacCracken, P. R. Mastrandrea and L. L. White (eds), Cambridge University Press, Cambridge, UK, and New York, NY, USA.

Jahn M., Jayamaha B., Mulhern W. S., Ross D. E., Rose M. A. and Treverton G. F. (2018). Global Food System Stability and Risk: At the Nexus of Defense and Development. Thomson Reuters, USA.

Johnson L. (2008). Geographic Information Systems in water resources engineering. CRC Press, New York, USA.

Kendon E. J., Roberts N. M., Fowler H. J., Roberts M. J., Chan S. C. and Senior C. A. (2014). Heavier summer downpours with climate change revealed by weather forecast resolution model. *Nature Climate Change*, **4**(7), 570–576.

Khazaei B., Khatami S., Alemohammad S.- H., Rashidi L., Changshan W., Madani K., Kalantari Z., Destouni G. and AghaKouchak A. (2019). Climatic or regionally induced by humans? Tracing hydro-climatic and land-use changes to better understand the lake Urmia tragedy. *Journal of Hydrology*, **569**(2019), 203–217.

Kumar D. (2002). Remote Sensing applications to water resources. World Scientific Publishers, Singapore.

Lau W. and Kim K. M. (2012). The 2010 Pakistan flood and Russian wave: teleconnection of hydrometeorological extremes. *Journal of Hydrometeorology*, **13**(1), 392–403.

Leonard M., Westra S., Phatak A., Lambert M., van den Hurk B., McInnes K., Risbey J., Schuster S., Jakob D. and Stafford-Smith M. (2014). A compound event framework for understanding extreme impacts. *Wiley Interdisciplinary Reviews: Climate Change*, **5**(1), 113–128.

Loaiciga H. (2001). Flood damages in changing flood plain: a forensic-hydrologic case study. *The American Water Resources Association*, **37**(2), 467–478.

Mazdiyasni O. and AghaKouchak A. (2015). Substantial increase in concurrent droughts and heatwaves in the United States. *Proceedings of the National Academy of Sciences*, **112** (37), 11484–11489.

Mehrian M. R., Hernandez R. P., Yavari A.-R., Faryadi S. and Salehi E. (2016). Investigating the causality of changes in the landscape pattern of lake Urmia basin, Iran using remote sensing and time series analysis. *Environmental Monitoring Assessment*, **188**(462), 1–13.

Mehran A., AghaKouchak A., Nakhjiri N., Stewardson M. J., Peel M. C., Phillips T. J., Wada Y. and Ravalico J. K. (2017). Compounding impacts of human-induced water stress 491 and climate change on water availability. *Scientific Reports*, **7**(1), 6282. doi: 10. 1038/s41598-017-492 06765-0.

Mendoza M. and Schwarze R. (2019). Sequential disaster forensics: a case study on direct and socio-economic impacts. *Sustainability*, **11**(21), 5898.

Moftakhari H. R., AghaKouchak A., Sanders B. F., Feldman D. L., Sweet W., Matthew R. A. and Luke A. (2015). Increased nuisance flooding along the coasts of the united 495 states due to sea level rise: past and future. *Geophysical Research Letters*, **42**(22), 9846–9852.

Moftakhari H. R., Salvadori G., AghaKouchak A., Sanders B. F. and Matthew R. A. (2017). Compounding effects of sea level rise and fluvial flooding. *Proceedings of the National Academy of Sciences*, **114**(37), 9785–9790.

Musselman K. N., Lehner F., Ikeda K., Clark M. P. and Prein A. F. (2018). Projected increases and shifts in rain-on-snow flood risk over western North America. *Nature Climate Change*, **8**, 808–812.

Namias J. (1955). Some meteorological aspects of drought, with special reference to the summers of 1952–54 over the United States. *Monthly Weather Review*, **83**, 199–205.

Nelsen R. B. (2003). Properties and applications of copulas: A brief survey. In: Proceedings of the First Brazilian Conference on Statistical Modeling in Insurance and Finance, J. Dhaene, N. Kolev and P. A. Morettin (eds), University of Sao Paulo, Sao Paulo, Brazil, pp. 10–28.

Nelsen R. B. (2007). An Introduction to Copulas. Springer, New York.

Nigam S. (2003). Teleconnections. In: Encyclopedia of Atmospheric Sciences, James Holton Judith Curry (eds). Elsevier, USA.

Noon R. (2001). Forensic engineering investigation. CRC-Press, Boca Raton.

Orlowsky B. and Seneviratne S. I. (2012). Global changes in extremes events: regional and seasonal dimension. *Climatic Change*, **110**(3), 669–696. doi: 10.1007/s10584-011-0122-9.

Power M. E., Bouma-Gregson K., Higgins P. and Carlson S. M. (2015). The thirsty Eel: summer and winter flow thresholds that tilt the Eel river of Northwestern California from Salmon-supporting to cyanobacterially degraded states. *Copeia*, **2015**(1), 200–211, doi: 10.1643/CE-14-086.

Ramirez A. I. and Herrera A. (2016). Forensic hydrology. In: Forensic Analysis – From Death to Justice, B. Suresh Shetty, (ed.), IntechOpen, London, UK.

Raymond C., M. Horton R., Zscheischler J., Martius O., AghaKouchak A., Balch J., G. Brown S., J. Camargo S., Hess J., Kornhuber K., Oppenheimer M., C. Ruane A., Wahl T. and White K. (2020). Understanding and managing connected extreme events. *Nature Climate Change*, **10**(7), 611–621, doi: 10.1038/s41558-020-0790-4.

Regonda S. K., Rajagopalan B., Clark M. and Pitlick J. (2005). Seasonal cycle shifts in hydroclimatology over the western United States. *Journal of Climate*, **18**(2), 372–384.

Renard B. and Lang M. (2007). Use of a Gaussian copula for multivariate extreme value analysis: some case studies in hydrology. *Advances in Water Resources*, **30**(4), 897–912.

Sadegh M., Ragno E. and AghaKouchak A. (2017). Multivariate copula analysis toolbox (MvCAT): describing dependence and underlying uncertainty using a Bayesian framework. *Water Resources Research*, **53**(6), 5166–5183.

Sadegh M., Moftakhari H., Gupta H., Rango E., Mazdiyasni O., Sanders B., Matthew R. and AghaKouchak A. (2018). Multi-hazard scenarios for analysis compound extreme events., *Geophysical Research Letters*, **45**(11), 5470–5480.

Schwalm C., Anderegg W., Michalak A., Fisher J., Biondi F., Koch G., Livak M., Ogle K., Shaw J., Wolf A., Huntzinger D., Schaefer K., Cook R., Wei Y., Fang Y., Hayes D., Huang M., Jain A. and Tian H. (2017). Global patterns of drought recovery. *Nature*, **548**(7666), 202–205.

Seneviratne S. I., Lüthi D., Litchi M. and Schär C. (2006). Land–atmosphere coupling and climate change in Europe. *Nature*, **443**(7108), 205–209.

Seneviratne S. I., Nicholls N., Easterling D., Goodess C. M., Kanae S., Kossin J., Luo Y., Marengo J., McInnes K., Rahimi M., Reichstein M., Sorteberg A., Vera C. and Zhang X. (2012). Changes in climate extremes and their impacts on the natural physical environment. In: IPCC 2012: Managing the Risks of Extreme Events and Disasters to Advance Climate Change Adaptation A Special Report of Working Groups I and II of the Intergovernmental Panel on Climate Change (IPCC), C. B. Field, V. Barros, T. F. Stocker, D. Qin, D. J. Dokken, K. L. Ebi, M. D. Mastrandrea, K. J. Mach, G.-K. Plattner, S. K. Allen, M. Tignor and P. M. Midgley (eds), Cambridge University Press, Cambridge, UK, and New York, NY, USA, pp. 109–230.

Sheffield J. and Wood E. F. (2008). Global trends and variability in soil moisture and drought characteristics, 1950–2000, from observation-driven simulations of the terrestrial hydrologic cycle. *Journal of Climate*, **21**(3), 432–458.

Shukla S., Safeeq M., AghaKouchak A., Guan K. and Funk C. (2015). Temperature impacts on the water year 2014 drought in California.*Geophysical Research Letters*, **42**(11), 4384–4393.

Sklar M. (1959). Fonctions de Repartition a n Dimensions Et Leurs Marges. Univ. Paris 8, Saint-Denis, France. [in French].

Su H., Yang Z.-L., Dickinson R. E. and Wei J. (2014). Spring soil moisture-precipitation feedback in the Southern Great Plains: How is it related to large-scale atmospheric conditions? *Geophysical Research Letters*, **41**, 1283–1289.

Trenberth K. E., Dai A., Rasmussen R. M. and Parsons D. B. (2003). The changing character of precipitation. *Bulletin of the American Meteorological Society*, **84**(9), 1205–1218.

Vahedifard F., AghaKouchak A. and Jafari N. H. (2016). Compound hazards yield Louisiana flood. *Science*, **353**(6306), 1374–1374.

Valdez L. D., Shekhtman L., La Rocca C. E., Zhang X., Buldyrev S. V., Trunfio P. A., Braunstein L. A. and Havlin S. (2020). Cascading failures in complex networks. *Journal of Complex Network*, **8**(2).

Van den Honert R. C. and Mcaneney J. (2011). The 2011 Brisbane floods: causes, impacts and implications. *Water*, **3**(4), 1149–1173.

Van Loon A., Glesson T., Clark J., Van Dijk A., Stahl K., Hannaford J., Di Baldassarre G., Teuling A., Tallaksen L., Uijlenhoet R., Hannah D., Sheffield J., Svoboda M., Verbeiren B., Wagener T., Rangecroft S., Wanders N and Van H. (2016). Drought in the anthropocene. *Nature*, **9**(2), 89–91.

Vormoor K., Lawrence D., Schlichting L., Wilson D. and Wong W. K. (2016). Evidence for changes in the magnitude and frequency of observed rainfall vs. Snowmelt driven floods in Norway. *Journal of Hydrology*, **538**, 33–48.

Wahl T., Jain S., Bender J., Meyers S. D and Luther M. E. (2015). Increasing risk of compound flooding from storm surge and rainfall for major us cities. *Nature Climate Change*, **5**(12), 1093–1097.

Wang W., Ertsen M., Svoboda M. and Hafeez M. (2016), Propagation of drought: from meteorological drought to agricultural and hydrological drought. *Advances in Meteorology*, **2016**(1–2), 1–5.

Westra S., Alexander L. V. and Zwiers F. W. (2013). Global increasing trends in annual maximum daily precipitation. *Journal of Climate*, **26**(11), 3904–3918.

Wilhite D. (2000). Drought as a natural hazard: concepts and definitions. *Drought: Glob, Assess*, **1**, 193–218.

Witte J- P.M., Zaadnoordijk W. J. and Buyse J. J. (2019). Forensic hydrology reveals why groundwater tables in the province of Noord Brabant (the Netherlands) dropped more than expected. *Water*, **11**(478), 1–14.

Zargar A., Sadigh R., Naser B. and Khan F. (2011). A review of drought indices. *Environmental Reviews*, **19**(1), 333–349.

Zhang L. and Singh V. P. (2019). Copulas and their Applications in Water Resources Engineering. Cambridge University Press, New York.

Zscheischler J., Westra S., Van den Hurk B., Seneviratne S., Ward P., Pitman A., AghaKouchak A., Bresch D., Leonard M., Wahl T. and Zhang X. (2018). Future climate risk from compound events. *Nature Climate Change*, **8**(6), 469–477.

Chapter 15

Citizen science

Hossein Rezaei[1], *Omid Bozorg-Haddad*[2*] *and Vijay P. Singh*[3]

[1]*M.Sc. graduate, Department of Irrigation & Reclamation Engineering, Faculty of Agricultural Engineering & Technology, College of Agriculture & Natural Resources, University of Tehran, Karaj, Tehran, 31587-77871, Iran. E-mail: HosseinRezaie18@ut.ac.ir*

[2]*Distinguished Professor, Department of Irrigation & Reclamation Engineering, Faculty of Agricultural Engineering & Technology, College of Agriculture & Natural Resources, University of Tehran, Karaj, Tehran, 31587-77871, Iran. E-mail: OBHaddad@ut.ac.ir*

[3]*Distinguished Professor, Regents Professor, Caroline & William N. Lehrer Distinguished Chair in Water Engineering, Department of Biological and Agricultural Engineering & Zachry Department of Civil & Environmental Engineering, Texas A&M University, 321 Scoates Hall, 2117 TAMU, College Station, Texas 77843-2117, USA. E-mail: VSingh@tamu.edu*

**Corresponding author.*

ABSTRACT

Today, developments in science and technology have made people more sensitive to their surroundings, as well as to the problems and challenges that surround them or threaten their lives in the future. This change in the way people look at the problems and big challenges of life today has led them to increasingly use science and technology. Scientific and technological progress has enabled humans to solve challenges that were once thought insoluble. These problems often involve three main limiting factors: cost, time and the need to collect lots of data. Today, in most countries, a solution called 'citizen science' is used to solve these big problems and projects. Indeed, the main foundation of citizen

Editor: Omid Bozorg-Haddad
doi: 10.2166/9781789062144_0315

science is that it is used to conduct extensive experiments by 'citizen scientists' in areas such as sampling and other time-consuming and costly activities. This chapter explains how, after an initial research plan is presented and has been evaluated, the decision is made by scientists whether or not to employ citizen scientists. In the next stages, described below, scientists attract volunteer citizens, then carefully plan the details of the subject and the relevant training and education needed to carry out the various stages of the research proposed. After completing these training courses, citizens scientists enter the stage of conducting experiments and collecting the required data. Finally, scientists analyze the information collected by scientists. The chapter describes this complete process, and supplies suitable examples.

Keywords: Citizen science, Water resources, Management.

15.1 INTRODUCTION

Research studies at vast scales need lots of data and information, and in most cases these data should be monitored and collected in a case study, over a long period of time. There are many approaches to monitoring data and collecting information. Typically, these are all very costly and time-consuming but, by contrast, one approach – citizen science – takes less time and money to monitor data and collect information. Citizen science, also called community science, is a practice of monitoring data and collecting information with the collaboration of the public (ordinary citizens, amateur scientists, students etc.), and uses these data in scientific research on a vast scale. It can be divided into three categories, based on the type of participation of the program: contributive (in which the public just monitors data and collects information), collaborative (the public also analyses the monitored data), and co-creative (where the public participates in all steps of the research study program). Citizen science can help members of a community to participate in research studies and learn a lot about science and their environment as a result. In other words, this approach is democratic science in action. Citizen science can also increase scientific literacy, public participation in local issues, and social capital, and has many benefits for governments and natural ecosystems that have been monitored.

Democratization in science, especially in environmental science, is a relatively new concept based on increasing public knowledge and expertise about environmental science. Before the professionalization of science in the 19th century, the only endeavors to promote science came from what we today call citizen scientists. Thus, citizen science is not a new concept and has a long life in human history. Some historic examples of citizen science include fossil finders and amateur geologists who helped to discover new species of extinct creatures that lived on the earth, and French wine growers who have accurately recorded the exact date of their grape harvest for over 640 years.

In recent years, the application of citizen science to monitor data and collect information in biology, social science, zoology, astronomy, environmental science, watershed planning, for example, has increased amazingly, so that we can perhaps refer to the last decade as the peak of citizen science. Because of the availability of digital tools and the internet, public participation in large scale projects has become easier than ever before. Also, with these kinds of facilities, the number of people who can participate in these projects and research studies is now huge. For instance, the use of local citizens in ecological research projects can help scientists to collect information over a vast area in just a short time, making monitoring data and collecting information in large-scale projects so much easier.

In research projects involving citizen science, the rule of citizens is not limited to a little part of the project. For instance, scientists can provide other sample papers and research to the citizen scientists in order to accustom them to the nature of research projects, and also fully use the research done by citizen scientists, under supervision. At the end of the research, if a scientific paper is being published in a journal, the authors must acknowledge the participation of all citizen scientists in the research project.

One of the most important factors that has promoted data gathering by the citizen science approach is the development of modern technology, and the low cost of using tools such as personal computers, smart phones and the internet – all of which have increased the speed and accuracy of collecting and sharing data. After monitoring data by the citizen science approach, scientists must validate and verify the data by analytical methods to prepare it for analysis. The results from previous research show that the data gathered from the citizen science approach are accurate and effective for use in large-scale research.

15.2 LITERATURE REVIEW

15.2.1 Water resources management

Williams (2002) examined community participation in protection and management of internal water resources. This research was applied in four case studies, including in Lake Washington and Mono Lake in the USA, the Aral Sea in Central Asia, and Lake Pedder located in Australia. The results in the Lake Washington and Mono Lake case studies were successful, whilst those in the Aral Sea and Lake Pedder were not, and demonstrate that citizens must be educated about the environment from childhood, and that education must be continued throughout life. They concluded that public actions caused success in Washington and Mono Lake and in the other lakes, and when citizens' plans are not considered and are not monitored, they are the reasons of failure.

Buytaert *et al.* (2016) investigated the application of citizen science in water resources management under the control and monitoring of government. New cost-effective technologies are employed by governments to monitor hydrologic systems and water resources evaluation than ever before. This process involves changing data collection from being done by major official institutions, by legally mandated supervising officials or water companies for example, to the community (citizens). This kind of monitoring work can increase the volume of science produced about water resources and the application of citizen science in water resources management.

Minkman *et al.* (2016) applied the citizen science approach in water management by using smart phone technology. The case study of this research is in the Netherlands. In this research, scientists faced a lack of awareness amongst citizens and other physical challenges. The proposed strategy to confront these challenges was to increase participation and citizen science by using smart phones. Results concluded that this kind of measurement could help scientists increase channel operation and control their forecasting model.

Kinchy (2017) used citizen science and democracy to monitor water resources in a case study located in northeastern USA. Buytaert *et al.* (2016) applied citizen science in water resources management and hydrology. In this research, they examined the opportunities of science production, management of ecosystem services, and sustainable development in four case studies in a mountain area in Peru. Results demonstrated that using new technologies such as cheap, strong, and automatic sensors increased the number of citizens who took part in the projects and scientists combined this data with strong environmental models to create visual methods.

Minkman *et al.* (2017) investigated the different visions held by practitioners about the application of citizen science in water management. In this research, they concluded that there were three kinds of visions: (1) citizen science as a potential solution that can achieve different goals, and as a result, encourage citizens' participation in collection and analysis of data; (2) citizen science to be used to access additional data like photos; and (3) citizen science, as an additional advantage, can give scientific awareness to citizens.

Paul and Buytaert (2018) applied citizen science and integrated water resources management to manage water resources. Their results show that the use of new technologies with citizen science could be very helpful in multi-purpose management of water resources. This method allowed the formation of a cheap and responsive network to examine different aspects of the water cycle in integrated water management.

15.2.2 Water quality

Thornton and Leahy (2012) used citizen science to create a database of groundwater quality and quantity, etc., for which 40 citizen scientists (students, state and local government employees, college faculties, etc.) volunteered to participate. This project was applied in five US states, and results showed that issues of trust were largely a question interpersonal trust.

Minkman (2015) applied citizen science to measure the quality of water using cellphones in the Netherlands. Results demonstrated that motivation was one of the major reasons for success in the research. The results of a survey from citizens determined that protecting drinking water was important and that citizen science progressed with new technologies.

Fox *et al.* (2016) used citizen science to examine general health and personal wells, using the research to help educate citizens. Huddart *et al.* (2016) used citizen science to identify pollution as part of an environmental reconstruction assessment. Citizens played their rule as environmental guards but scientists needed more effective social relationships with citizens to have effective environmental monitoring. Results demonstrated that citizen scientists' cooperation with scientists has a more effective role rather other factors, to help discover solutions for future problems such as climate change. They also demonstrated that the evaluation approach seeks to shift citizenship science to a stakeholder participation process.

Zheng *et al.* (2017) examined the application of citizen science in monitoring the quality of surface water by using social networks in a case study in China. They concluded that citizen scientists were more willing to send their photos as unknown citizens. Results showed that water quality data collected by citizen scientists were relatively valid and if there was an economic motivation to the project, more citizens participated in a short time. Farnham *et al.* (2017) applied citizen science to make a database of the density of bacteria in waterways. They used citizen science data to make a database of bacteria density in groundwater in New York, USA, and results showed that sensors can't help scientists collect and analyze data automatically, so using a citizen science approach can be very helpful. Hyder *et al.* (2017) used citizen science to monitor and measure contamination at a small scale on the coast of England. Results of this research discovered that the use of citizen science at a small scale was useful. McGoff *et al.* (2017) used citizen science and professional researchers to find a clean water habitat in London, England. They concluded that citizen science at a large scale could be very successful in measuring contamination in the basin. Jollymore *et al.* (2017) investigated the application of citizen science in water quality projects to examine the nature of data collected by citizens and to deal with traditional methods to find the nature of data deviation. Scientists discovered six points from this research and concluded that the association approach, like the citizen science approach, was very basic to achieving scientific and social goals. Muenich *et al.* (2016) examined the effect of citizen science in monitoring the water quality of a river in the USA. They concluded that data collected by a citizen science approach could be very useful in watershed management groups of the basin. Because watershed management groups exist around the world and often work as non-profit organizations, cooperation of citizens with this kind of organization not only increases the volume of data collected by citizens but also increases citizens' education and participation in environmental protection.

Shupe (2017) applied citizen science to examine the water quality of a river in Vancouver, British Columbia, Canada. Results demonstrated that due to the population growth, collection of data with a citizen science approach to monitor and model the effect of water quality in that area was essential. Barrows *et al.* (2018) used citizen science to measure the amount of microplastics (particles less than 5 mm) in river water in the Galattin basin in the USA. In this research, 120 persons participated as citizen scientists and monitored data from 72 positions in the basin, and 714 microplastic samples were gathered from the basin. Results showed that the citizen science approach was useful for this research and can be helpful for gathering microplastics.

Brouwer *et al.* (2018) monitored the quality of urban water and microbiological sustainability by using the citizen science approach. This research project was applied in Amsterdam, the Netherlands. Citizen scientists took samples from urban drinking water in their homes, which were then examined and 10 000 species of bacteria were discovered. Result determined that self-confidence of citizen scientists after participation in this research increased, and concluded that using citizen science increased awareness of consumers in the community. Chmielewski (2018) examined the combined use of citizen science and GIS software in the evaluation of contamination in a case study in Lublin, Poland.

Esmaeilian *et al.* (2018) used citizen science to improve university students' experiences in design engineering construction and sustainable educational development. Eastoe and Clark (2018) used citizen science to realize water resources in rural areas, in Cascabel, Arizona, USA. Herman-Mercer *et al.* (2018) applied citizen science to examine the quality of water quality data in the Yukon River basin in Canada and Alaska. Results showed that local network data had high quality and accuracy, and with consistent protocols and training, they could continue to collect accurate and reliable data. Valois *et al.* (2019) used citizen science to monitor urban water quality in Atyaro in New Zealand. Results demonstrated that if citizens were educated well, they could collect reliable data.

15.2.3 Evaluation of citizen science

Branchini *et al.* (2015) investigated citizen science and its impact on general education about the environment in a Red Sea case study. Results demonstrated that citizen science projects had an important value in education. Chase and Levine (2018) used citizen science to examine the potential of environmental monitoring programs in a case study in the USA, finding that, in order to maximize their potential for impact on environmental attitudes and decision making, they should cooperate with volunteer citizens. Follett and Strezov (2015) examined analyses of citizen science-based research finding papers published in different kinds of journals, showing that citizen science is being used in many different science fields. Hollow *et al.* (2015) used citizen science to develop management policies for koala bears in South Australia. Results demonstrated that citizen science in early steps of policy development was very useful. Alender (2016) examined the motivation of citizens to participate in citizen science projects in a case study in the USA and results showed that two factors were very important for the motivation of citizens: that they should learn and share experiences with others, and that scientists should try to engage them in activities needing a wide range of experience and skills. Jalbert and Kinchy (2016) examined the power of citizen science and environmental tools, and investigated which method had more power and influence. This research was applied in a case study in Pennsylvania, USA. They concluded that the application of automatic technologies in collecting qualitative environmental data were not as good as general participation in environmental science and politics. Kosmala *et al.* (2016) examined the quality of citizen science data and concluded that there were many duties and systems that showed volunteers could generate accurate and quality data, and that the accuracy in easy tasks and with experienced volunteers was very high.

Keeler *et al.* (2015) used citizen science to examine the relationship between clean water and recreational demand for visiting lakes in the Minneapolis area, USA, and showed that clean water encouraged more visitors to a lake and that visitors were also willing to pay more for a clear lake. Scientists use photographs to understand the impact of water quality on recreational goals. Wehn and Evers (2015) investigated using citizen science to manage local flood risks. The case studies of this research were located in England and the Netherlands. Result showed that integrating urban observation with an organizational structure helped to reinforce flood risk management and citizen participation. They also found that promoting citizen science needs cooperation between officials and citizens.

Little *et al.* (2016) used citizen science to monitor groundwater levels in an area in Alberta, Canada. In this research, citizens measured the level of water in their groundwater wells and uploaded data to a web-based portal. Due to the good use made of the data by the government, local municipalities, and watershed management and planning officials, the community monitoring of groundwater was very efficient and affordable.

Hadj-Hammou *et al.* (2017) used citizen science to evaluate the oversight and monitoring of urban organizations. This case study was conducted in France. They concluded that the basin-based approach in the management of freshwater resources is a good foundation for governance and the central water ecosystem. By creating stakeholder engagement in catchments, different goals and citizen oversight efforts can work better. In addition, river basin management programs can accommodate an extensive database and use an extensive citizen science network to enter information. Efforts to collect data can be encouraged by agencies and also improved to indicate where monitoring is needed. Shuker *et al.* (2017) applied citizen science and a MoRPh survey to survey physical habitats and assess the pressure applied to rivers by humans. The indicators calculated by the MoRPh survey, using the plaster flow test, can distinguish lots of difference in hydraulic, sediment, and vegetation habitat across rivers.

Starkey *et al.* (2017) used citizen science to show the value of this approach in modeling and used it to gain the characteristics of a basin. The case study of this research was of the

Haltwhistle Burn catchment, a tributary of the River Tyne, in the northeast of England. This research showed that the citizen science approach helped local communities to learn about the local water environment, association between citizen science and professional stakeholders was effective and increased monitoring efficiency in basin management.

Tipaldo and Allamano (2017) used citizen science to monitor rainfall that combined water science and sociology. They concluded that the internet and its application (intelligent sensor, network and derivatives) received more attention and enthusiasm in public debate. Collaborative practices can actually lead to more uncertainty than traditional initiatives, which is mainly due to the lack of a standard measurement tool. A multidisciplinary approach based on community monitoring and trustworthy users should prevent the most adverse side effects of citizen science. How much we trust in numbers, and how much is an epistemological knot at the intersection of science and sociology, is still largely unimportant.

Lamine *et al.* (2018) investigated the application of citizen science to monitoring groups of fish in the Tunisian Sea. Results demonstrated an increased interaction between citizens and the scientific community, and an increased environmental awareness and stakeholder interference in water resources management. Mercer (2018) researched the meaning of citizen science in a case study in North America. He concluded that citizen science projects could be undertaken with low cost because citizens typically cared about their civic duty. In addition, this kind of project needs time, resources, monitoring, promotion, employment, education, collecting and validating of data.

Paul *et al.* (2018) used citizen science to collect voluntary geographic data in the field of hydrologic risks. They concluded that citizen participation in collecting data increased multidirectional decentralized data, multi-risk management, and local flexibility. However, they believed that the future of citizen science was not only in collecting information but in information processing and across the whole research stage.

Weeser *et al.* (2018) examined the results of increased motivation of participants citizen science in poor countries. This research was applied in Kenya. They concluded that giving feedback to citizen scientists was essential and causes an increase in their contribution to the project. Results showed that it was possible to interact with the citizen scientists to ensure that their data from monitoring of water level were reliable, compatible and of high quality.

15.3 DEFINITIONS AND TERMS

15.3.1 Citizen science

Citizen science is defined as cooperation and participation of citizens (ordinary people) in research and scientific studies being done by scientists or researchers in order to increase scientific knowledge in the community. Citizens cooperate and participate in citizen science projects, such as monitoring and collecting of data, and other collection programs. Typically, citizens take part in citizen science projects voluntarily and without any payment.

15.3.2 Citizen scientists

Citizen scientists are the group of citizens who participate in large-scale scientific research, not as laboratory test subjects or volunteers but by doing an experiment themselves, making observations, monitoring data and applying their minds to the project. Citizens help researchers do large-scale research at low cost and in quick time. Every person in the community can take part in research as citizen scientists, for example, students, amateur scientists, educators, etc.

15.3.3 Validation

In general, validation means actions that examine the accuracy and validity of something. In citizen science, validation is defined as comparing the data monitored by citizen scientists and real world data by using different statistical methods; this comparison is made by scientists or researchers to ensure the correctness of the data. At this stage, any inaccurate and inappropriate data will be remove from the database. In citizen science, scientists cannot use data in the next processing stage before validating it.

15.4 FUNDAMENTALS AND LOGIC OF CITIZEN SCIENCE

Because of the developments in science in the last decades, scientists and researchers in various fields understand that, in order to discover other examples of interest, they must examine vast areas of their communities. For example, to examine the impact of global warming on biodiversity across a continent or across the whole world, examining the impact of a rule or decision in all parts of a community is needed. One of the greatest challenges in these projects is the lack of data and the need to monitor and collect data over vast areas or across a large targeted community. Monitoring and collecting data for these projects across vast areas, because of the time-consuming and costly nature of the work involved, can in most cases not be done successfully, or employers (government, big companies etc.) will change their minds about doing such projects. Therefore, scientists have now turned to a new approach to monitor and collect data over vast areas, in a less time-consuming and less expensive way than previously possible.

Citizen science is an approach that helps scientists undertake their projects without caring about anything else. In this approach, scientists use citizens to monitor and collect data for other stages of processing and analysis. For the first time, scientists used citizen science to count the number of birds at Christmas by the National Audubon Society in 1900. Frank Chapman, birdwatcher and one of the Desert Guide's authors, suggested that hunters counted the number of birds instead of hunting them in the New Year holidays. At that time, 27 birdwatchers supported this idea. The foundations of birdwatching were further established in 1932 when British amateur bird watchers were encouraged to protect nature and improve ecology science. As a result of this work, the number of data collected by the National Biodiversity Network was more than 31 million, including details of 27,000 species of different plants and animals (Stewart, 1954).

The citizen science approach is based on the direct involvement of citizens in projects or research such as monitoring and collecting data, processing and analysis of data, and participation at all stages of a project. This approach, like any other, has advantages and disadvantages. The advantages include:

(1) it is a cost effective approach, which does not employ government, association or business staff;
(2) it is often a quicker approach;

(3) citizens can collect more data over a larger area;
(4) scientists can teach citizens and encourage them to care about their community and environment;
(5) it fosters excellent engagement of citizens in projects; and
(6) scientists can trust data collected by the citizen science approach.

Disadvantages of the citizen science approach are:

(1) it is much more effective in simple rather than difficult or complicated projects;
(2) sometimes, there are tensions between the motivations of organizers and citizens (Rotman *et al.*, 2012);
(3) data gathering is outside of scientists' direct control; and
(4) sometimes, volunteers need to get paid, especially in long-term monitoring projects.

Today, scientists use new technologies such as the internet and social networks, etc., to call for volunteers to take part in citizen science projects and research. Citizens are then selected from the volunteers, based on requirements, and they then form a targeted community as citizen scientists to help scientists in different stages of research and projects. After being chosen, researchers explain the research goals and duties, at each stage, to the citizen scientists. In the next stage, scientists provide any necessary training needed for the project. Then the citizen scientists start their work based on the goals defined for them by the scientists – such as monitoring data, collecting information, etc. It is worth mentioning that, in cases where the results of any research begin to help citizens solve their own community's problems and concerns, the relationship and cooperation between scientists and citizen scientists often improve and, consequently, the quality of citizen scientists' work is often higher than before in a way that allows scientists to do the rest of their research with even more accuracy and quality.

Different important factors related to the citizen science approach are explained below:

- **training of citizen scientists.** One of the most important aspects in a citizen science project is the citizen scientists' training to do their duties correctly and accurately. If there is sufficient training of good quality, the citizen scientists can do their duties expertly and accurately, and consequently data collected for the next processing stages are more accurate and the research is better. One of the common duties of citizen scientists is to monitor and collect data, and because this stage is the basis of other stages, so scientists must educate citizen scientists to do this stage very accurately.
- **using appropriate equipment.** Since equipment can often do some tasks without errors and more accurately than human beings, it is better to provide citizen scientists with appropriate equipment to do some of their duties, where possible.
- **explaining the relevance and benefits of the research.** If scientists can convince citizens that the results of their research could solve problems or improve their standards of living and, as a result, provide them with a better life in some way, it can create a positive sense of duty and responsibility in the citizen scientists to do their duties better, more accurately and to a high quality.
- **building a sense of community.** Discovering, building or creating a culturally or socially relevant aspect to the research can create more participation of citizen scientists in scientific projects. Because of the sense of responsibility and social duty of citizens for

their community, they are more willing to take part in citizen science projects and undertake them to a better quality.

- **importance and necessity of citizen science.** The development of science and technology in modern society often increases citizens' concerns because they are faced with new problems and challenges. In order to solve this, scientists should examine the scale and range of problems proposed for experiments, or for collecting data and discussing results. With this focus, the quality and quantity of data can be made higher, and scientists can investigate a problem better and present more comprehensive solutions. Three major limiting factors of large-scale projects are that they are time-consuming, require a vast area to be checked, and need reliable data to be collected. These limiting often cause projects to be stopped, despite their importance. The new approach of citizen science can solve some of these problems and, scientists have concluded, it is often the best approach to overcome the above mentioned limitations.

15.5 METHODS AND FLOWCHART

The citizen science approach is presented in Figure 15.1. As can be seen in the figure, scientists should first present a research plan based on the problem(s) being addressed, then determine the importance of the research and the aspects which could involve citizen scientists. Then, the methods and equipment required for the research have to be chosen. At the next stage, scientists call for volunteer citizens to cooperate in the research project, and then select eligible citizens to work with them. Scientists then educate the volunteer citizen scientists about their duties throughout the research, including explaining the research, the importance of doing it correctly, the application of research results to improve citizens' living standards (where appropriate), and the methods of monitoring data and collecting information. After data is collected by citizen scientists, it is validated and verified by scientists and then used processed and analyzed. At the end of the research, the role of citizen scientists in the research is acknowledged, and scientists share the results with them.

15.6 EXAMPLES

15.6.1 Example 1: measuring surface water temperatures in Sweden

Weyhenmeyer *et al.* (2017) applied citizen science to measure the surface water temperature of lakes, ponds, beaches, and rivers. For this purpose, scientists trained 3500 high school students as citizen scientists in Sweden. The researchers used the data collected by students to calculate the concentration of carbon dioxide and methane in the air by the thermal flux created between air and water, based on temperature differences. High concentrations of the gases cause an increase in global temperature. The results demonstrated that, as the temperature differences between air and water increased, the thermal flux increased too, and consequently the concentration of carbon dioxide and methane increased, and vice versa. The difference of temperature between air and water had a linear relation with the concentration of gases in the air. In that way, if their difference decreased, the concentration of carbon dioxide and methane decreased too, and

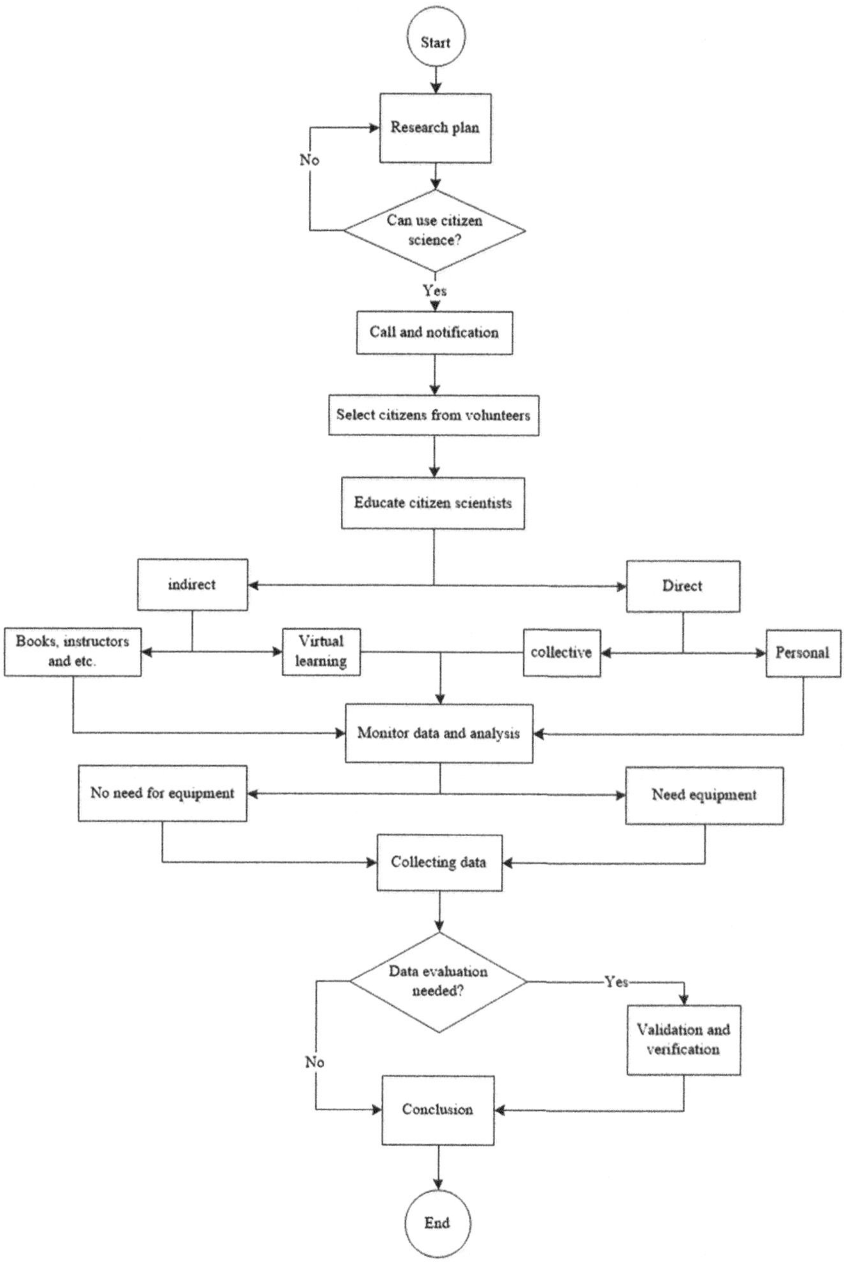

Figure 15.1 A flowchart of the citizen science approach.

as a result the global temperature decreased. They concluded that data collected by the citizen science approach were accurate and encouraged students living beside lakes and ponds to become interested in scientific research. Research organizations try very hard to guide the public in science's direction by using members of the public in research, and increasing general awareness of scientific knowledge. Scientists in this particular research used the citizen science approach to understand the transfer of temporal and spatial changes in the potential transfer of carbon dioxide and methane greenhouse gases from inland waters to the atmosphere.

Scientists in partnership with high school students and their teachers were able to create a suitable team to monitor and collect the temperature data of air and water over a vast geographical area. In this citizen science project, 14–16 year old students were chosen specifically to encourage them to continue their education in natural sciences. First, scientists defined citizen science and then explained the project in the format of a short video, sharing it through web pages and social networks. The initial plan was for 100 classrooms all over Sweden to take part but, because of the unexpected interest, the number of classrooms was increased to 240. Then, training packages included field protocols, thermometers, sampling pipes, and an instructions for teachers were supplied. The researchers requested teachers and students to select a water collection site near their school. Students then started to measure the temperature of water at 0.5 m depth from the surface and air temperature at 1.5 m above the sampling site, based on the instructions and protocols. At every sampling site, students also recorded the water color, temperature, pH and also observed the local climate. In addition, students supplied photographs of the site and sample containers.

When the project started, the researchers were unsure of what results to expect but, given the teachers' keen interested in participation in the research, the scientists were eager to continue the project. Based on the teachers' evaluation throughout the project, students increased their knowledge of natural sciences and discovered a clear relation between mathematics, physics, chemistry, geography, and biology. Scientists concluded that this citizen science project was very successful and caused students to increase their knowledge of natural sciences and understand natural processes better. It also enabled scientists to have access to very useful data collected by the citizen scientists. The scientists concluded that temperature data collected by the citizen science approach is very cheap, easy to collect, and reliable.

15.6.2 Example 2: monitoring water quality in the USA

Alender (2016) examined the influence of citizen scientists' motivation when engaged in water quality monitoring projects, in a case study in the USA. In general, a citizen science project can have several overlapping goals:

(1) provide results for scientific research, such as by collecting data;
(2) provide results for citizen scientists (participant in program), such as teaching of new skills and giving new knowledge; and
(3) provide results for social-ecological systems, such as conservation, stewardship, and policy.

Volunteers are the most important aspect of citizen science projects and they help organizations to collect data in a cost-effective approach. Understanding of volunteer motivation helps project managers with recruitment and retention. Scientists must pay attention to ensure that volunteers are not merely free labor. Previous research has demonstrated that many volunteers have complex motivations. Motivations can be put in six categories: (1) values: concern for others; (2) understanding; (3) social; (4) career; (5) ego protective: escape from negative feelings; and (6) ego enhancement: personal growth. In order for volunteers to recognize their accomplishments, financial and human resources are required to train, recruit, supervise, and retain volunteers (Jacobson *et al.*, 2012). According to Conrad and Hilchey (2011), in applying citizen science, scientists running a project are faced with two challenges: program organization and achieving a set of accurate data. For Dickinson and Bonney (2012), the main challenges in citizen science projects are the budget and voluntary employment; funding is needed to recruit volunteers and, typically, volunteer participation is needed to attract funding.

Understanding citizen scientists and their motivation to participate in a project is very important for scientists. They may be motivated to take part to understand how the results can be used to manage the ecosystem, or to see them in scientific publications. For instances, in the research by Roggenbuck *et al.* (2001), volunteers said that they were keen to get feedback and understand how the results were used by government agencies. The results in this case demonstrated that data collected by citizen scientists were as good as data collected by experts. Results of this kind of voluntary researchers (citizen science) are influenced by two factors: quality and quantity of participation. For this reason, understanding citizens' motivation for participating in projects can help scientists and managers make better decisions when organizing them.

15.7 SUMMARY

Although, volunteer motivation has been studied in different fields of sciences but a few of them have been done on citizen science projects. Today, citizen science projects increase because the popularity of this kind of project increases too. This research demonstrates that volunteer motivation for participation in citizen science research on monitoring water quality is very similar to volunteer participation in environmental research. In citizen science, age and the number of volunteers are very important and younger volunteers have more motivation than elders. For this reason, coordinators must try to define some projects that are suitable for a wide range of experience and expertise. Creating an attitude of social learning and regimentation of experienced with younger volunteers lead to more project efficiency.

Today, we hear a lot of good news about promotion and brilliant results of citizen science research. This promotion of citizen science is because of doing different researches accurately, so for progress in different fields of science we should use the results of research done before. One of the crises that cause many challenges for new modern world is water resources management problems that scientists should always investigate in large-scale areas. One of the approaches that can help scientists solve water resources management problems is citizen science. In order to use citizen science in real projects government should create suitable situations, including cultural, political, economic, social, software and hardware infrastructure etc. Today more than any time in history, our societies need to solve their water resources problems, and accurate data collection with lack of information and equipment is very valuable. For this reason, citizen science projects can help scientists understand and solve water resources problems.

Citizen science is not only cost-effective and low time consuming but also can motivate citizens with a sense of responsibility and cooperation to solve the societies' problems and make better future for citizens and their children.

REFERENCES

Alender B. (2016). Understanding volunteer motivations to participate in citizen science projects: a deeper look at water quality monitoring. *Journal of Science Communication*, **15**(3), A04.

Barrows A. P., Christiansen K. S., Bode E. T. and Hoellein T. J. (2018). A watershed-scale, citizen science approach to quantifying microplastic concentration in a mixed land-use river. *Water Research*, **147**, 382–392.

Branchini S., Meschini M., Covi C., Piccinetti C., Zaccanti F. and Goffredo S. (2015). Participating in a citizen science monitoring program: implications for environmental education. *PloS One*, **10**(7), e0131812.

Brouwer S., van der Wielen P., Schriks M., Claassen M. and Frijns J. (2018). Public participation in science: the future and value of citizen science in the drinking water research. *Water*, **10**(3), 284.

Buytaert W., Dewulf A., De Bièvre B., Clark J. and Hannah D. M. (2016). Citizen science for water resources management: toward polycentric monitoring and governance? *Journal of Water Resources Planning and Management*, **142**(4), 01816002.

Chase S. K. and Levine A. (2018). Citizen science: exploring the potential of natural resource monitoring programs to influence environmental attitudes and behaviors. *Conservation Letters*, **11**(2), e12382.

Chmielewski S., Samulowska M., Lupa M., Lee D. J. and Zagajewski B. (2018). Citizen science and WebGIS for outdoor advertisement visual pollution assessment. *Computers, Environment and Urban Systems*, **67**, 97–109.

Conrad C. C. and Hilchey K. G. (2011). A review of citizen science and community-based environmental monitoring: issues and opportunities. *Environmental Monitoring and Assessment*, **176**(1–4), 273–291.

Dickinson J. L. and Bonney R. (eds). (2012). Citizen Science: Public Participation in Environmental Research. Cornell University Press, Ithaca, NY, USA.

Eastoe C. J. and Clark B. (2018). Understanding the water resources of a small rural community: citizen science in Cascabel, Arizona. *Journal of Contemporary Water Research & Education*, **164**(1), 19–40.

Esmaeilian B., Rust M., Gopalakrishnan P. K. and Behdad S. (2018). Use of citizen science to improve student experience in engineering design, manufacturing and sustainability education. *Procedia Manufacturing*, **26**, 1361–1368.

Farnham D. J., Gibson R. A., Hsueh D. Y., McGillis W. R., Culligan P. J., Zain N. and Buchanan R. (2017). Citizen science-based water quality monitoring: Constructing a large database to characterize the impacts of combined sewer overflow in New York City. *Science of the Total Environment*, **580**, 168–177.

Follett R. and Strezov V. (2015). An analysis of citizen science based research: usage and publication patterns. *PloS One*, **10**(11), e0143687.

Fox M. A., Nachman K. E., Anderson B., Lam J. and Resnick B. (2016). Meeting the public health challenge of protecting private wells: proceedings and recommendations from an expert panel workshop. *Science of the Total Environment*, **554**, 113–118.

Hadj-Hammou J., Loiselle S., Ophof D. and Thornhill I. (2017). Getting the full picture: assessing the complementarity of citizen science and agency monitoring data. *PloS One*, **12**(12), e0188507.

Herman-Mercer N., Antweiler R., Wilson N., Mutter E., Toohey R. and Schuster P. (2018). Data quality from a community-based, water-quality monitoring project in the Yukon River Basin. *Citizen Science: Theory and Practice*, **3**(2).

Hollow B., Roetman P. E., Walter M. and Daniels C. B. (2015). Citizen science for policy development: The case of koala management in South Australia. *Environmental Science & Policy*, **47**, 126–136.

Huddart J. E., Thompson M. S., Woodward G. and Brooks S. J. (2016). Citizen science: from detecting pollution to evaluating ecological restoration. *Wiley Interdisciplinary Reviews: Water*, **3**(3), 287–300.

Hyder K., Wright S., Kirby M. and Brant J. (2017). The role of citizen science in monitoring small-scale pollution events. *Marine Pollution Bulletin*, **120**(1–2), 51–57.

Jacobson S. K., Carlton J. S. and Monroe M. C. (2012). Motivation and satisfaction of volunteers at a Florida natural resource agency. *Journal of Park and Recreation Administration*, **30**(1), 51–67.

Jalbert K. and Kinchy A. J. (2016). Sense and influence: environmental monitoring tools and the power of citizen science. *Journal of Environmental Policy & Planning*, **18**(3), 379–397.

Jollymore A., Haines M. J., Satterfield T. and Johnson M. S. (2017). Citizen science for water quality monitoring: data implications of citizen perspectives. *Journal of Environmental Management*, **200**, 456–467.

Keeler B. L., Wood S. A., Polasky S., Kling C., Filstrup C. T. and Downing J. A. (2015). Recreational demand for clean water: evidence from geotagged photographs by visitors to lakes. *Frontiers in Ecology and the Environment*, **13**(2), 76–81.

Kinchy A. (2017). Citizen science and democracy: participatory water monitoring in the Marcellus shale fracking boom. *Science as Culture*, **26**(1), 88–110.

Kosmala M., Wiggins A., Swanson A. and Simmons B. (2016). Assessing data quality in citizen science. *Frontiers in Ecology and the Environment*, **14**(10), 551–560.

Lamine E. B., Di Franco A., Romdhane M. S. and Francour P. (2018). Can citizen science contribute to fish assemblages monitoring in understudied areas? The case study of Tunisian marine protected areas. *Estuarine, Coastal and Shelf Science*, **200**, 420–427.

Little K. E., Hayashi M. and Liang S. (2016). Community-based groundwater monitoring network using a citizen-science approach. *Groundwater*, **54**(3), 317–324.

McGoff E., Dunn F., Cachazo L. M., Williams P., Biggs J., Nicolet P. and Ewald N. C. (2017). Finding clean water habitats in urban landscapes: professional researcher vs citizen science approaches. *Science of the Total Environment*, **581**, 105–116.

Mercer K. L. (2018). Citizen Science. *Journal – American Water Works Association*, **110**(6), 2–2.

Minkman E. (2015). Citizen Science in Water Quality Monitoring. MSc thesis, Department of Civil Engineering, TU Delft, Delft, Netherlands.

Minkman E., Rutten M. M. and van der Sanden M. C. (2016). Acceptance of mobile technology for citizen science in water resource management. *Journal of Irrigation and Drainage Engineering*, **143**(3), B6016001.

Minkman E., Sanden M. V. D. and Rutten M. (2017). Practitioners' viewpoints on citizen science in water management: a case study in Dutch regional water resource management. *Hydrology and Earth System Sciences*, **21**(1), 153–167.

Muenich R., Peel S., Bowling L., Haas M., Turco R., Frankenberger J. and Chaubey I. (2016). The Wabash sampling blitz: a study on the effectiveness of citizen science. *Citizen Science: Theory and Practice*, **1**(1).

Paul J. D. and Buytaert W. (2018). Citizen science and Low-cost sensors for integrated Water resources management. *Advances in Chemical Pollution, Environmental Management and Protection*, **3**, 1–34.

Paul J. D., Buytaert W., Allen S., Ballesteros-Cánovas J. A., Bhusal J., Cieslik K., Clark J., Dugar S., Hannah D. M. and Stoffel M. (2018). Citizen science for hydrological risk reduction and resilience building. *Wiley Interdisciplinary Reviews: Water*, **5**(1), e1262.

Roggenbuck J. W., Haas S. C., Hall T. and Hull R. B. (2001). Motivation, Retention, and Program Recommendations of Save Our Streams Volunteers. Report No. SR19-2001. College of Natural Resources and Environment, Virginia Institute and State University, Virginia, USA.

Rotman D., Preece J., Hammock J., Procita K., Hansen D., Parr C., Lewis D. and Jacobs D. (2012). Dynamic changes in motivation in collaborative citizen-science projects. CSCW '12: Proceedings of the ACM 2012 Conference on Computer Supported Cooperative Work, Association for Computing Machinery, New York, USA, pp. 217–226. See: https://doi.org/10.1145/2145204.2145238 (accessed 21 May 2021).

Shuker L. J., Gurnell A. M., Wharton G., Gurnell D. J., England J., Finn Leeming B. F. and Beach E. (2017). MoRPh: a citizen science tool for monitoring and appraising physical habitat changes in rivers. *Water and Environment Journal*, **31**(3), 418–424.

Shupe S. M. (2017). High resolution stream water quality assessment in the Vancouver, British Columbia region: a citizen science study. *Science of the Total Environment*, **603**, 745–759.

Starkey E., Parkin G., Birkinshaw S., Large A., Quinn P. and Gibson C. (2017). Demonstrating the value of community-based ('citizen science') observations for catchment modelling and characterisation. *Journal of Hydrology*, **548**, 801–817.

Stewart P. A. (1954). The value of the Christmas bird counts. *The Wilson Bulletin*, **66**(3), 184–195.

Thornton T. and Leahy J. (2012). Trust in citizen science research: a case study of the groundwater education through water evaluation & testing program 1. *Journal of the American Water Resources Association*, **48**(5), 1032–1040.

Tipaldo G. and Allamano P. (2017). Citizen science and community-based rain monitoring initiatives: an interdisciplinary approach across sociology and water science. *Wiley Interdisciplinary Reviews: Water*, **4**(2), e1200.

Valois A., Davies-Colley R., Storey R., Wright-Stow A., Stott R., Kin E. and van Hunen S. (2019). Volunteer monitoring as a focus for community engagement in water management in Aotearoa-New Zealand: review and prospects. *Water Supply*, **19**(3), 671–680.

Weeser B., Kroese J. S., Jacobs S., Njue N., Kemboi Z., Ran A., Rufino M. and Breuer L. (2018). Citizen science pioneers in Kenya – a crowd-sourced approach for hydrological monitoring. *Science of the Total Environment*, **631**, 1590–1599.

Wehn U. and Evers J. (2015). The social innovation potential of ICT-enabled citizen observatories to increase eParticipation in local flood risk management. *Technology in Society*, **42**, 187–198.

Weyhenmeyer G. A., Mackay M., Stockwell J. D., Thiery W., Grossart H.-P., Augusto-Silva P. B., Baulch H. M., de Eyto E., Hejzlar J., Kangur K., Kirillin G., Pierson D. C., Rusak J. A., Sadro S. and Woolway R. I. (2017). Citizen science shows systematic changes in the temperature difference between air and inland waters with global warming. *Scientific Reports*, **7**, 43890.

Williams W. D. (2002). Community participation in conserving and managing inland waters. *Aquatic Conservation: Marine and Freshwater Ecosystems*, **12**(3), 315–326.

Zheng H., Hong Y., Long D. and Jing H. (2017). Monitoring surface water quality using social media in the context of citizen science. *Hydrology and Earth System Sciences*, **21**(2), 949–961.

Index

IWA Publishing's authorised EU representative for General Product Safety Regulations is Diane D'Arras, 15 rue Duret, 75116 Paris, France, e-mail: safety@iwap.co.uk.

Printed and bound by CPI Group (UK) Ltd, Croydon, CR0 4YY
16/06/2026
02140403-0001